MIG-29 FLIGHT MANUAL

Also by Alan Wise

JET AGE FLIGHT HELMETS

MiG PILOT SURVIVAL

MIG-29 FLIGHT MANUAL

DECLASSIFIED

ALAN R. WISE

Schiffer Military History
Atglen, PA

Acknowledgements

I would like to thank Ruth Shablak of ScantekDesign.com for the expert scanning and redefining critical to the creation of this book. Thanks to David Shablak of Digital Spider Design.com for the cover art. Also many thanks to Rob Young, Captain Richard Neal-USAF, Master Sergeant Grant Weaver-USAF, David Johnston, Dennis Desmond, and the United States Air Force.

Book design by Robert Biondi.
Front cover design by David Shablak.

Library of Congress Catalog Number: 2001090575.

Printed in China.
ISBN: 0-7643-1389-4

We are always looking for people to write books on new and related subjects. If you have an idea for a book, please contact us at the address below.

Published by Schiffer Publishing Ltd.
4880 Lower Valley Road
Atglen, PA 19310
Phone: (610) 593-1777
FAX: (610) 593-2002
E-mail: Schifferbk@aol.com.
Visit our web site at: www.schifferbooks.com
Please write for a free catalog.
This book may be purchased from the publisher.
Please include $3.95 postage.
Try your bookstore first.

In Europe, Schiffer books are distributed by:
Bushwood Books
6 Marksbury Ave.
Kew Gardens
Surrey TW9 4JF
England
Phone: 44 (0)208 392-8585
FAX: 44 (0)208 392-9876
E-mail: Bushwd@aol.com.
Free postage in the UK. Europe: air mail at cost.
Try your bookstore first.

Contents

Foreword

The Mikoyan MiG-29 (NATO-designated *Fulcrum A*) is acknowledged as the finest lightweight, multi-role fighter ever produced in Russia. It was designed to operate under wartime conditions and fight head-to-head with Western combat aircraft such as the U.S. F-16 Falcon, F/A-18 Hornet, French Mirage 2000, Rafale, and Swedish JAS-39 Grippen. First flown in 1977, the MiG-29 became operational in 1983. Roughly the size of the F/A-18 Hornet, the Mach 2.3 capable Fulcrum has also demonstrated many extraordinary capabilities and set several world records.

One of these records includes topping the Lockheed SR-71 Blackbird in altitude. The Russians have long considered high altitude flight very important. To demonstrate this, on April 26, 1995, a MiG-29, with standard RD-33 engines, was piloted by Mikoyan Chief Test Pilot Roman Taskaev to a ratified world altitude record of 27,460m (90,092 ft.). The Federation Aeronautique Internationale (FAI) record classification is for a non-payload, turbojet, land-based aircraft weighing between 12,000 and 16,000kg. and was previously held by the SR-71 at 25,929.031m (85,069 ft.) set on July 28, 1976. The absolute altitude record for an air-breathing aircraft is 37,650 m (123,524 ft.) set in 1977 by a Ye-155M – a MiG-25/31 research aircraft.

The MiG-29 Fulcrum A is presently in service with the Russian Air Force, as well as the air forces of Georgia, Ukraine, Moldova, India, Germany, Yugoslavia, Serbia, Iraq, Czech Republic, Slovakia, Syria, Poland, Malaysia, Hungary, and Yemen.

While economics and political realities have prevented the MiG-29 from becoming an even greater worldwide success, such as the legendary F-16 Falcon, the aircraft itself possesses all the necessary ingredients for such status. It is a remarkable aircraft, and one that NATO pilots are pleased to have never encountered during the Cold War.

This opinion became especially true following German reunification when former East German MiG-29's were integrated into the German Air Force. It was only then the MiG-29's deadly capabilities were fully revealed.

Quoting *Military Technology Magazine,* 1995, referring to the facing off of German Air Force MiG-29's and Dutch, U.S., and Belgian F-16's: "From the technological and, what matters more, the operational points of view, there can be no doubt whatsoever that the main lesson stemming from this Tactical Air Meet has been the final confirmation of the overwhelming superiority of the Russian air-to-air weapon system comprising the R-73 short-range missile interfaced with the helmet-mounted sight, as installed on the MiG-29. Mock F-16/MiG-29 dogfight after mock dogfight have shown that anyone venturing within visual range (3-5km) of an R-73/HMS fitted MiG-29, is basically mincemeat, in that as soon as the MiG-29 pilot can see the F-16, he can lock onto it and launch." Mock engagements with USAF F-15C fighters had similar results. As expected by Western analysts, at medium and long ranges, the weapons systems were more equal.

With a qualified pilot at the controls, the MiG-29 has proven to be a rugged and reliable fighter capable of exceptional combat performance. Upgraded with the Vympel R-73M short-range missile, which has a 90° seeker head and is capable of 40-50g turns, with a host of counter-countermeasures, the MiG-29 will continue to be a factor in future conflicts, both as ally or foe. Understanding this aircraft is, therefore, an interesting exercise in foresight.

Introduction

Modern fighter aircraft are marvels of technology. They intrigue us with their awesome performance, impressive weapon systems, and unimaginable costs. For decades, Russian aircraft have carried the added aura of Soviet secrecy; shadowy, mysterious, and dangerous. With the veil now lifted, there have been many impressive surprises. Perhaps none greater than the MiG-29 Fulcrum A. Looking beyond the specifications, and past the first-hand Western pilot impressions, however, what is it like to fly this complex combat machine? What are the Fulcrum pilots' instructions?

It is with this in mind that *MiG-29 Flight Manual: Declassified* was created – to provide a rare and unprecedented glimpse inside a top Russian aircraft. Using information that only a few years ago was highly classified, this translated text presents an extraordinary look at the capabilities and complexities of one of the very best fighters of the current age.

The text, illustrations, and tables in this book were taken directly from the actual Russian flight manual produced for the Mikoyan MiG-29 multirole fighter. Editing considerations were made regarding translation, and the level of technical detail. Many terms. phrases, and sentences where left in Russian sentence structure to provide "character" to the text. In addition, we used the Library of Congress method of translating the spelling. However, the overall presentation of the book took precedence over perfect English.

So, strap yourself into the KM-36DM ejection seat, put on your ZSh-7M or GSh-6AA helmet, and read carefully . . .

Section One

GENERAL INFORMATION ON AIRCRAFT

AIRCRAFT MiG-29 FLIGHT MANUAL

The present Flight Manual is effective for the MiG-29 aircraft of all series and contains the basic instructions to be obligatory followed by the pilot in operation and piloting technique of the aircraft as well as the recommendations on actions to be taken in case of emergency in flight. The Flight Manual comprises the following sections:

Introduction

Section 1. General Information on Aircraft
Section 2. Operating Limitations
Section 3. Preparation for Flight
Section 4. Performance of Flight
Section 5. Combat Employment
Section 6. Emergencies in Flight
Section 7. Flight Performance and Peculiarities of Aircraft
Section 8. Operation of Aircraft Systems and Equipment
Section 9. Amendments

In the present Flight Manual, everywhere, if not specified otherwise, the term "speed" implies the indicated airspeed, the term "altitude" implies the indicated altitude whereas the terms "Mach number", "angle of attack" and "gload" imply their values against the respective indicators.

PURPOSE OF AIRCRAFT

The MiG-29 aircraft is actually a lightweight frontline fighter intended for conducting air combat at the middle and short ranges. The aircraft may be also employed for attacking the ground (surface) targets of the enemy.

DESIGN FEATURES

Structurally, the MiG-29 aircraft has an integrated configuration.

The integrated configuration provides for a single lifting surface comprising the loadcarrying airframe smoothly coupled with the wing through the area of a heavily developed extension.

The aircraft structure consists of a profiled loadcarrying airframe developed in length and span, a tapered wing, horizontal and twinfin vertical tails.

The airframe is made in the form of the wing profile section with a low aspect ratio, whose outlines enclose cabin and engine nacelles. Such a construction of the airframe with use of highly swept extensions makes it possible to obtain the high lift qualities of the aircraft and to increase the net volumes, with the airframe and wing root relative thickness being small. The airframe comprises the centre wing section, the nose section with extensions, side and tail sections.

The wing is trapezoidal in plan, having the contour made irregular in space, with the wing leading edge swept through 42°. The andedral is minus 3°. The highlift devices include the deflectable leading edges, singleslotted flaps, and ailerons.

The horizontal tail is an allmovable stabilizer, whereas the vertical tail comprises two fins set at 6°, on which the rudders are secured.

The aircraft power plant consists of two engines (RD33) with common aircraft accessory gear box, two variable air intakes and two fullyvariable nozzles.

The aircraft general view and basic geometrical dimensions are shown in Figs 1 to 4.

BASIC SPECIFICATIONS

Maximum takeoff mass .... 18,000 kg
Maximum level flight speed:
near ground (V ind) .... 1500 km/h
at altitude (V true) .... 2400 km/h

Maximum levelflight Mach number 2.3
Service ceiling (with 800kg fuel remainder) 18,000m
Maximum gload 9.0
Maximum operational range of flight CXf = 0.785):
near ground (H = 200 m, M = 0.5) .... 1430 km
at altitude (H = 12,000 to 13,000 m, M = 0.8 without drop fuel
ferry range (during cruising climb flight with ventral drop fuel tank attached) .... 2100 km

Acceleration time at altitude of 1000 m:
from 600 to 1100 km/h 13.5s
from 1100 to 1300 km/h 8.7s
Takeoff run length .... 600 to 700 m
Landing roll length .... 650 to 750 m
Unstick speed .... 260 to 280 km/h
Landing speed .... 250 to 260 km/h

The aircraft may be operated from the concrete runway of the third class.

FLIGHT RANGE AND ENDURANCE CHARACTERISTICS

The flight range and endurance characteristics depend on the available fuel reserve, flight mass, airspeed and altitude. The MiG-29 aircraft reaches the maximum operational range of the flight at an altitude of 13,000 m and N = 0.8 and near the ground, at M = 0.5.

The aircraft capabilities in flight range and endurance may be calculated with use of Tables 1 to 3; in this case, the following should be subtracted from the total fuel reserve:

- the amount of fuel consumed during the engines operation on the ground (starting, runup, taxiing) 75 kg;
- the amount of fuel consumed at takeoff 50 kg;
- the 7% enroute fuel reserve 250 kg;
- the amount of fuel consumed for executing the approach:
- the visual circling approach 200 kg (4 min);
- the straightin approach from H = 5000 m 330 kg (12 min, 35 a);
- the approach from the estimated line from E = 2000 m 160 kg (5 min, 15 a);
- the approach with execution of two 1800 turns 355 kg (9 min, 25 a).;
- the approach with use of the navigation system and turn Point 135 kg (4 min, 15s).

The time, distance and amount of fuel consumed in climbing at the maximum power setting with allowance made for the acceleration of the airspeed up to V true = 900 km/h and the further climb at this airspeed (the aircraft is armed with two missiles R60 and two launchers API27) are indicated in Table 1.

The time, distance and amount of fuel consumed in gliding to an altitude of 500 m at the idle power setting and an airspeed of 500 km/h (the aircraft is armed with two missiles R60 and two launchers API27) are indicated in Table 2.

The fuel consumption per kilometer, when the aircraft is armed with two missiles R60 and two launchers ARI27, with the aircraft mean mass being 13,000 kg, are indicated in Table 3.

Table 1

Altitude, m	Time, min-s	Distance, km	Fuel consumption, kg
1000	1-00	10	115
3000	1-30	15	170
5000	2-00	20	230
7000	2-30	30	280
9000	4-00	50	340
11,000	5-45	75	420
13,000	11-30	160	600

Table 2

Altitude, m	Time, min-s	Distance, km	Fuel consumption, kg
13,00	8-00	100	120
11,000	6-35	80	100
9000	5-25	60	80
7000	4-25	50	60
5000	3-00	35	40
3000	1-50	15	20
1000	0-20	5	5

Table 3

Altitude, m	Mach number				
	0.5	0.6	0.7	0.8	0.9
0	4.33	4.77	5.38	6.11	7.6
1000	3.95	4.32	4.79	5.46	6.76
3000	3.35	3.47	3.8	4.29	5.27
5000	2.96	2.88	3.05	3.39	4.11
7000	2.69	2.43	2.46	2.64	3.14
9000	2.93	2.25	2.12	2.18	2.53
11,000	-	2.36	1.93	1.84	2.06
13,000	-	-	-	1.75	1.82

Section Two

OPERATING LIMITATIONS

LIMITATIONS IN AIRSPEED, MACH NUMBER AND GLOADS

The maximum permissible airspeeds, Mach numbers and gloads are given in Table 4 (for or the symmetrical external stores), and the other limitations, in Table 5.

LIMITATIONS IN AIRSPEED, MACH NUMBER AND G-LOADS

The maximum permissible airspeeds, Mach numbers and g-loads are given in Table 4 (for the symmetrical external stores), and the other limitations, in Table 5.

Table 4

Parameter	Without external stores and with missiles	With empty ventral drop fuel tank	With ventral drop fuel tank filled partially or to full capacity	With rockets and 4 aerial bombs АБСП	With small load pack КМГУ
Airspeed, km/h	1500	1000	1000		900 at H = 5000 m 1100 at H = 5000 m
Mach number	2.35	0.9	0.9	0.8	1.5
G-load at m_{acft}, kg:	14,200	14,200	18,480	Takeoff	
at M < 0.85	+9.0 to -3.0	+9.0 to -3.0		+5.0 to -2.5	+5.0 to -1.0
at M > 0.85	+7.0 to -1.5	+7.0 to -1.5			
at M < 0.9			+4.0 to -1.5		

CAUTION

1. WHEN THE MASSES EXCEED INDICATED VALUES, THE GLOADS ARE DECREASED BY 1.0.

2. PROCEEDING FROM THE STABLE OPERATION OF THE POWER PLANT, DO NOT CREATE THE NEGATIVE GLOADS AT MACH NUMBERS EXCEEDING 2.0.

3. PROCEEDING FROM THE STRENGTH OF THE VERTICAL TAIL, NEVER OVERRIDE THE PEDALS (RIGID SPRING) ADDITIONAL ARTIFICIAL FEEL AT MACH NUMBERS EXCEEDING 0.85. THE TIME OF DEFLECTION TO FULL TRAVEL AT MACH NUMBERS BELOW 0.85 AND TO THE RIGID SPRING AT MACH NUMBERS EXCEEDING 0.85 SHOULD BE 1 S MINIMUM.

Table 5

No	Limitation	Reason for limitation
1.	Aircraft Maximum takeoff mass is 18,480 kg	Landing gear strength considerations
2.	Maximum permissible airspeed during retraction, extension of landing gear and flight with extended landing gear is 700 km/h (at airspeed of 500 to 700 km/h, slipping should not exceed two diameters of ball)	Landing gear panel strength considerations
	Flight with wing leading edges extended, at ≤ 0.95	Wing leading edge strength considerations

Table

No.	Limitation	Reason for limitation
4	Maximum angle of attack of aircraft carrying no external stores, but missiles, is limited by setting stall limitation system to 26°, with wing leading edges extended and to 15°, with wing leading edges retracted Maximum permissible angle of attack of aircraft with any variants of asymmetrical external stores (except for one missile P-60 or aerial bomb АБСП-100), irrespective of wing leading edge position, is 13° Maximum angle of attack of aircraft carrying aerial bombs АБСП, small load pack КМГУ with wing leading edges extended, is 19° Maximum angle of attack of aircraft carrying aerial bombs АБСП, small load pack КМГУ with wing leading edges retracted, carrying rockets and drop fuel tanks, irrespective of wing leading edges position, with failed leading edges and disengaged (failed) automatic flight control system, is 13° Maximum permissible angle of attack of aircraft with all armament variants, set to takeoff-landing configuration, is 15° Maximum permissible angle of attack of aircraft carrying no external stores and carrying missiles, with wing leading edges extended and one engine running, is 19°	Stability and controllab__ity considerations
5	Takeoff and landing of aircraft is cleared provided cross-wind component does not exceed 15 m/s	Proceeding from rudder effectiveness to eliminate turn during slipping, and landing gear strength considerations
6	Unstick ground speed should not exceed 375 km/h, landing speed should not exceed 330 km/h	Landing gear wheel strength considerations
7	Normal landing mass of aircraft is 14,200 kg. Limit landing mass of aircraft is 15,760 kg. Number of landings of aircraft weighing 14,200 to 15,760 kg should not exceed 3 % of total number of landings. After landing of aircraft weighing 14,200 to 15,760 kg, perform inspection and check retraction of landing gear	
8	Maximum permissible airspeed during deployment of brake chute is 310 km/h	Brake chute attachment strength considerations
9	Wheel brake application initiation speed during landing roll without employment of brake chute is not more than 215 km/h. If necessary, it is allowed to apply brakes at any speed	Brake kinetic energy considerations
	Power Plant	
10	Maximum permissible exhaust gas temperature: - up to value marked with yellow notch on exhaust gas temperature indicator during engine starting on ground and in flight; - 500 °C, with engines running on ground at IDLE; - up to value marked with red sector on exhaust gas temperature indicator at all steady power settings of engine; - not more than 3 s within red sector during acceleration up to maximum and reheat power settings, engagement and disengagement of afterburner.	Turbine blade strength and reliable cooling considerations

Table

No.	Limitation	Reason for limitation
	CAUTION. OPERATION OF ENGINE AT POWER SETTING WITHIN RED SECTOR OF EXHAUST GAS TEMPERATURE INDICATOR IS PROHIBITED	
11	Maximum speed of high-pressure rotor should not exceed 103 % at maximum and reheat power settings	Turbine blade strength considerations
12	Difference in engine high-pressure rotor speed at maximum and reheat power settings should not exceed 4 %	Engine individual peculiarities considerations (on set-up and adjustment)
13	Short-time (up to 3 s) rise (drop) of high-pressure rotor speed during acceleration, throttling, engagement and disengagement of afterburner - up to 3 %	Engine strength and stability considerations
14	Duration of close-to-zero (±0.2) g-loads should not exceed 10 s at maximum power setting and 5 s at reheat power setting and duration of negative g-loads should not exceed 15 s at maximum power setting and 5 s at reheat power setting. Zero g-loads (weightlessness) should not exceed 2 s in duration as transient ones. Negative and close-to-zero g-loads may repeat not more than five times during flight at interval of at least 30 s	To ensure continuous supply of fuel to engine and to restore oil pressure in engine and aircraft accessory gear box
15	Time of engine continuous operation: - in flight at all power settings - without limitations; - on ground: at idle power setting - not more than 15 min, at maximum and reheat power settings - not more than 10 min; at other power settings - without limitations.	Turbine blade reliable cooling considerations
16	Operation of engines at reheat power setting when 550 KG REMAIN (ОСТАЛОСЬ 550 КГ) signal lights up IS PROHIBITED	To ensure fuel supply to engines
17	When air intake upper inlet is used, flight at M > 0.8 is prohibited	Power plant stability considerations
	Life-Support, Emergency Escape and Rescue Equipment	
18	Safe emergency escape of pilot from aircraft is ensured: (a) during takeoff run and landing roll at airspeed of at least 75 km/h; (b) in level flight: at airspeed of not more than 950 km/h without limitations in altitude; at airspeed from 950 km/h to airspeed not exceeding 1200 km/h at altitudes not below 40 m over terrain; at airspeed from 1200 to 1300 km/h at altitudes not below 80 m over terrain;	To ensure time required for operation of parachute system elements and effect of airstream on pilot considerations

Section Three

PREPARATION FOR FLIGHT

1. Before the flight, the pilot should receive the report from the aircraft technician on the aircraft readiness (fueling, charging) for flight in compliance with the flight mission assigned, on the last operations performed aboard the aircraft, on the magnetic declination set on the selector, on departure airfield latitude set on the latitude updating panel, on type of the navigation computer, on programs introduced into the navigation computer and number of the program selected for the flight.

2. During visual inspection of the aircraft, check to see that:

• the blanking and slip covers are removed;
• the airframe, landing gear and antenna radomes are free from external damage;
• the maintenance hatch doors are closed;
no traces of leakage of hydraulic mixture and fuel are available on the outer surface;
• the external stores are suspended in compliance with the flight mission assigned and make certain that they are free from external damage;
• the chocks are reliably installed under the main landing gear wheels.

PILOT'S ACTIONS BEFORE CLIMBING INTO CABIN

3. Before climbing into the cabin, the pilot should give the aircraft technician a command to remove all the safety pins, except for the safety pins installed on the ejection handle and canopy emergency jettison handle and make certain that:

• no foreign objects are available in the cabin;

CAUTION. IT IS PROHIBITED TO PLACE THE NONSTANDARD PADS AND OTHER OBJECTS ON THE SEAT.

the ejection handle is placed in the initial position and the ground safety pin is installed therein (on aircraft equipped with a broken ejection handle the ground safety pin is installed therein; the ejection handle is thrown forward);

• the foot restraint lines are secured in the springloaded holders on the instrument board;
• the safety pins are installed in the canopy emergency jettison handle;
• the movable canopy interlock line is coupled with the snap book of the seat interlocking mechanism and locked.

PILOT'S ACTIONS AFTER CLIMBING INTO CABIN

4. After climbing into the cabin, it is necessary to:

• put on and tightly fit the parachute harness system; perform the above actions independently or with the aid of a technician;
• couple the special outfit connectors with the seat;
• when the life belt is used, make certain that the straps of the parachute harness system do not cover (overlap) the floats and inflation plugs;
• check the waist harness restraint for proper functioning with the aid of the WAIST RESTRAINT handle, tightly become pressed against the seat back and lock the waist harness; the adjusting buckles of the leg straps should be between the body and the ejection handle. Do not release the waist harness during the entire flight.

• check the shoulder harness for proper functioning with the aid of the SHOULDER LOCK) handle having placed the latter to the STRAPS UNLOCKED position.

5. Make certain that the equipment and system controls are set to the initial position (from left to right):

• the head of the automatic pressure control unit is set to the MIN position when antig suit PKK-1 is used in flight or to the MAX position when the highaltitude partial-pressure suit is used in flight;
• the pilot's oxygen supply valve is open;
• the HOT COLD knob controlling the temperature in the ventilated suit is placed in the middle position;
• the SUIT VENT knob is placed in the OPEN position when the ventilated suit is used in flight;
• the additional oxygen feed knob on remote control unit DI7 is set to the MIX position, whereas the emergency oxygen feed and helmet ventilation knobs are set to the OFF position;
• the cabin emergency depressurization handle is set to the PRESS position and locked;
• the APY (automatic transmission ratio controller) selector switch is set to the AUTO position and locked;
• the WSM, EMERG OFF switch is turned off and locked;
• the EMERG PUMP UNIT switch is turned off and locked;

On the radio set control panel, the operating channel is selected, the volume control knob is fully turned clockwise, the emergency reception signal monitoring switches are turned off and noise suppressor (US) is cut in;

• the throttle control levers are set to the STOP position;
• the REHEAT EMERG OFF switch is set to the REHEAT position and locked;
• the AIR RELIGHT switches are turned off and locked;
• the GEN DRIVE EMERG OFF switch is turned off and locked;
• the RAMP EMERG RETR selector switches are locked in the neutral position;
• the SHUTOFF VALVES selector switches are set to the OPEN position and locked;
• the EXITING selector switch is set to the neutral position and locked;
• the EMERG RELEASE button is capped;
• the OUTER INNER selector switch is set in the OUTER position;
• the LTS selector switch is placed in the OFF position;
• the LG handle is placed in the EXTENDED position;
• the MASTER OFF switch is in the OFF position; EMERG LG handle is sunk (forward) all • the way home and locked;
• the SET COURSE selector switch is set to the AUTO position;
• the NOSE WHEEL BRAKE lever is placed in the ON position;
• the clock reads the exact time and is prepared for the assigned flight mission;
• the shutters (flags) on the ramp position indicator are set to the 0% marks (divisions);
• the cabin air temperature selector reads 15 to 20°C;
• the PITOT (MM) lever is set to the MAIN position;
• the CABIN HEAT selector switch is set to the AUTO position;
• the PITOT GLASS HEAT switch is turned off; on the navigation system control panel, • the CHANNELS RB WPT AFLD, COURSE selector switches are set to the MAIN, AUTO SHORAN, AFLD positions, respectively, in compliance with the take off course, and the LANDING switch is turned off; the OPERATION PREPARATION

• selector switch located on the right console is set to the PREPARATION position;
• the function selector switch located on the CO69 ATC transponder is set to the position corresponding to the assigned mission;
• the AM FM selector switch is set to the AM position and locked;

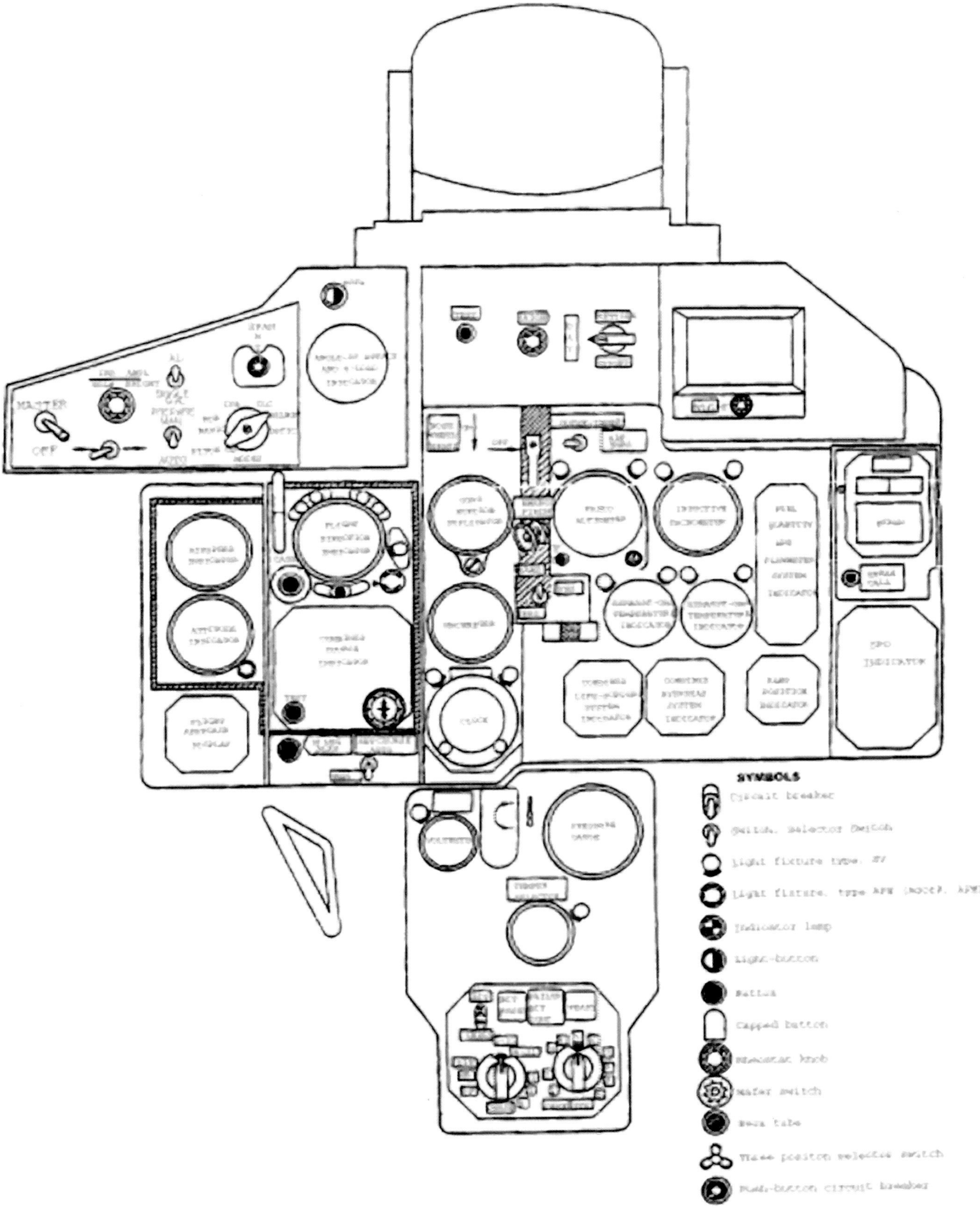
MASTER
OFF
AUTO
RPM
INDICATOR
SYMBOLS
Circuit breaker
Switch, Selector Switch
Indicator lamp
Light-button
Capped button
Wafer switch
Three position selector switch
Push-button circuit breaker

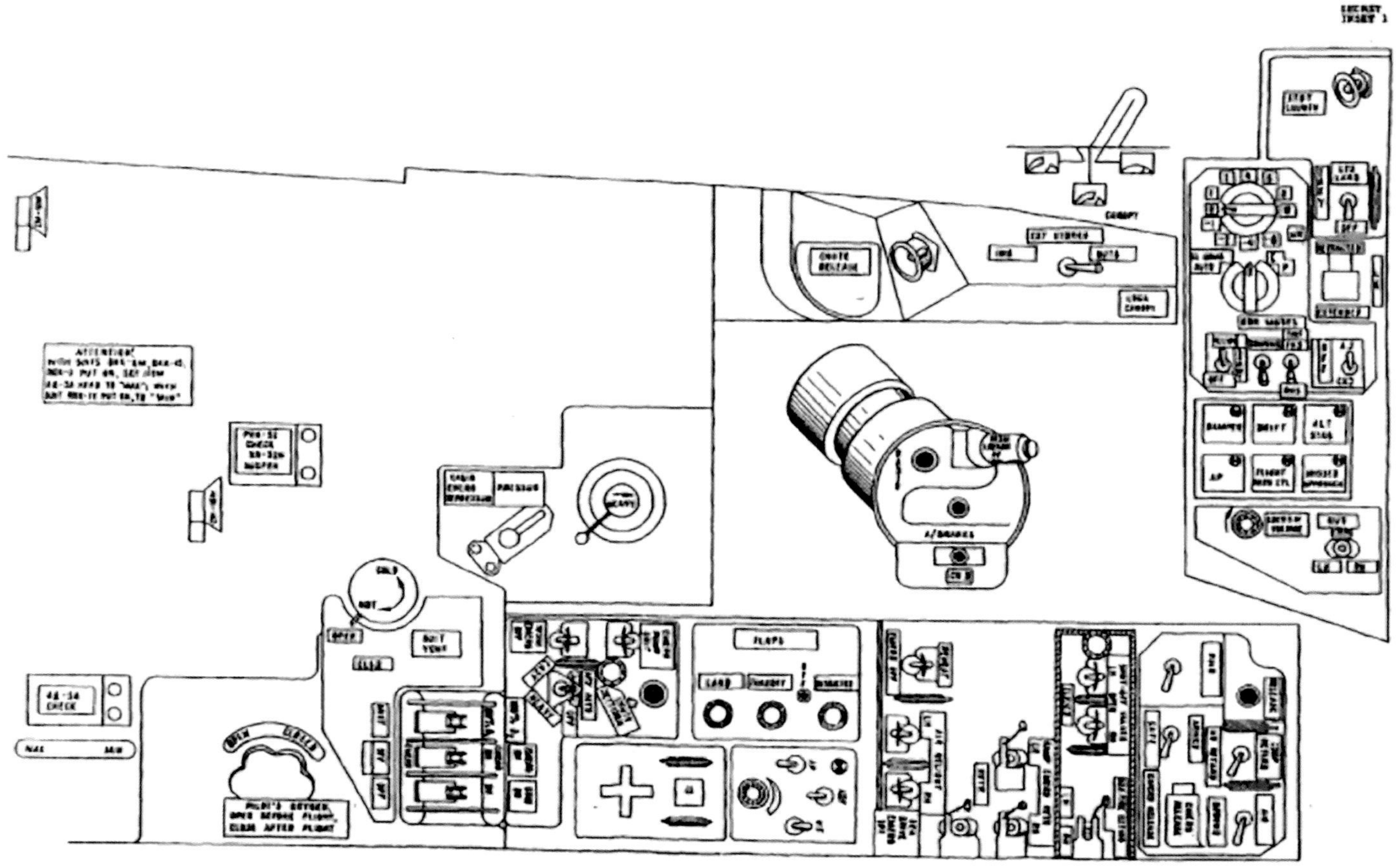

On the radar illumination warning station panel: the VOLUME knob is fully turned clockwise, the SCAN OFF selector switch is set to the SCAN position, the BRIGHTNESS switch is turned off;

- the COOL valve is set to the CANOPY position;
- on unit *INN,* set the pilot's call sign number;

On the automatic direction finder panel: the departure airfield outer homing radio station channel number is selected with the aid of the ADF CHANNELS selector switch, the VOICE CW and ADF ANT selector switches are set to the VOICE and ADF positions, respectively, the VOLUME knob is fully turned clockwise;

- on the IFF transponder panels: the DISTRESS switch is turned off and locked, the code selector switch is set to the assigned code position;
- the necessary radio data are set with the aid of the wafer switches on the guidance equipment control panel;
- the PWR SYS START selector switch is set to the START position and clamped;
- the START selector switch is set in the START TWO position;
- the CABIN SUPPLY valve is set to the OPEN position;
- all other switches located on the right vertical and horizontal control panels are turned off.

The arrangement of the controls, instruments and annunciators in the cabin is shown in Figs 1 to 3 (pages 17-19).

CHECK OF AIRCRAFT SYSTEMS AND EQUIPMENT

6. To avoid discharge of the airborne storage batteries, check the aircraft systems and equipment for proper functioning only with the use of the ground AC and DC power supply sources. During autonomous starting of the engines, check the systems for proper functioning after starting the engines.

Check the airborne storage batteries under load for voltage, for which purpose turn on the BAT GND SUP FUEL PUMP, INV switches. The voltage measured against the voltmeter should be at least 22 V. Turn off the switches in the reverse order.

Give the flight technician a command to connect the ground power supply sources and turn on the BAT GND SUP switch. The voltage should be within 27.5 to 28.5 V when the airfield power unit is operative or 22 V when said unit is inoperative.

Note. If the KSTs critical mode warning lightbutton flickers when the BAT GND SUP switch is turned on, press this lightbutton to turn off the latter.

7. After connecting the power supply source, make certain that:

- the seat is adjusted to your height (adjust the seat to your stature, if required);
- the flight indication system (annunciators and lamps) is serviceable by depressing the LAMP TEST button;
- the amount of oxygen reaches (100+10)% against the IKSh combined lifesupport system indicator;

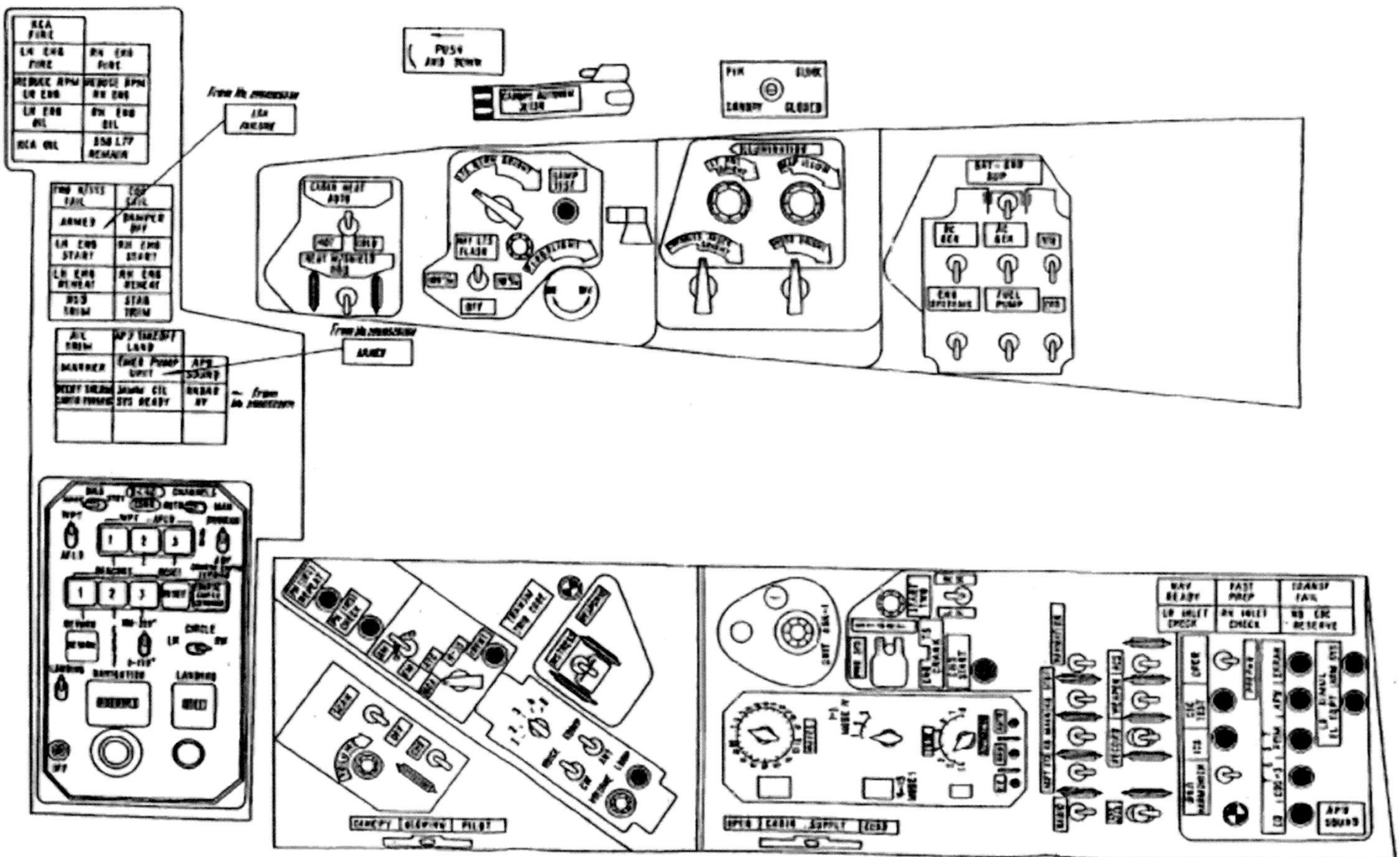

CABIN RIGHT-HAND CONSOLE

- the amount of fuel against the ISTR indicator is in compliance with the fuelling;
- the pressure in the main and emergency air systems is within the values corresponding to the green sectors on the IKG combined hydrogas system indicator;
- the pressure in the hydraulic systems is at division vacuum of the IKD combined hydrogas system indicator.

8. Check the braking system for proper functioning. When the brake lever is fully pressed, the pressure in the system should be within 6 to 8 kgf/cm 2 and when the lineup brake is applied, the pressure should be within 10 to 12 kgf/cm. When the brake lever is released, the pressure should drop to zero.

9. Check the oxygen equipment set for proper functioning in the following sequence:

- put on the oxygen mask;
- make two or three deep inhalations and exhale; breathing should be easy and the combined life-support system indicator index should be motionless or move insignificantly; set the additional oxygen feed knob to the 100% 02 position and make two or three deep inhalations and exhalations. The oxygen flow indicator index should move up during exhalation and move down during inhalation;
- set the additional oxygen feed knob to the MIX position;
- set the emergency oxygen feed knob to the ON position; in this case, the oxygen is continuously flowing to the oxygen mask;
- disconnect the emergency oxygen feed line; in this case, the oxygen should not be fed to the oxygen mask;
- set the helmet ventilation unit valve handle to the ON position and make certain that the helmet ventilation unit functions normally, which is indicated by continuous flow of oxygen to the oxygen mask;
- switch off the helmet ventilation unit and make certain that no oxygen is applied there.
- Perform the functional check of the oxygen equipment set under the excessive pressure before flying to an altitude over 12,000 m, only if the highaltitude pressure suit is put on and connected, in the following sequence:

  - set the additional and emergency oxygen feed knobs to the 100% 02 and ON positions, respectively, and make certain that the oxygen pressure does not rise in the oxygen mask for 15 to 20 s against the pressure differential scale on the IKSh combined lifesupport system indicator;
  - depress (to be done by the technician) the oxygen continuous feed manual initiation button on the oxygen demand regulator and keep this button pressed throughout the check period;
  - after a considerable amount of oxygen has been supplied (the pressure) exerted by the suit on the pilot's body starts decreasing), stop the hole on the oxygen demand regulator and create the excessive pressure of oxygen in the mask up to 0.05 kgf/cm against the AP cab scale of the IKSh combined lifesupport system indicator;

- make several inhalations and exhalations; if the oxygen feed index moves down during inhalation and returns to the initial position during exhalation, and the highaltitude pressure suit snugly fits the body (without any painful perceptions), the oxygen set is considered to be serviceable;
- release the oxygen continuous feed manual initiation button and open the hole on the oxygen demand regulator;
- set the additional and emergency oxygen feed knobs to the MIX and OFF positions, respectively.

The pressure rise in the oxygen mask when the hole on the oxygen demand regulator is open or absence of the pressure in the highaltitude partial-pressure suit, with the pressure available in the oxygen mask, as well as the failure to create an excessive pressure of 0.05 kgf/cm 2 in the oxygen mask with the hole closed, is indicative of the oxygen equipment set failure.

In this case, the flight is Prohibited.

11. Make sure that the OPERATION PREPARATION selector switch is set to the PREPARATION position. Turn on the RADIO, ACFT SYST, KB MAIN, KB STBY. NAVIGATION switches.

Make sure that the navigation system is engaged referring to illumination of light buttons No. I in the WPT AFLD and BEACONS rows, disappearance of the KC (compass system) flag on the combined course indicator and dead state of the CAGE lamp on the flight director indicator.

Perform the fast or navigational preparation of the navigation system.

In the process of the fast preparation, proceed as follows:

- make sure that the WPT AFLD selector switch is set to the AFLD position and depress the lightbutton in the WPT AFLD row, corresponding to the number of the departure airfield (if this lightbutton is dead);
- 30 to 40 s after turning on the KB MAIN and KB STBY switches, simultaneously depress and keep depressed for 10 to 15 s the M HDG SLAVE button with the magnetic declination value set on the magnetic declination selector or the ICS (initial course setting) button with the aircraft ramp heading set on the selector and the COURSE CMPTR/ZEROING lightbutton.

Note. Upon depression of the ICS button, do not fail to set the departure airfield magnetic heading on the magnetic declination selector and depress the M HDG

SLAVE button for 10 to 15 s.

Check follow-up of the aircraft ramp heading by the combined course indicator; set the OPERATION PREPARATION selector switch to the OPERATION position; on a lapse of 3 min, check changeover of the navigation system to the operating conditions ref erring to lightup of the FAST PREPARE annunciator, ground static angles in roll and pitch read off the flight director indicator, desired heading and selected airfield (WPT) range read off the combined course indicator.

Whenever necessary, the navigation system fast preparation may be performed after start of one engine (the second engine should be started upon the FAST PREPARE annunciator illumination) or after start of two engines.

Taxiing out may be executed after illumination of the FAST PREPARE annunciator.

Note. The FAST PREPARE annunciator goes dead at takeoff.

As soon as the navigation system navigational preparation is over, the OPERATION PREPARATION selector switch should be set to the OPERATION position upon illumination of the NAV READY annunciator, but not earlier than 15 min after turningon of the KB MAIN and KB STBY switches; as a result, the NAV READY annunciator extinguishes.

If the navigational preparation is performed by the ground crew, the OPERATION PREPARATION selector switch should be thrown to the OPERATION position by the pilot before the engine start.

On the aircraft equipped with the inertial direct directional/vertical gyro systems updated by means of the initial setting filters, the navigational preparation may be performed with the engine (engines) started; in this case, the NAVIGATION, KB MAIN, KB STBY switches should. be turned on after setting of the engine (engines) at IDLE. When the preparation is executed with one engine started, the second engine should be started upon throwing of the OPERATION PREPARATION selector switch to the OPERATION position.

To specify the aircraft t ramp heading by *reference to the landmark 'in tie process of the fast or navigational preparation of the navigation system, before shifting the OPERATION PREPARATION selector switch to the OPERATION position, do not fail to:

- turn on the ACS switch (during the fast preparation the NAVIGATION and ACS switches should be turned on 2 to 3 min before turning on the KB MAIN, KB STBY switches);
- depress the lightbutton (1 to 3) in the WPT AFLD row or the RESET button corresponding to the selected landmark number;
- as soon an the integrated display system is warmed up and the windshield indicator shows some Indication, manipulate the control button to align the movable reticule with the landmark image;
- depress the WSM, LOCK ON PF button;
- check specification of the heading by the combined course indicator.

Notes:

1. If the aircraft are equipped with computers TsVI and TsVI-M, prior to specifying the heading by reference to the landmark, set the WPT AFLD selector switch to the AFLD position.

2. Specification of the aircraft ramp heading by reference to the landmark is performed to ensure the assigned accuracy of the selfcontained deadreckoning modes.

In the process of the aircraft preparation in the metalconcrete shelters on the airfield programmed as AFLD 1, the ramp heading should be introduced by the ground crew. In this case, simultaneous depression of the 9 HDG SLAVE button and COURSE CMPTR/ZEROING lightbutton on a lapse of 30 to 40 s after turning on the KB MAIN and KB STBY switches should not be performed.

CAUTION. FLIGHTS FROM THE UNPROGRAMMED AIRFIELD CAN BE EXECUTED ONLY DURING THE NAVIGATION SYSTEM FAST PREPARATION AM WITH THE "CHANNELS" SELECTOR SWITCH SET TO THE "MANUAL" POSITION.

Drop of the mains voltage below 22 V after accomplishment of the navigation system preparation can result in shorttime appearance of the CS, DCA, DKE flags on the combined course indicator, extinguishing of the lightbuttons on the control panel and FAST PREPARE annunciator, changeover of the AFLD 2 or AFLD 3 display to the AFLD I display and illumination of the NAVIG COUP signal on the EKRAN display system during starting of the engines, cutout of the ground power supply source and cutin of the highpower loads. The system serviceability becomes restored as soon as the mains voltage reaches a value exceeding 22 V. Changeover to the display relative to the AFLD 2 or AFLD 3, if necessary, is effected through depression of the respective lightbuttons in the WPT AFLD row.

If the aircraft are equipped with computers TsVI, the flight from the airfield programmed as AFLD 2 or AFLD 3 can be executed only during the fast preparation. In this case, perform preparation proceeding as follows:

• set a grid magnetic declination relative the true meridian of the airfield programmed as AFLD I on the magnetic declination selector;
• on a lapse of 30 to 40 s after turning on the KB MAIN and KB STBY switches, simultaneously depress and keep depressed the M HDG SLAVE button and COURSE CMPTR/ZEROING light button until the ‘ combined course indicator follows up the reduced ramp heading relative to the true meridian of the airfield programmed as AFLD 1;
• set the OPERATION PREPARATION selector switch to the OPERATION position;
• set the departure airfield magnetic declination on the magnetic declination selector;
• as soon as the FAST PREPARE annunciator lights up, set the WPT AFLD selector switch to the AFLD position and, while the AFLD I lightbutton is on, check illumination of the D < 40 Annunciator and followup of the 0±2.5.km range by the combined course indicator range counter;
• upon starting the engines and cutting out the ground power supply source, depress the COURSE CMPTR/ZEROING lightbutton, and, while keeping it depressed, push the light button with the departure airfield number for 1 to 2 s;
• upon releasing the buttons, depress the M HDG SLAVE button for 10 to 15 s;
• make sure that the D < 40 annunciator lights up and that the combined course indicator follows up the 0+2.5km range,
• refer to the combined course indicator to check the value of the ramp heading relative to the departure airfield true meridian supplied by the inertial directional/vertical gyro system, having set the CHANNELS selector switch to the MAN and by the navigation computer, having set the selector switch to the AUTO position;
• make sure that the preset course and range values to the programmed airfields and way points are followed up correctly.

Further in the text, the sequence of the Pilot's actions to be performed is specified for the navigation system completed with computer TsVIM. The text also presents the features which make the navigation system operation different when completed with computers TsVI and TsVI-M.

12. Perform functional check of:

(a) the automatic direction finder by monitoring the call signs and by referring to the reading of the radio station relative bearing pointer of the combined course indicator on the outer homing radio station and inner homing radio station with the OUTER INNER selector switch set to the respective positions;

(b) functioning of the turn indicator of instrument DA200; when some portion of the instrument board is depressed, the pointer should deflect to the opposite side; when it is released, the pointer should return to the zero position;

(c) set an altitude of 0m on the altitude indicator; in this case, the pressure read by the indicator may differ from that at the runway level by no more than 2mm Hg at the reported atmospheric pressure at the airfield level equal to 760mm Hg and exceeding this value and by no more than 3mm Hg at the reported atmospheric pressure at the airfield level below 760 mm Hg;

(d) check the Mach number and true airspeed readings; the Mach number should range from 0.2 to 0.29, and the true airspeed, from 200 to 350 km/h;

(e) functioning of the EKRAN system by successive illumination of the SELFTEST EKRAN READY annunciators 15 s after depression of the EKRAN CALL button.

Note. when the navigation system is not serviceable, the FAIL annunciator illuminates and the EKRAN FAIL frame is displayed or the SELFTEST and MAN READY frames are displayed in the distorted form (or no printing takes place);

(f) the standby inertial direction/vertical reference system by reference to the readings of the combined course and flight director indicators, with the IVG selector switch set to the STBY position;

(g) the IFF equipment by illumination of the TRANSM TRIG CODE annunciator and by periodical illumination of the RESPONSE lamp;

(h) the aircraft transponder by illumination of the TRANSPONDER (CO) lamp when the TRANSPONDER CHECK button is pressed;

(i) the communication radio set on the required channels;

(j) the airborne command radio link equipment and warning station, if necessary.

ENGINE STARTING

13. It is allowed to start the engines both from the ground and airborne power supply sources.

Prior to starting the engines, proceed as follows:

give a command to the aircraft technician to close the canopy. Following this command, the technician removes the ground safety pins and shows them to the pilot;

set the ejection handle to the flight position, for which purpose, depress the side triggers on the handle and shift the handle to the vertical position;

check the canopy for proper closing by reference to the canopy closed position match marks alignment provided on the placard and on the canopy handle, by fadeout of the LOCK CANOPY annunciator, and after starting the engines, by absence of the LOCK CANOPY signal in the EKRAN system and the "Lock canopy" voice information when the IRS REPEAT button is depressed. On the aircraft provided with the canopy lock closed position mechanical indicator, canopy closing is additionally checked by reference to smooth retraction of the mechanical indicator into the subcanopy panel;

check for ease of the throttle control lever travel, check the throttle control levers for locking in the IDLE, MAX MIN REHEAT and FULL REHEAT (30 positions, and set the throttle control levers to the STOP position;

on receiving the engine starting permission, give the "Get ready for starting the engine" command to the aircraft technician over the aircraft intercom system and on receiving the technician's "Ready for starting" reply, proceed to starting the engines.

14. The main method of starting the engines is starting the engines when the START selector switch is set to the START TWO position.

For executing this starting, proceed as follows:

turn on the DC GEN, AC GEN, AC INV, ENG SYSTEMS, FUEL PUMP, ASS, RECORD switches;
turn off the RADIO switch;
set both the throttle control levers to the IDLE position;
depress the GND START button.

The engines (the RH engine first) should automatically accelerate to the idle power and the LH ENG START and RH ENG START annunciators should extinguish.

The idle speed amounts to 58 to 72% at the ambient air temperature of +40°C.

Perform the separate starting in the LH RH engine sequence.

CAUTION. IT IS PROHIBITED TO DEPRESS REPEATEDLY THE "GND START" BUTTON DURING THE ENGINE STARTING AND WHEN THE ENGINES ARE RUNNING.

15. For starting the engine, set the START selector switch to the LH ENG position and the LH engine throttle control lever to the IDLE position and depress the START button. Start the RH engine not earlier than 40 s after the LH engine has accelerated to the idle power.

16. In the process of the engine starting, check:

illumination of the LH ENG START (RH ENG START) lamp;
rise in the engine speed and temperature;
pressure in the main and booster hydraulic systems;
closing of the axial entry (inlet) by reference to shift of the shutters of the front panels to the TO position on the ramp position indicator and by absence of the LEFT (RIGHT) AIR INTK signals on the EKRAN annunciator;

Note. When the front shutter is shifted from "0% to the TO position, the LH (RH) INLET annunciator may flicker;

- absence of the START TURB CRIT CONDITION signal;
- fadeout of the LH ENG START (RH ENG START) lamps when the engine speed gains 34 to 36%.

17. Discontinue the starting by shifting the throttle control lever to the STOP position the following cases:

when the LH ENG START (RH ENG START) lamp remains dead;
when the START TURB CRIT CONDITIONS, KCA VIBRATION , LH (RH) VIBRATION, KCA OIL or LH (RH) ENG 'OIL signal lights UP;

- when the engine speed and exhaust gas temperature do not increase on a lapse of 10 s after depressing the GND START button;
- in case of the engine speed holdup at the exhaust gas temperature increase;
- when the exhaust gas temperature exceeds the tolerable value;
- during simultaneous acceleration of the rotors of both engines.

It is allowed to restart the engines after finding out and eliminating the cause of failed starting.

18. When the engine starting is discontinued because of the fuel ignition failure, before restarting the engine, crank the latter in the following sequence:

- make certain that the throttle control levers are set to the STOP position and the engine rotors are motionless;
- set the PWR SYS .START EKG CRANK TS CRANK selector switch to the ENG CRANK position, the START selector switch to the LH ENG (RH ENG) position;
- press the GND START button.

While cranking, the highpressure rotor speed should be at least 20%.
The interval between the successive startings or crankings should be at least three minutes.

CHECK OF AIRCRAFT SYSTEMS WITH ENGINES RUNNING

19. After one of the engines has accelerated to the idle speed, check the pressure in the main and booster hydraulic systems, the functioning of the trim actuators as follows below:

(a) for checking the longitudinal control, shift the stabilizer trim actuator slider first forward till the STAB annunciator extinguishes and then, backward till the STAB TRIM annunciator illuminates. The relieved control stick should deflect in the direction of the slider displacement;

(b) for checking the lateral control, shift the roll trim actuator to the right till the AIL TRIM annunciator extinguishes and then, to the left till the AIL TRIM annunciator illuminates. The relieved control stick should deflect in the direction of the slider displacement;

(c) for checking the functioning of the rudder, shift the RUD TRIM PTO selector switch to the right till the RUD TRIM annunciator extinguishes and then, to the left till the RUD TRIM annunciator illuminates. The relieved pedals should move in the direction of the selector switch shift.

The check over, give the command to the aircraft technician to check the neutral position of the stabilizer, ailerons and rudders.

CAUTION. TO AVOID THE ONSET OF THE ADDITIONAL FORCES ON THE CONTROL STICK AIM PEDALS AT TAKEOFF, IT IS NECESSARY TO SET THE TRIM ACTUATOR TO THE NEUTRAL POSITION IN THE DIRECTIONS MENTIONED ABOVE.

20. Turn on the AFCS switch and check the automatic flight control system for proper functioning. After turning on the AFCS switch, the builtin testcheck circuit is actuated. During the check, the aircraft controls are deflected and the DAMP lightbutton flickers on the automatic flight control system panel.

When the automatic flight control system is serviceable, on a lapse of 1.5 to 2 minute DAMPER mode is automatically selected. The DAMPER OFF annunciator goes out. Set the trim actuator knob to the neutral position. If the automatic flight control system failure is detected, the testcheck ceases and all the lightbutton except for the MISSED APPROACH lightbutton, flicker.

To start the testcheck repeatedly, set the control stick to the neutral position and depress the AFCS MODE OFF red button on the control stick for a short period of time.

Notes: 1. During the testcheck the aircraft controls should be relieved from forces.

2. During the testcheck all the lightbuttons except for the MISSED APPROACH lightbutton on the automatic flight control system panel may illuminate for a short period of time. In this case, the testcheck does not cease.

21. After the both engines have accelerated to the idle speed, give the command to the aircraft technician to turn off the ground power supply, check the generators for being put into operation the DC GEN and AC GEN signals should be displayed on the EKRAN annunciator, whereas the aircraft electrical system voltage should be within 28 to 29 V as read by the voltmeter. On the aircraft carrying the drop fuel tank, with the engines running at a speed less than 85 %, the information reporting and EKRAN systems may deliver the information on the absence of the fuel consumption from the drop fuel tank. After starting the second engine and turning off the ground power supply, cut out the ASS switch located on the right vertical panel and then, turn on this switch again. During the internal start-

ing, after the second engine has accelerated to the idle speed, perform the same operation.

22. Check the functioning of the aircraft control system, for which purpose proceed as follows:

- smoothly shift the control stick to the extreme positions three or four times. The control stick should move smoothly, without creeping and jerking, the trim actuator neutral position signals should be displayed;
- make certain that the forces originated by the artificial feel units are percept and the relieved control sticks and pedals return to the neutral position.

23. Check the hydraulic systems for proper functioning, for which purpose make. several quick and full movements of the control stick over the diagonal; in this case, the pressure in the main and booster hydraulic systems should not drop below the 0 m division on the EKD combined hydrogas system indicator.

24. Check the air brakes for proper functioning by extending and retracting them and check their actuation by reference to the Hn LG and flaps position indicator (with the landing gear retraction simulation button depressed).

25. Check the functioning of the wing flops and leading edges extension and retraction system visually or by reference to the LG and flaps position indicator.

26. Check the serviceability of the automatic transmission ratio controller by reference to illumination of the APY TAKEOFF LANDING annunciator and absence of the APY signal on the EKRAN Annunciator.

27. Check the stall limitation system for serviceability by absence of the COC FAIL and NO COC RESERVE signals.

28. Check the air intake ramp control system for proper functioning, for which purpose proceed as follows:

- make certain that the shutters on the ramp position indicator are matched with the CH D mark, the warning signals are not displayed on the EKRAN annunciator and the LE INLET TEST and RH INLET TEST annunciators are dead.

Note. If the LEFT (RIGHT) AIR INTK signal is displayed on the EKRAN annunciator or the LH (RH) INLET TEST annunciator lights up, shut down the engines and find out the trouble.

- increase the engine speed up to 80 to 90% for 1 or 2 s. In this case, the LH INLET TEST and RH INLET TEST annunciators should illuminate;
- decrease the engine speed the annunciators should extinguish.

CAUTION. IF THE "LH (RH) INLET TEST" ANNUNCIATORS ILLUMINATE AT THE IDLE POWER SETTING OR ARE DEAD AT THE ENGINE SPEED OF 80 TO 90% OR IF THE "LEFT (RIGHT) AIR INTK" SIGNAL IS DISPLAYED, SHUT OFF THE ENGINES AND FIND OUT THE TROUBLES.

29. Check for feed of air to the high altitude partial-pressure suit (VKK) or to the antig suit (PKK) with the aid of the automatic pressure controller. When depressing smoothly the pressure controller head, the legs and belly should be smoothly and uniformly compressed. With the decrease in depression on the automatic pressure controller head, the compression decreases, respectively, and when the head is released, the compression ceases at all.

30. When the ventilated suit or the marine survival set (VMSK) is used, check the functioning of the ventilation system of the special outfit, for which purpose proceed as follows:

- open the SUIT VENT valve;
- make certain that the air is fed into the suit space;
- regulate the temperature and select the required rate of air ventilation.

31. Prior to taxiing out, look through all the commands available on the EKRAN annunciator panel by repeatedly depressing the EKRAN CALL button.

Section Four

PERFORMANCE OF FLIGHT

PREPARATION TO TAXI OUT AND TAXIING

32. On completion of the aircraft systems check, turn on the RADIO switch, request the flight control officer for the taxiout clearance and on receiving the latter, depress the braking lever and give the command to the aircraft technician to remove the chokes from under the aircraft wheels.

33. Before the taxi, make sure that the brakes function normally. For this purpose increase the engine speed up to 70% to 80% the aircraft in this case should hold in place; make certain that there are no obstacles in the taxiway, throttle back the engines and request the aircraft technician for the taxiout clearance. Upon receiving the taxi-out clearance (at the same time the aircraft technician must show the removed launcher safety pins to the pilot), check once more for absence of the obstacles, release the braking lever and start taxiing. The aircraft response to the crosswind during taxi is counteracted by the leg. While in taxi, for shifting the nose wheel steering mechanism to the large steering angles, it is necessary to depress the WSM LOCKON PF button on the throttle control lever. The sharp braking results in smooth, but considerable lowering of the aircraft nose section. The taxi speed should ensure the safety in moving over the airfield and preclude any collision with the obstacles. In this case, do not turn the aircraft in place and do not brake sharply one of the wheels.

CAUTION: SHOULD THE MAIN BRAKING SYSTEM FAIL, STOP MOVING, USING THE EMERGENCY BRAKING SYSTEM FOR BRAKING THE AIRCRAFT AND TAXI OFF THE RUNWAY. SHUT DOWN THE ENGINES, IF REQUIRED.

In the process of taxi over the straight way and when making the turns, pay attention to the proper functioning of the navigation system (to the indication of the present heading, roll, pitch, change of range and presence of updating).

CIRCLING FLIGHT

34. Before you taxi out to the runway, lock the shoulder straps in the restrained position, turn on, if necessary, the HEAT: W/SHIELD, switch, request the clearance for taxi to the runway. On receiving the clearance, taxi out to the runway and align the aircraft with the centre line of the runway, smoothly braking the aircraft, with the pedals set to the neutral position.

35. Before the takeoff, proceed as follows:

- make certain that the APY TAKEOFF LANDING annunciator illuminates;
- extend the flaps, check the extension of the wing flaps and leading edges;
check the readings of the combined course indicators for correspondence to the position of the aircraft on the runway;
- check for absence of the PRIORITY (signal on the EKRAN annunciator (if the signals are available, take a decision to take off);
- check for the absence of the information from the information reporting system, with the engines running at the speed of 85% and over;
- request the takeoff clearance.

In case of necessity, specify the present heading by reference to the runway heading proceeding in the following sequence:

- make sure that the WPT AFLD selector switch is in the AFLD position and the lightbutton corresponding to the departure airfield number is depressed;
- set the COURSE selector switch on the navigation system control panel to the position corresponding to the runway heading for taking off;
- proceeding in the RETICLE mode by reference to the windshield indicator, and using the nose wheel steering mechanism, make sure that the aircraft is positioned along the runway centre line and depress the COURSE CMPTR/ZEROING lightbutton;
- check the course readings followed by the combined course indicator and integrated display system.

Note. Computers TsVI and TsVI-M have no provisions for the mode of checking the runway heading.

CAUTION:
1. IF THE PRESENT HEADING DIFFERS FROM THE RUNWAY COURSE By 50 OR MORE, DO NOT TAKE OFF, TAXI OUT AND FIND OUT THE TROUBLE.

2. PERFORM ALL LOW FLIGHTS WITH THE AUTOMATIC FLIGHT CONTROL SYSTEM SWITCHED INTO THE "DAMPER" MODE.

3. SELECT THE LIMIT ALTITUDE RECOVERY" MODE ONLY AFTER CLIMBING OVER THE ASSIGNED LIMIT ALTITUDE.

36. Perform the takeoff, with the flaps extended, as a rule, at the maximum power setting. On receiving the takeoff clearance, switch on the flight time counter on the aircraft clock, and, holding the aircraft in position by the lineup brakes, smoothly shift the throttle control levers to the MAXIMUM position. After the engines have accelerated to maximum speed or if the wheels start skidding, release the braking lever and start the takeoff run.

Check the operation of the engine at the beginning of the takeoff run. While moving off from the rest, the aircraft tends to nose up to some extent and till the air intakes are changed over to the axial inlet, the aircraft continues moving in this position.

During the takeoff run, hold the control stick in the neutral position. Maintain the direction by deflecting the rudders. During the takeoff run, the air intakes changeover to the axial inlet is perceived by the pilot by the increment in the longitudinal acceleration and lowering of the aircraft nose.

At this point, the aircraft takeoff run becomes more steady. At an airspeed of 230 to 250 km/h, pulling the control stick backward, lift the nose wheels to a pitch angle of 10 to 11 against the flight director indicator.

The aircraft unsticking practically coincides with the moment at which the takeoff angle is created and the aircraft becomes airborne at an airspeed of 260 to 280 km/h, depending on the takeoff mass of the aircraft and the nose wheel lift rate. As soon as the aircraft becomes airborne, depart it from the ground, maintaining the pitch angle to be approximately equal to the unstick angle; at this point, the aircraft should be stable, and slight corrective forces should be applied to the control stick.

If necessary, it is allowed to perform the takeoff in the reheat power setting.

Switch on the afterburner simultaneously with the release of the braking lever, check the afterburner for being switched on by the jerk caused by the longitudinal acceleration, by change in the sound of the engine operation and by illumination of the LE EKG REHEAT and RE ENG REHEAT annunciators.

Since the takeoff run is short in time (7 to 8s), start pulling the control stick to lift the nose wheels at an airspeed of 150 to 160 km/h. The nose wheels are lifted at an airspeed of about 180 km/h, whereas the unsticking takes place at an airspeed of 230 to 250 km/h, coinciding with the moment of creation of the takeoff angle. The takeoff run is 250 to 350 m. The negative pitching moment resulted from the transition of the air intakes to the axial inlet (the system fails to succeed in actuation during the takeoff run) is counteracted by pulling the control stick to increase the angle of climb.

At an altitude of 10 to 15m, set the landing gear control to the RETRACTED position, make certain that the landing gear is completely retracted by reference to the LG and flaps position indicator, by restoration of the pressure in the hydraulic systems as read against the KKr combined hydrogas system indicator and by the absence of the UPPER INLET signal on the EKRAN annunciator.

If the landing gear has failed to be completely retracted the LG intermediate position lamp lights up on the LG and flaps position indicator, it is necessary to keep the landing gear control valve in the RETRACTED position, to decrease the airspeed to 400 km/h and check for the complete retraction of the landing gear.

Climb to an altitude of 100 m, retract the flaps, check for retraction of the wing flaps and leading edges against the LG and flaps position indicator.

Set an airspeed of 450 to 500 km/h and execute the further circling flight at this altitude. At an altitude of 150 to 200 m, bring the aircraft into a turn at a bank of 35 to 45° . Perform the turns on the crosswind and downwind legs in the close succession.

On accomplishing the turn on the downwind leg, it is necessary to check the pressure in the hydraulic systems, the air pressure in the main and emergency air systems and engagement of the nose wheel brake.

When flying abeam the outer homing radio station, extend the landing gear, the aircraft should be stable in extending the landing gear and the retrimming is practically not perceived. Check the landing gear for proper extension by reference to the indicator lamps located on the LG and flaps position indicator and by restoration of the pressure in the main hydraulic system, make certain that the wheels are not braked.

Extend the flaps, for which purpose depress the flaps extension button and, keeping the hand on the flaps control board, check the extension of the wing flaps and leading edges by reference to the LG and flaps position indicator. During extension of the wing flaps and leading edges, the retrimming moment is insignificant and is easily counteracted by the control stick.

CAUTION: IN CASE OF THE ENERGETIC BANKING OF THE AIRCRAFT AT THE MOMENT OF THE FLAPS EXTENSION, IMMEDIATELY RETRACT THE FLAPS AND PERFORM THE LANDING WITH THE FLAPS RETRACTED.

Before executing the turn on the base leg, set an airspeed of 400 km/h, check the shoulder straps for proper locking, report the extension of the landing gear to the flight control officer and request the landing clearance.

Start turning on the base leg at a radio station relative bearing of 250°(110).

Accomplish the 100 to 110d turn in the horizontal plane at a bank of 35 to 45°.

After recovery from the turn, bring the aircraft into descent at a vertical speed of 3 to 5 m/s and proceed flying, maintaining the radio station relative bearing to be equal to 350° (100) before starting the turn on the final.

Execute the turn on the final at an airspeed of at least 350 km/h at a bank of 30 to 35°. The altitude for recovery from the turn is at least 200 m. Maintain the runway alignment by varying the bank up to 45°.

On completing the turn on the final, fly over the outer homing radio station at an altitude of 200 m and airspeed of 320 to 330 km/h and bring the aircraft into descent at a vertical speed of 3 to 5 m/s so that the glide slope is directed to the flareout initiation point located at a distance of 50 to 70 m from the runway approach end. Execute the gliding with the gradual decrease in the airspeed up to 280 to 290 km/h by the moment of the flareout initiation. The altitude of flying over the inner homing radio station, with the glidepath slope normally maintained (without sinking) amounts to 50 m.

With the airspeed decreasing, the angle of pitch increases in the amount perceptible for the pilot. The pull forces are applied to the control stick. The runway is visible all time both when gliding and flaring out. The retrinmings developed in correcting the estimation as a result of change in the engine power should be counteractedby the counter-roll stick.

M an altitude of 30 to 20 m, after ascertaining that the estimation has been made from the glance to the ground to the left, forward correctly and checking the airspeed, shift and towards the aircraft descent.

From an altitude of 10 to 8 m, begin flaring out the aircraft by gradually pulling the control stick till the touchdown. Decelerate the engines after flaring out the aircraft or during the second stage (half) of the flareout. During the aircraft flareout, the control stick travel is considerable from the neutral position to almost the extreme backward position.

The touchdown is performed gently at an airspeed of 250 to 260 km/h if the lending mass of the aircraft is normal and the landing angle of attack is about 110. The aircraft nosedown tendency at the moment of the touchdown is counteracted by the respective application of the control stick backward.

When landing the aircraft having the maximum landing mass, the gliding and landing speeds increase by 10 to 15 km/h.

37. When landing the aircraft at the airspeed higher than that required for the given landing mass, the aircraft may bounce. In this case, it is necessary to proceed as follows:

- gradually pushing the control stick forward, cease the further separation of the aircraft from the runway;
- pulling the control stick backward by applying the force thereto corresponding to the aircraft descent, create the aircraft landing position;
- during the repeated touchdown of the aircraft, depress the brake chute release button at once to preclude development of the progressive bouncing.

As soon as the landing roll becomes steady, lower the nose wheels, shift the glance forward, deploy the brake chute and start braking the wheels by smoothly depressing the braking lever; in this case, the landing roll will amount to 600 to 700 m. For reducing the landing roll, the braking lever may be fully applied since the beginning of braking. With the crosswind component not exceeding 6 m/s, it is allowed to deploy the broke chute before lowering the nose wheels.

38. In the case the aircraft is landed without deployment of the brake chute, hold the aircraft on two main wheels before lowering the nose wheels, after which push the control stick and at an airspeed of not more then 215 km/h, start braking the wheels. In this case, the landing roll will amount to about 1300 m. If necessary, start braking at an airspeed of more than 215 km/h.

The brake chute should be obligatorily deployed in the following cases:

- when the aircraft is landed just after the takeoff with complete amount of fuel;
- with four aerial bombs AB500;
- on the short or wet runways;
- when the aircraft is landed with the flaps retracted;
- when the automatic transmission ratio controller is set to the small arm position;
- in case of abortion of the takeoff after the nose wheel liftoff.
- During the landing roll, and the aircraft is stable, showing no tendency to yaw and turn, use the steerable nose wheels.
- Hold the direction with ding roll, it is prohibited to select
- For precluding the sharp dashes during the Ion wheels is the large angles of nose wheel turn till the end of the roll If one of the sharply braked after depressing the braking lever, release the letter, immediately deploy, the broke chute and again begin braking the wheels smoothly. Increasing the pressure in the brakes with the decrease in the airspeed, depending on the remaining length of the runway. If the aircraft brakes fail to operate when the braking lever is depressed (the main braking system has foiled to operate), deploy the broke chute if the letter has not been deployed, shut off the engines and use the emergency braking system. In this case, it is necessary to tear off the lock-

ing device on the lever, take up the free travel of the lever and apply the brakes by smoothly pulling the control stick backward, in conformity with the landing roll speed and the remaining length of the runway. If the skidding occurs, push the control stick forward to some extent, restore the direction of the roll with the aid of the pedals and increase the pressure further on at the smaller airspeed.

39. When in taxi to off the runway, retract the flaps, turn off the HEAT: W/ SHIELD, switch and check the closing of the axial inlets of the air intakes by reference to the position of the shutters on the ramp position indicator.

After stopping, release the brake chute in the assigned place. If the axial entry of any air intake is not closed, shut down the both engine, the further taxi is prohibited.

At the high temperature of ambient air, it is allowed to taxi the aircraft after landing, with the canopy partially open.

40. After taxi the aircraft to the parking area (the area where the engines are shut down) shut down the engines in the following sequence:

- turn off all the switches, except for the ENG SYSTEM BAT GXD SUP RECORD switches;
- set the throttle control levers to the STOP position and switch off the flight time counter;
- make certain that the axial inlet is open by reference to the ramp position indicator;
- turn off the ENG SYSTEMS, BAT GND SUP, RECORD switches.

CAUTION: THE ENGINES (ENGINE) SHOULD NOT BE USED IF THEY ARE SHUT DOWN BY CLOSING THE SHUTOFF VALVE.

After a taxi to the parking area, opening the canopy and installing the ground safety Pin into the ejection handle, hinge the latter forward, for which purpose depress the side triggers.

41. Tow the aircraft about the airfield, with the canopy closed or partially open, at a speed not exceeding 20 km/h on the concrete pavement and at a speed not exceeding 10 km/h on the ground. When the aircraft is towed at the night time, the navigation lights should be switched on. While towing the aircraft, the pilot or technician should take a seat in the aircraft cabin and be ready for application of the brakes during the towing truck halts.

GO AROUND

42. The goaround is allowed to be performed from the altitudes down to the flareout initiation altitude. However, for the safety purpose, it is recommended to take a goGround decision at an altitude of at least 50 m. When going around at an airspeed of 300 to 330 km/h, with the engine speed increasing at the rate of acceleration, the energetic pitching moment develops, which requires the application of the control stick and the trimout of the control stick forces with the aid of the trimming mechanism.

For going around, proceed as follows:

- shift the throttle control levers to the MAXIMUM position without changing the angle of gliding. Counteract the developed pitching moment with the aid of the control stick
- decrease the angle of gliding with the increase in the airspeed so as to bring the aircraft into the level flight at an airspeed of 320 to 330 km/h;
- bring the aircraft into climb with the gradual increase in the airspeed;
- retract the landing gear at the altitude not below 10 to 15 m;
- retract the flaps at the altitude not below 100 m;
- perform the further flight at an airspeed of 450 to 500 km/h.
- If it is late to take a goaround decision, increase the engine speed to maximum and proceed with landing till the landing gear touches the runway with the subsequent unsticking from the ground after gaining the maximum speed.

LANDING WITH ONE ENGINE RUNNING

43. The landing of the aircraft with one engine running presents no special problems in the piloting technique.

In this case, the descent and landing practically do not differ from those executed under the normal conditions. The slippingtowards the running engine is counteracted by deflecting the aircraft control stick and pedals.

With one engine running, the goaround is allowed to be performed from any altitude down to the runway touchdown; in this case, it is recommended not to delay in counteracting the turning moment when the engine gains the maximum power.

LANDING WITH FLAPS RETRACTED

44. It is allowed to land the aircraft with the flaps retracted, only in the case it is impossible to extend them (in case of failure of the main hydraulic system) or in case of the aircraft banking during the flaps extension.

When performing the landing approach, execute the turn on the base leg through an angle of 100 to 110° at a radio station relative bearing of 240° (120°).

When performing the approach, turns and gliding, maintain the airspeed by 20 to 25 km/h more then that assigned for the normal landing.

Select the flareout point 200 to 250 m away from the runway approach end; in this case, fly over the inner homing radio station at an altitude of 40 m.

Perform flareout at the normal rate, throttle down the engines only after the aircraft approaches the runway.

Taking into account the elevation landing speed, execute the landing approach with the flaps retracted at the minimum fuel remainder.

During the landing roll, after lowering the nose wheel, deploy the brake chute at a speed not exceeding 310 km/h and fully depress the braking lever. In this case, the landing roll will be about 1000 m.

After landing the aircraft with the flaps retracted, examine the landing gear.

TAKEOFF AND LANDING AT CROSSWIND

45. The crosswind takeoff and landing are allowed if the crosswind component does not exceed 15 m/s. The crosswind causes the Aircraft to perform turning into the wind and banking downwind.

At the takeoff, eliminate the aircraft turning with the aid of the pedals, whereas the banking is counteracted by applying the control stick against the wind. With the increase in the airspeed during the takeoff run, the required deflections of the pedals and control stick decrease to some extent. Lift the nose wheel. through the smaller value, up to 100, against the flight director indicator so that the aircraft unstick takes place at an airspeed increased by 10 to 15 km/h. Avoid unsticking from one wheel or the aircraft banking at the moment of unsticking.

When landing the aircraft, counteract the aircraft drift by selecting the appropriate course. At the moment of touchdown, to preclude the wheel skidding, push the pedal in the drift direction and maintain the direction along the runway axis.

For improving the aircraft stability during the landing roll, lower the nose wheals, with the pedals set to the neutral position, and proceed with braking the wheels; this done, deploy the brake chute. The landing roll with the crosswind component being 15 m/s increases by 20% with respect to the landing roll without the crosswind effect.

TOUCH AND GO FLIGHT

46. It is allowed to perform the next takeoff on the aircraft carrying no external ,stores directly after lending (the touchandgo takeoff). In this case, the first lending should be performed with the fuel remainder not exceeding 2500 kg, and the last takeoff, with the fuel remainder being not less than 700 kg.

Request the touchandgo takeoff clearance when flying the aircraft over the outer homing radio station.

The landing procedure till the touchdown is ordinary.

Upon the touchdown of the runway, make certain that the aircraft is stable during the landing roll on two main wheels, shift the glance forward and without lowering the nose wheels, accelerate the engines to the maximum speed for 2 or 3 s and execute the takeoff, avoiding the excessive lift of the nose wheels.

- The total landing roll and takeoff run distance amounts to 700 M.
- The fuel consumption during one touchandgo flight amounts to:
- 230 kg, with the landing gear retracted after the takeoff;
- 330 kg, during the circling flight, with the landing gear extended.
- The maximum number of touchandgo flights should not exceed four.

EXECUTION OF TWO SUCCESSIVE FLIGHTS

47. When executing two successive flights, perform the first landing when the fuel remainder is not more than 2500 kg, without the external stores, and not more than 1200 kg when the aircraft carries external stores 25 indicated in Table 8.

Perform the first landing without deployment of the brake chute during the landing roll and since the beginning of the braking at an airspeed of not more than 215 km/h.

Prior to taxiing out to the runway for the next takeoff, check the air pressure in the system, which should not be below the value indicated by the yellow sector on the EKD combined hydrogas system indicator.

After accomplishment of the second takeoff, execute the flight for 5 to 6 minutes with the landing gear extended for cooling off the wheels, after which the landing gear' may be retracted.

FLIGHTS WITH AIRCRAFT PARTIALLY FUELED

48. It is allowed to perform the shorttime flights with the aircraft partially fuelled.

Before the flight with the aircraft partially fueled, it is necessary to make the fuelofflight calculation.

AIR NAVIGATION

EXECUTION OF ENROUTE FLIGHT

49. Perform the enroute flight with the use of the navigation system in the manual control mode; in this case, it is allowed to use the STABILIZATION and ALTITUDE STABILIZATION modes of the automatic flight control system.

The accuracy' of the enroute flight is dependent on the presence of the radio correction (updating) at which it amounts to 0.4% of the distance to the radio beacon and during the autonomous flight, it depends on the time required for preparation of the system on the ground: during the fast preparation of the system, the accuracy in computing the distance amounts to 4% of the distance flown and daring the navigation preparation, 8 km for one hour of the flight.

In case of the departure from any airfield, the latter should be programmed as AFLD. In all other cases, perform the enroute flight in the manual control mode.

After the takeoff, set the WPT AFLD. selector switch to the WPT position, depress the button of the respective waypoint, match the pointer of the selected course with the upper fixed index on the combined course indicator and proceed with the flying of the aircraft, constantly keeping the course selector pointer matched with this index. Perform the climb and maintain the altitude according to the assigned flight mission.

For ensuring the enroute flight to the programmed radio beacon as to the waypoint, depress the RESET lightbutton and the lightbutton with the desired radio beacon number. The radio correction is ensured from the radio beacon whose number lights up on the navigation system control panel. To change over to the flight to the WPT (AFLD), depress the lightbutton in the WPT AFLD row.

Note. Computer LIBY makes no provision for the mode of flying to the radio beacon as to the WPT, so depression of the RESET lightbutton results in the failure of the navigational information generation which can be restored by repeated depression of the respective lightbutton in the WPT AFLD row.

For ensuring the radio correction in flight, depress the BEACONS respective lightbutton. Check the transmission of the radio correction signals by reference to illumination of the CORR. lamp.

For ensuring the correction by reference to the visual landmarks programmed as a waypoint or airfield, when approaching the landmark, depress the COURSE CMPTR/ZEROING lightbutton (in this case, the distance readings on the combined course indicator will vary within ±2.5 km) and release it at the moment the aircraft flies over the landmark.

When approaching the waypoint and when the D <40 lamp illuminates, depress the lightbutton of the next waypoint, the course selector pointer will show the direction to this waypoint. Turning the aircraft, match the fixed index with the course selector pointer and proceed with flying to the next waypoint.

If the lightbutton of the next waypoint has not been depressed when the D < 40 lamp illuminates, the course selector pointer reads the course to the previous waypoint and is fixed when entering the zone with a radius of 4 km (6 km for computers TsVI and TsVI-M around the waypoint. The course selector pointer will show the course to the flown waypoint after the recovery of the aircraft from the indicated zone,

When executing the enroute flight with the use of the navigation system, the following is checked against the combined course indicator, flight director indicator and the integrated display system:

the selected course, distance to the waypoint and azimuth of the correction radio beacon against the combined course indicator and integrated display system, in this case, the deviation bars on the flight director indicator are within the circle, whereas the integrated display system circle, within the centre of the crosshairs.

Besides, to periodically check the present heading and the aircraft fix with respect to the radio beacon, it is necessary to:

- use the NAVIGATION and LANDING selector switches to set the channels of the selected navigation and landing radio beacons;
- set the CHANNELS selector switch to the KANUAL position.
- Check the readings of the azimuth and distance to the radio beacon against the combined course indicator, after which proceed as follows:
- set the channel selector switch to the AUTO position;
- depress the lightbutton of the required radio beacon;
- proceed with flying with the use of the navigation system.

EXECUTION OF FLIGHT TO PROGRAMMED AIRFIELD

50. It is allowed to execute the flight to any of the three programmed airfields from any point of the route. This flight is ensured by the navigation system in the RETURN mode.

For executing the flight in this mode, proceed as follows:

- turn on the RETURN lightbutton. The AFLD lightbutton will come on;
- depress the respective WPT AFLD lightbutton;
- depress the BEACONS lightbutton of the selected radio beacon;
- check the COURSE selector switch for proper setting in compliance with the runway landing heading;
- check to see that the CIRCLE selector switch is set in accordance with the traffic circle.

Note. Depressing only the RETURN lightbutton ensures the approach to the airfield programmed for the AFLD 1 lightbutton.

When the RETURN mode is selected, displayed on the combined course indicator are the selected course to the intermediate approach initiation point, distance to the landing airfield and airfield azimuth.

The flight in the RETURN mode is allowed during the manual control with the use of the AFCS STABILIZATION mode.

The RETURN mode may be executed in the vertical plane a ‘ long the breakthrough flight path or at the altitude assigned from the command post. In this case, the indicator follows up the selected course to the intermediate approach zone.

Execute the flying, matching the course selector pointer with the upper fixed index on the combined course indicator. If the aircraft flies over the cruising altitude (13,000 m), descend the aircraft so as to gain the cruising altitude at a distance of 120 to 150 km from the airfield and proceed with Plying at this altitude.

After intersecting the breakthrough flight path, keep the horizontal position bar on the flight director indicator within the circle, which will correspond to the glide slope angle of about 7°. If the glide slope bar on the flight director indicator assumes the lower stop position, it is recommended to increase the vertical speed and descend the aircraft till the entry into the descent glide slope so that the horizontal position bar is within the circle.

Check the flight in the RETURN mode by illumination of the RETURN lightbutton and CORR lamp. When the CORR lamp extinguishes, the combined course indicator will display the course not to the intermediate approach point, but to the airfield. In this case, it is allowed to execute the approach by reference to the radio station relative bearing or by the commands delivered from the command post (groundcontrolled approach).

For disengaging the RETURN mode, if necessary, depress the RETURN lightbutton once more.

EXECUTION OF FLIGHT TO UNPROGRAMMED AIRFIELD

51. For executing the flight to the unprogrammed airfield, it is necessary to proceed as follows:

- set the CHANNELS selector switch to the MANUAL position;
- use the NAVIGATION and LANDING selector switches to select
- the channels of the navigation and landing radio beacons;
- set the flight course to the airfield on the combined course indicator.
- Perform the flight with the selected course till the entry into the coverage zone of the airfield navigation radio beacon.

LANDING APPROACHES WITH USE OF LANDING CONTROL SYSTEMS

LANDING APPROACHES WITH USE OF NAVIGATION SYSTEM

52. It is allowed to execute the landing approaches with the use of the navigation system by intercepting the initial point of turn (IPT) of the programmed airfield by resorting to the "return" and "missed approach" methods or by using the navigation system on the final after approaching the unprogrammed airfield runway alignment.

LANDING APPROACHES BY INTERCEPTING INITIAL POINT OF TURN TO PROGRAMMED AIRFIELD

53. Prior to intercepting the initial point of turn to the programmed airfield, check to see that:

- the CORR lamp comes on, the RETURN lightbutton is depressed and alive;
- the WPT AFLD selector switch is placed in the AFLD position;
- the respective lightbutton of the WPT AFLD row is depressed and alive;
- the COURSE selector switch is placed in the position corresponding to the runway landing heading;
- the CIRCLE selector switch is placed in the position corresponding to the traffic circuit.

Besides, when intercepting the initial point of turn after the takeoff or goaround by the "missed approach" method after climbing to an altitude of 200 m and beginning of execution of the turn at a bank of up to 30° proceed as follows:

- switch off the AFCS automatic and director flight control modes (if they have been switched on) by depressing the AFCS MODE OFF button for a short time;
- depress the MISSED APPROACH lightbutton and check it for cutin by reference to the illumination of this lightbutton;
- check the followup of the selected course pointer on the combined course indicator for execution of the turn on downwind leg.

54. After reaching an altitude of 600 m, perform the flight to the initial point of turn at an airspeed of 450 to 500 km/h.

Perform the turn to the runway landing heading by reference to the deflection of the selected course pointer and the vertical position bar on the combined course indicator at banks of up to 30°.

When flying a 30° flight leg, after rolling into the final, extend the landing gear, flaps and proceed with the further level flight at an airspeed of 400 km/h till the aircraft intercepts the equisignal zone of the glide slope beacon (at a distance of 12 to 14 km at an altitude of 600 m).

When the aircraft enters the LANDING mode automatic engagement zone and the difference between the present and runway landing headings is less than 60° (less than. 45° for computers TsVI and TsVI-M, the following takes place:

- followup of the runway landing heading by the selected course pointer of the combined course indicator;
- changeover to the relative bearing indication by the radio station relative bearing pointer from the automatic direction finder;
- disappearance of the LOC and GS flags on the combined course indicator and appearance of the localizer and glide slope indexes on the integrated display system (with the landing and radio beacon groups (PRMG) operating);
- going dead of the CORR annunciator. The RETURN lightbutton continues glowing.

For forced selection of the LANDING model proceed as follows:

- select (check for selection of) the channels of the navigation and lending radio beacons with the aid of the NAVIGATION and LANDING selector switches;
- use the setting knob to set the true landing beading;
- turn on the LANDING toggle switch while executing the turn on final.

When the LANDING mode is selected, the pilot receives the following information:

- the course and glide slope deviation bars on the combined course indicator show the position of the equisignal zones of the course and glide slope with respect to the aircraft;
- the large circle on the integrated display system screen displays the departure of the aircraft from the equisignal zones of the course and glide slope;
- the distance to the runway is displayed on the combined course Indicator and on the integrated display system screen;
- the selected course pointer on the combined course indicator shows the lauding heading.

The lauding approach by rolling out of the initial point of turn may be executed in the automatic, director or manual control mode and by reference to the radio beacon relative bearing.

Landing Approach in Automatic Control Mode

55. Select the automatic control mode from the director control one at speeds less than 400 km/h.

For accomplishing the landing approach in the automatic control mode after entry into the coverage zone of the course and glide slope radio beacons, with the flags on the combined course indicator disappeared and the position bars on the combined course indicator operating, depress the FL FLIGHT PATH CTL button, make certain that the indicator lamps of the FLIGHT PATH CTL and ALT STAB button light up, flags LOC (K) and P (T) disappear on the flight director indicator, by maneuvering the aircraft get the command pointers on the flight director indicator within the circle and depress the AP (Arl) command button. Make certain that the indicator lamp of the AP button is illuminated, release the automatic flight control system mode trigger and do not intervene with the control stick motions.

The landing approach in the automatic control mode may be selected from the ALTITUDE STABILIZATION or STABILIZATION mode by depressing the FLIGHT PATH CTL button after entry into the coverage zone of the course and glide slope radio beacons both: with the position bars on the combined course indicator set to zero in the COMBINED CONTROL mode or without zeroing of the position bars.

The process of landing approach in the automatic control mode is accompanied by insignificant trimout of the aircraft control stick.

Before interception of the glide slope equisignal zone, the pressure altitude is stabilized in the longitudinal channel. In this case, one or two cycles of longperiod longitudinal oscillations with the altitude variation up to 50 m may occur.

When the pilot intervenes with the aircraft control (when the control stick trigger is depressed), the automatic control mode is switched off, the indicator lamps of the ALT. STAB. and AP buttons flicker and the indicator lamps of the DAMP and FLIGHT PATH CTL buttons remain alive. After releasing the control stick, the automatic control mode is switched on again, and the indicator lamps of the ALT STAB and AP buttons continue glowing constantly.

When the aircraft intercepts the glide slope equisignal zone, it is automatically changed over to descent, the Indicator lamp of the ALT. STAB. button extinguishes. The aircraft and try into the equisignal zones of the course and glide slope radio beacons in the automatic control mode may be performed with a certain over control. In this case, transient processes terminate before entering the outer homing station zone.

In the process of the entire lending approach in the automatic control mode, timely correct the engine power setting in compliance with the assigned flying regime, avoiding the airspeed loss. Maintain the values of the airspeed and altitude of the flight versus the distance to the runway as indicated in Table 6.

Table 6

Parameter	Distance, km			
	14 to 12	9 to 7	4	1
Altitude, m	600	400	200	50
Airspeed, km/h	400	370 to 360	330 to 320	310 to 300

At on altitude of 50 m, switch off the automatic control mode by depressing the AFCS MODE OFF button located on the control stick for a abort time, visually verify the landing approach and its calculation, and land the aircraft.

Check the landing approach in the automatic control mode against the command pointers on the flight director indicator and the command circle on the integrated display system screen, which may deflect to some extent with respect to the middle position as well as against the readings of the automatic direction finder, the deviation bars on the combined course and flight director indicators, smaller circle on the integrated display system screen, altimeter, rateofclimb indicator and the distance flown counter.

The automatic control mode may be switched off by one of the two methods:

- by depressing for a short time the AFCS MODE OFF red button located on the control stick;
- by selecting the leveling mode with the aid of the LEVELING lightbutton located on the control stick.

CAUTION: IF THE READINGS OF THE COMMAND POINTERS ON THE FLIGHT DIRECTOR INDICATOR (THE COMMAND CIRCLE ON THE INTEGRATED DISPLAY SYSTEM SCREEN) DO NOT CORRESPOND TO THE POSITION OF THE RESPECTIVE POSITION BARS ON THE COMBINED COURSE INDICATOR DURING THE FLIGHT IN THE EQUISIGNAL ZONES OF THE COURSE AND GLIDE SLOPE RADIO BEACONS (EXCLUDING THE ZONE ENTRY PROCESS) AND ALSO IN CASE OF THE STEADY REFLECTION OF THE COMMAND POINTERS FROM THE CIRCLE (THE DEFLECTION OF THE COMMAND CIRCLE FROM THE CROSSHAIRS), IMMEDIATELY SWITCH OFF THE AUTOMATIC CONTROL MODE BY DEPRESSING THE "AFCS MODE OFF RED BUTTON LOCATED ON THE CONTROL STICK AND PROCEED WITH THE FURTHER FLIGHT IN THE MANUAL CONTROL MODE.

Landing Approach in Director Control Mode

56. For executing the landing approach in the director control mode after intercepting the coverage zone of the course and glide slope radio beacons, with the flags disappeared, course and glide slope deviation bars on the combined course indicator operating, depress the FLIGHT PATH CTL button, if the previous flight has been accomplished without selection of the AP (stabilization) mode. If the AP (stabilization) mode has been selected, first switch off the mode with the aid of the red button located on the control stick and then, select the landing approach director control mode by depressing the FLIGHT PATH CTL button. When the FLIGHT PATH CTL button is depressed and the indicator lamp of the AP button is alive

(the AFCS stabilization mode is switched on) the automatic control mode will get switched on.

The selection of the director control mode is indicated by disappearance of the LOC and G/S flags and illumination of the indicator lamp of the FLIGHT PATH CTL button.

For additionally checking the landing approach in the director control mode, use the automatic direction finder. For this Purpose, when the aircraft intercepts the runway landing heading, set the RB selector switch to the ADF Position.

When the director control mode is used, the aircraft is piloted manually. When flying the aircraft along the flight path, keep the command pointers on the flight director indicator within the circle (keep the command circle of the integrated display system (IDS) screen at the crosshairs). The deflection of the command pointers indicates the direction of deflection of the control stick in roll and pitch for the optimum arrival at the assigned flight path.

Till the moment the aircraft intercepts the equisignal zone of the glide slope radio beacon, the horizontal position bar on the flight director indicator and the command circle on the integrated display system screen deliver the commands to eliminate the deviations from the flight altitude at which the director control mode has been selected and the indicator lamp of the ALT STAB button lights up on the AFCS control panel. . After intercepting the equisignal zone of the glide slope (the distance is 12 to 14 km), the altitude stabilization mode is switched off (the indicator lamp of the ALT STAB button extinguishes) the command pointer on the flight director indicator and the command circle on the integrated display system screen deliver the commands to eliminate the deviations from the equisignal zone of the glide slope radio beacon.

Check the landing approach in the director control mode by reference to the deviation bars on the combined course and flight director indicators, smaller circle on the IDS screen, altimeter, rateofclimb indicator and the distance flown counter.

CAUTION: 1. WHEN THE GLIDE SLOPE FLAG APPEARS ON THE COMBINED COURSE INDICATOR THE HORIZONTAL POSITION COMMAND POINTER ON THE FLIGHT DIRECTOR INDICATOR WILL DELIVER THE COMMAND TO RETAIN THE FLIGHT ALTITUDE WHICH HAVE BEEN GAINED AT THE MOMENT OF THE FAILURE. IN THIS CASE, IT IS NECESSARY TO EXECUTE THE DESCENT IN ACCORDANCE WITH THE DISTANCE TO THE RUNWAY.

2. BANK THE AIRCRAFT IN CONTROLLING THE AIRCRAFT ON THE RUNWAY LANDING HEADING AT A RATE OF 5 TO 8° PER SECOND FOR ME CORRECT REPRODUCTION OF THE REQUIRED COMMANDS BY TEE COURSE COMMAND POINTER.

Landing Approach in Manual Control Mode

57. The landing approach manual control mode is used when it is assigned or in case of failure of the automatic and director control modes of the automatic flight control system.

When the LANDING mode is automatically selected or is switched on manually with the aid of the LANDING selector switch at the end of the turn to the runway landing heading, it is necessary to check appearance of the flags of the course and glide slope channels and after interception of the runway landing heading set the RB selector switch to the ADF position for additionally checking the landing approach against the automatic direction finder.

As soon as the flags on the combined course Indicator appear, the glide slope deviation bar deflects, indicating the position of the equisignal zone of the glide slope radio beacon, whereas the course deviation bar deflects towards the equisignal zone of the course radio beacon.

While intercepting the runway landing heading, it is necessary to determine the aircraft attitude with respect to the runway alignment, make the necessary corrective turns to enter the equisignal zone of the course radio beacon, having set the course deviation bar on the combined course indicator within the circle and proceed with flying to the field at an altitude of 600 m.

Check the arrival of the aircraft at the equisignal zone of the glide slope radio beacon by reference to the approach of the glide slope deviation bar on the combined course indicator to the centre.

After intercepting the equisignal zone of the glide slope, bring the aircraft into descent, keeping the glide slope deviation bar on the flight director indicator within the circle.

Maintain the altitude and airspeed of the flight depending on the distance to the runway.

Landing Approach with the Use of Radio Beacon Relative Bearing

58. It is allowed to execute the landing approach to the programmed airfield by reference to the radio beacon relative bearing in case of the failure (cutoff) of the landing radio beacon group of the landing control system and when the navigation beacons of the system are arranged at a distance of 300 m, from the runway axis.

CAUTION: TO AVOID THE BANK ANGLE OVERRUN, SET THE "CHANNELS AUTO MAN" SELECTOR SWITCH OF THE NAVIGATION SYSTEM CONTROL PANEL TO THE "MAN" POSITION ONLY WITH THE AFCS AUTOMATIC MODES DISENGAGED OR WITH THE CONTROL STICK TRIGGER DEPRESSED.

The initial point of turn is intercepted by the aircraft in the RETURN or MISSED APPROACH mode.

Before beginning the turn to the runway landing heading (prior to intercepting the. initial point of turn), proceed as follows:

- set the CHANNELS selector switch to the MAN. position;
- use the NAVIGATION selector switch to select the channel of the landing airfield radio navigation beacon;
- set the landing heading on the combined course indicator.

Begin the turn to the runway landing heading by deflection of the selected course pointer on the combined course indicator and the vertical position bar on the flight director indicator. While making the turn, control the landing approach by varying the bank angle.

Perform the flight to a distance of 12 to 14 km at an altitude of 600 m and then, bring the aircraft into descent according to the assigned mode.

While executing the landing approach, select the course and radio beacon relative bearing ensuring the aircraft flight within the runway alignment.

Execute the instrument flying till the aircraft arrives at the area under the clouds at an altitude of 300 m and a distance of 6 km to the runway, after which change over to the visual flying, verify the landing approach and its calculation and land the aircraft.

The altitude and airspeed of the flight versus the distance to the runway are similar to those used in executing the landing approach with the landing control system fully operating.

LANDING APPROACH TO UNPROGRAMMED AIRFIELD WITH USE OF NAVIGATION SYSTEM

59. When performing the landing approach to the unprogrammed airfield, it is allowed to use the navigation system in the LANDING mode after the aircraft has intercepted the runway alignment.

For switching on the LANDING mode, proceed as follows:

- set the CHANNELS selector switch to the MAN. position;
- use the NAVIGATION and LANDING selector switches to select the channels of the navigation and landing radio beacons of the landing airfield;
- set the SET COURSE selector switch to the MAN . position;
- set the landing heading on the combined course indicator;
- when on the runway (landing) heading, turn on the LANDING switch at an altitude of 600 m and distance of up to 35 km from the runway.

After disappearance of the localizer flog on the combined course indicator, proceed as follows:

- select the director control mode by depressing the FLIGHT PATH CTL button;
- maneuvering the aircraft, set the command pointers within the circle;
- switch on (if necessary or according to the mission assigned) the automatic control mode by depressing the AP button.

LANDING APPROACH WITH USE OF AUTOMATIC DIRECTION FINDER

60. It is allowed to perform the landing approach with the use of the automatic direction finder, using the following methods:

- the straightin landing approach
- the landing approach from the estimated line;
- the landing approach by making two 180° turns.

For executing the landing approach with the use of the automatic direction finder, proceed as follows:

- set the RB selector switch to the ADF position;
- set the SET COURSE selector switch to the MO. position;
- set the runway landing heading on the combined course indicator.

Execute the landing approach with the use of the automatic direction finder by reference to the data from the groundcontrol approach and marker receiver.

Straight In Landing Approach

61. When performing the straightin landing approach after flying over the homing radio station, execute the flight at an airspeed of 500 km/h till the interception of the runway landing-heading.

The initial point of turn may be intercepted by turning the aircraft through the estimated angle after flying over the homing radio station or by intercepting the heading reverse to the landing one, by turning the aircraft through 180° or making the circle over the homing radio station.

The moment of flying over the homing radio station is determined by the change in the readings of the t automatic direction finder through 180°, and at the altitudes of up to 2000 m this moment is also determined by the operation of the audio and light signaling systems of the marker receiver. The automatic direction finder pointer changes its position through 180° after flying over the homing radio station at the distance equal to 1.5 to 2 flight altitudes.

When flying over the homing radio station, start the stopwatch, make a corrective turn to the course, with the turnaway angle being taken into consideration (to the course reverse to the landing heading) and proceed flying on this course to the initial point of turn. On expiration of the estimated time, start turning to the landing heading at the initial bank of 30°. During the second half of the turn, verify the runway alignment approach by varying the amount of bank.

When flying a 30° flight leg, after intercepting the landing heading, extend the landing gear, and flaps and bring the aircraft into the descent if the transitional and vertical speeds versus the flight altitude and distance to the runway (Table 7).

Table 7

Flight altitude at descent, m	Flight speed, km/h	Vertical speed, m/s	Approximate distance to runway, km
Up to 2000	450	40	30 to 28
Up to 1000	450	15	21 to 20
Up to 600	400	8 to 6	14 to 12
Up to 400	370 to 360	5 to 3	9 to 8
Up to 200	330 to 320	5 to 3	4

Decrease the airspeed gradually, avoiding the airspeeds less than specified ones at the distance check points.

While descending, select the drift lead so that the aircraft descends within the runway alignment zone.

At an altitude of 200 m, bring the aircraft into the level flight till the aircraft arrives at the outer homing radio station zone.

On passing the locator middle marker (after receiving the light and audio signals from the marker receiver), proceed as follows:

- bring the aircraft into descent at a vertical speed of 3 to 5 m/s;
- check the changeover of the automatic direction finder to the inner homing radio station by reference to the light signals and behavior of the pointer (if necessary, change over the automatic direction finder to the inner homing radio station manually);
- verify the drift lead;
- continue descending along the glide slope so as to fly over the inner homing station at an altitude of 50 m and at an airspeed of 310 to 300 km/h.

If the drift angle has not been selected or the wind "drift" has beer observed before arriving at the outer homing radio station zone, then, after passing the locator outer marker, proceed as follows:

- make a corrective turn to the lending heading;
- bring the aircraft into descent;
- check the changeover of the automatic direction finder to the inner homing radio station (change over the automatic direction finder to the inner homing radio station);
- make a corrective turn of the aircraft to the inner homing radio station;
- proceed with descending towards the inner homing radio station at the relative bearing equal to zero.,
- After passing the locator middle marker, make a corrective turn of the aircraft to the landing heading.

Proceed to the visual flight during the landing approach under the IFR conditions only when the runway is visually observed by the pilot flying the aircraft by reference to the instruments.

In the event of the stable visibility of the runway, visually verify the landing approach and its estimation (planning) and land the aircraft.

Landing Approach from Estimated Line

62. Intercept the landing heading by the commands and in the modes assigned by the flight control post.

After hitting the landing heading and receiving the "Mode" command from the flight control post (tower) act just as in case of the straightin landing approach.

Landing Approach by Making Two 180° Turns

63. For executing the landing approach by asking two 180° turns after going around, proceed as follows:

- make a corrective turn to the runway landing heading;
- retract the landing gear and flaps if they have been extended;
- turn on crosswind leg at an airspeed of 450 to 500 km/h at a bank of 30° climbing to an altitude of 600 m or to the altitude assigned for the given airfield (assigned by the flight control officer).

During the training flights, the flight time till the first 180° turn should amount to:

- one minute from the beginning of the takeoff run;
- one minute after flying over the locator middle marker during the goaround;
- one minute and 30 seconds after flyingover the locator outer marker during the goaround.

Execute the flight on the course reverse to the landing one with the drift angle taken into account. When flying abeam the outer homing radio station, start the stopwatch. On expiration of one minute and 20 seconds, at a relative bearing of 210° (150°), start making the second 180° turn at the initial bank of 30°; during the second half of the turn, verify the runway alignment approach by changing the amount of the bank.

After arriving at the final leg, extend the landing gear, flaps and decrease the airspeed to 400 km/h and proceed with the further flight just as during the straightin landing approach.

PECULIARITIES OF FLIGHT PERFORMANCE UNDER IFR CONDITIONS

64. As a rule, the flights under the IFR conditions are performed with the use of the automatic control modes of the automatic flight control system. In case of the failure of the automatic control system and for the training purpose, the manual control mode way be used.

When preparing for the flight in the IFR conditions, pay special attention to the serviceability of the flight director indicator, combined course indicator, combined instrument AA200, radio altimeter, automatic direction finder, navigation system, automatic flight control system and the windshield heating system. Set the limit altitude selector to the position corresponding to the minimum instrument flight altitude in the given area or in the landing airfield area during the transit flight.

Prior to taxiing out to the runway, switch on the windshield and pitot static tube heating systems.

Before the takeoff, check the readings of the navigation and flight instruments for compliance with the actual position of the aircraft an the runway.

Before entering the clouds, estimate the readings of the navigation and flight instruments according to the position of the aircraft with respect to the natural horizon (ground), check the AFCS mode selection for compliance with the mission assigned and proceed with instrument flying.

CAUTION: SELECT THE LIMIT ALTITUDE RECOVERY MODE AFTER CLIMBING THE ALTITUDE WHICH EXCEEDS THAT SET ON THE RADIO ALTIMETER BY 50 TO 100 M.

In case of the AFCS failure, immediately change over to the manual control mode. Proceed with the landing approach in the manual control mode, because due to the lack of time it is rather difficult to estimate the capability of the lauding approach in the director mode.

Before the lauding approach, set the LIMIT altitude on the radio altimeter by 15 m higher than the required one for timely reception of the LIMIT ALTITUDE voice information.

WARNING: 1. NEVER EXECUTE THE VISUAL SEARCH OF THE RUNWAY DURING THE IFR LANDING APPROACH.

2. IF THE RUNWAY THRESHOLD IS NOT VISUALLY OBSERVED BY THE PILOT AFTER INTERCEPTING THE INNER HOMING RADIO STATION, IT IS NECESSARY TO GO AROUND, REPORT TO THE FLIGHT CONTROL OFFICER AND FOLLOW HIS DIRECTIONS.

65. When the aircraft inadvertently enters the thick cumulus or thunderstorm clouds, it is necessary, without diverting the attention from the instrument flying, gradually decrease the angles of pitch and roll, reduce the airspeed to 500 600 km/h and take measures to recover from the cumulus (thunderstorm) clouds till the turn to the reverse course. Perform the turns at the severe turbulence in the clouds at the banks not exceeding 30° and a speed of 500 to 600 km/h with the energetic return of the aircraft to the assigned position in the case the aircraft deviates from its position due to the effect of the air streams.

66. Execute the taxiing and takeoff of the aircraft in the heavy icing conditions on the ground (the supercooled rain, etc.) without ranup of the engines.

Perform the flight in the icing zone at an indicated airspeed of at least 700 km/h or true airspeed of 800 to 900 km/h.

Perform the descent and climb in the flight regimes ensuring the minimum time of flying in the icing zone.

In case of the engine malfunction and when it is impossible to increase the airspeed or to recover from the icing zone, it is necessary to land the aircraft at the nearest airfield.

If the icing zone is at am altitude below 1000 m, it is advisable to execute the landing approach from the estimated line or to execute the straightin landing approach and extend the landing gear at a distance of 8 to 10 km short of the outer homing radio station.

Ice formation on the aircraft surface leading edges can cause:

- premature aerodynamic buffeting;
- decrease of the longitudinal stability margin by 1.5 to 2% MAC;
- decrease of the aileron rolling effectiveness at angles of attack exceeding 5°.

If the aircraft icing is detected, do not bring the aircraft to angles of attack exceeding 15° and do not execute abrupt maneuvers in yaw and bank. Due to possible overlapping of the wing flap slot with ice accretion, extend the flaps on the base leg.

Each time after flying in the icing zone, the pilot must order the aircraft technician to carry out the thorough inspection of the air intakes and compressor first stage blades. making the respective entries in the aircraft preparation log book

PECULIARITIES OF FLIGHT PERFORMANCE AT NIGHT

67. Fly the aircraft at night by reference to the flight and navigation instruments, irrespective of the night darkness and difficult weather condition. The night flight is considered to be an IFR one even if it is possible to determine the aircraft attitude by reference to the natural horizon and ground (the clear night, reference light marker, etc.).

PREPARATION FOR FLIGHT

68. Before taking seat in the cabin at the night time, give order to the aircraft technician to switch on the cabin lighting system, thoroughly check to see that the canopy and sight reflector are clean.

After climbing into the cabin, apart from the ordinary check of the equipment, it is also necessary to prepare the aircraft lighting equipment as follows below:

- adjust floodlight lamps
- adjust the light intensity of the indicator lamps, annunciators, consoles and instruments.

Note. The best (optimum) lighting of the cabin is gained by the combined use of the light fixtures employing the main and flood lights.

- check the map illumination light fixture for proper functioning;
- adjust the position of the shutters of all indicator lamps and annunciators;
- check with the assistance of the aircraft technician, the exterior signal lighting of the aircraft for proper functioning and use the NAV ITS selector switch to set the necessary mode of functioning;
- check to see that the taxi/land light boom is directed properly to the ground surface when the landing and taxi lamps are switched on;
- cut in the taxi lights.

Before taxiing out, proceed an follows:

- give the command to the aircraft technician to remove the chokes (by the navigation light flickering);
- designate the beginning of taxiing by switching over the lights from the landing position to the taxi one.

Perform the taxiing with the taxi lights switched on, making certain that the taxi light intensity is sufficient for taxiing the aircraft over the unit airfield.

Prior to taxiing out to the runway, switch on the pitotstatic tube and canopy windshield heater.

Before taking off, after taxiing out to the runway and setting the aircraft in alignment with the runway axis, switch off the taxi light and estimate the intensity of the cabin lighting.

TAKEOFF

69. The technique of the takeoff performance at night is similar to that in the daytime, but slightly differs in the distribution of the pilot's attention.

Maintain the direction of the aircraft at takeoff by reference to the runway lights. The aircraft approach to one of the sides from the middle of the runway is determined by the visible "decrease" in the distance between the lights of this side and "increase" in the distance between the lights of the opposite side.

The amount of the aircraft wheel lift is determined by the lights of the horizon, and if the latter are not available, by the runway leadout lights; in this case, the wheel will be lifted somewhat less with respect to the normal lift, but within the tolerable one (10 to 11°).

While unsticking, it is necessary to notice the angle of pitch and avoid banking. Upon steady separation of the aircraft from the ground (the runway edge lights start "going" under the aircraft), completely change over to the instrument flying, set the landing gear control valve to the retraction position and continue unsticking, maintaining the angle of pitch, avoiding the banking and keeping the airspeed and altitude of the flight within the assigned values.

FLIGHT

70. Fly the aircraft at night by reference to the instruments at all legs of the flight, constantly checking the altitude, especially during the flights at the low altitudes and approach. The visual determination of the distances and the spatial attitude by reference to the light markers at night may be erroneous and may be performed only on the base of the instrument flying.

When flying at night in the clouds, it is necessary to take into account the following peculiarities of the lighting situation:

- the presence of the screen formed by the navigation lights and periodical afterglows and darkening of the cabin, caused by the screen due to the irregular density of the clouds;
- the presence of the screen formed by the engines operating at the reheat power;
- the creation of the brightly illuminating sphere in launching the missiles.

In case of appearance of the heavy screen formed by the navigation lights, it is necessary to change over the lights from the flickering mode to the steady burning mode.

When flying at night, fully turn the rear view mirrors backward (completely retract them).

LANDING

71. The technique of the landing approach, landing approach planning and execution of the landing on the runway illuminated by the searchlights does not principally differ from that of the landing in the daytime, but is complicated to some extent because of the necessity to fly the aircraft by reference to the instruments and constantly check the flight altitude. The airspeeds at which the aircraft flies over the outer and inner homing radio stations and the airspeed at the which the aircraft starts flaring out should be increased by 5 to 10 km/h.

Switch on the landing lights at an altitude of 80 to 100 m (before flying over the inner homing radio station). The landing lights ensure the beginning of the ground observation from an altitude of 60 to 80 m.

CAUTION: IN CASE OF HAZE OR PRECIPITATION, THE SWITCHINGON OF THE LANDING LIGHTS RESULTS IN CREATION OF THE LIGHT SCREEN. IF THE SCREEN IS AVAILABLE, SWITCH OFF THE LANDING LIGHTS AND LAND THE AIRCRAFT WITHOUT THEIR APPLICATION.

When executing the landing approach, direct the glide slopes to the flareout initiation point illuminated by the landing searchlights and accomplish the flareout over the illuminated ground surface.

The execution of the landing on the runway which is not illuminated by the searchlights, with the landing lights switched on, is more complicated, therefore it calls for the particular attention on the part of the pilot in determining the flareout initiation altitude.

The glide slope should be directed under the runway approach end lights. It is recommended to shift the glance to the ground from an altitude of 20 to 30 m in case of the contrast illumination of the ground by the landing lights. On reaching the runway approach lights, start flaring out, from an altitude of 8 to 10 m, during which the pilot's glance should follow the light spot of the landing lights.

CAUTION: THE APPROACH TO THE FLAREOUT ALTITUDE AT THE GREAT ANGLES OF ATTACK (LOW AIRSPEEDS) AND THE HIGH FLAREOUT COMPLICATE THE VISUAL INSPECTION SINCE THEY WEAKEN THE GROUND ILLUMINATION INTENSITY BY THE LANDING LIGHTS.

During the touchdown and landing roll, the sequence of actions on the part of the pilot and distribution of his attention are similar to those performed in the daytime. After completing the landing roll, switch over the lights to the taxi lighting, taxi in and switch off the lights before shutting down the engines. The navigation lights remain alive till the onboard power is cut off.

MANEUVERING FLIGHTS

72. It is allowed to perform the elementary, advanced maneuvers and aerobatics within the operating angles of attack and gloads and within the wide range of altitudes and airspeeds of the flight. It is allowed to execute the maneuvers without employment of the trim tabs in all flying regimes. When accomplishing the vertical and horizontal maneuvers on the aircraft, the pilot may withstand the great gloads for a long time, which calls for the constant training on the part of the flying personnel to withstand these gloads.

When accomplishing the maneuver, the control stick pusher of the stall limitation system may actuate.

CAUTION:
1. NEVER OVERRIDE THE CONTROL STICK PUSH FORCE.

2. WITH THE CONTROL STICK COMPLETELY PULLED BACKWARD, THE PUSH FORCE APPLIED TO THE CONTROL STICK BY THE STALL LIMITATION SYSTEM IS CONSIDERABLY DECREASED.

3. ROLL OUT THE AIRCRAFT WHEN FLYING AT THE ANGLES OF ATTACK CLOSE TO ACTUATION OF THE STALL LIMITATION SYSTEM BY DEFLECTING THE CONTROL STICK TO ROLL OUT WITH SIMULTANEOUS DECREASE OF THE ANGLE OF ATTACK.

4. WHEN THE EMPTY BELLY DROP FUEL TANK IS ATTACHED, ALL FLIGHT MANEUVERS MAY BE EXECUTED.

Flying the aircraft at the low altitudes and at the reheat power settings calls for the constant check of the gloads and angles of attack since the execution of the maneuvers at the limit value of one parameter may result in exceeding the limit value or the other parameter due to the high power of the power plant.

During the initial stage of training, it is expedient to execute the maneuver at an altitude of 1000 to 5000 m, with the Gloads being within 5 to 6 g.

Prior to executing the maneuver, check to see that the automatic flight control is changed over to the DAMPER mode and the wing leading edges function normally.

360° TURN

73. It is allowed to perform a 360° turn It the reheat and nonreheat Power settings of the engines within the entire operating range of the altitudes and airspeeds. There is no difference in executing the lefthand and righthand 360° turns.

Prior to bringing the aircraft into the 360° turn set the assigned airspeed and then, enter the turn by the coordinated application of the control Stick and pedals, simultaneously increasing the engine thrust to the required level.

Check the correct accomplishment of the 360° turn by reference to the neutral horizon, rateofclimb indicator, altimeter, airspeed and gload indicators.

During the limit thrust 360° turn, it is necessary to maintain the airspeed by slightly varying the gload and hold the altitude by changing the bank.

Perform the 360° turn recovery by the coordinated application of the control stick and pedals, simultaneously decreasing the engine thrust at a rate ensuring maintaining of the airspeed.

The maximum angular rates of the steady turn of the aircraft are gained at the airspeeds close to those corresponding to the maximum gloads of the steady 360° turn.

It is allowed to execute the 360° accelerated (unsteady) turns within the subsonic airspeeds, not exceeding the tolerable angles of attack and maximum operational gloads.

The external stores suspended from the aircraft does not adversely affect the technique of execution of the 360° turn (Fig. 3).

DIVING

74. It is allowed to perform the diving at the angles of UP to 80°.

Bring the aircraft into a dive, depending on the previous maneuver and required (selected) angle of dive from the turn, with twohalf rolls, splitS maneuvers or a loop.

While diving from an altitude of 3000 to 5000 m, with the air brakes retracted, at the dive angles of 20 to 30°, the airspeed increment per 1000 m of the altitude lost amounts to:

- 130 to 140 km/h, with the engines running at the idle power;
- 250 to 270 km/h, with the engines running at the maximum power.

Recover the aircraft from the dive without banking and sideslipping by energetically creating the gload (for 2 or 3 s), simultaneously increasing the engine speed to provide the thrust required for executing the next maneuver. The dive recovery initiation airspeed, with the engines running at the idle power, should be less than the maximum permissible one for the given variant of the external stores by 50 to 100 km/h, and with the engines running at the maximum power, by 150 to 200 km/h.

WINGOVER

75. It is allowed to perform the wingover, depending on the engine power setting, the range of the altitudes and airspeeds indicated in Figs 10 and 10a. Before entering the wingover maneuver, set the selected engine power setting, air and altitude. Create an angle of pitch of 10 to 15°, then, coordinately deflecting the control
and pedals (for 2 or 3 s), turn the aircraft through 180° with respect to the long axis and, without fixing it in the inverted position, pulling the control stick to create the assigned angle of attack (gload) for 2 or 3 s, simultaneously set the required engine speed, if necessary. Further on, maintain the assigned angle of attack 8° (gload) till the aircraft is back into the level flight.

NORMAL LOOP

Enter the normal loop at the maximum or reheat power setting within the range of airspeeds specified in Figs 12, 13, proceeding from the conditions to ensure speed of at least 300 km/h at the top point of the maneuver.

Entering the normal loop, set the required (assigned) power of the engines, and altitude of the flight, then, pulling the control stick backward, create the assigned gload for 2 or 3 s. without surpassing the gload tolerable for the armament (external stores).

Upon the ascending portion of the loop, maintain the gload until the assigned angle of attack is reached. Further on, keep this angle of attack until the aircraft reaches the top point of the loop.

Upon passing the top point of the normal loop, set the required power setting for of the second half of the normal loop. Further on, proceed with flying as in the wingover.

For executing the normal loop at the maximum angle of attack and maximum gload, it is necessary to create the maximum permissible gload at the normal loop entry and keep it until the stall limitation system actuates.

Then, fly the aircraft close to the stall limitation system actuation boundary. When flying over the top point of the normal loop, set the power setting required for accomplishing the descending portion of the normal loop. Execute the second half of the normal loop at the angles close to the stall limitation system actuation boundary, until the limit gload is gained. Further on, maintain the gload till the aircraft is brought into the level flight.

When executing the normal loop at the stall limitation system actuation boundary the loop recovery altitude is close in its magnitude to the loop entry altitude. Therefore, it is necessary to select the entry minimum altitude for these conditions taking into account the possible errors made in the second half of the normal loop.

If the altitude at the top point of the normal loop is less than that required for executing the wingover, complete the maneuver with a halfloop. If the wing leading edges fail to extend and the stall limitation system is changed over to angle of attack of 15° at the ascending portion of the loop (halfloop, oblique loop), bring the aircraft into level flight by executing two halfrolls or complete the maneuver with halfloop. Further execution of flight maneuvers is not recommended.

HALF LOOP

77. It is allowed to execute the halfloop at the same power settings, altitudes and airspeeds as at those which are used for performing the normal loop. The procedure for execution of the maneuver and distribution of the pilot's attention is just the same as that employed during the execution of the first half of the normal loop.

When the aircraft approaches the top point of the maneuver (the aircraft nose "rests" on the horizon) at the airspeed of at least 300 km/h, execute the halfroll for 3 or 4 s by the coordinated application of the control stick and pedals.

If the airspeed proves to be less than 300 km/h when the aircraft approaches the top point of the maneuver, it is necessary, without decreasing the engine thrust, to continue executing the normal loop, check to see that the angles of attack are within the tolerable limits and the banking and sideslipping are not available. After lowering the aircraft nose and gaining an airspeed of 350 km/h, perform the halfroll at descent and bring the aircraft into the level flight.

OBLIQUE LOOP

78. It is allowed to execute the oblique loop at the same power settings, altitudes and airspeeds as at those which are used for performing the normal loop.

Prior to entering the oblique loop, select a landmark and set the assigned airspeed.

While executing the entry, create the assigned bank and, maintaining this bank, with the ailerons and pedals set to the neutral position, bring the aircraft into the oblique loop in the some way as into the normal loop.

While performing the oblique loop, the bank continuously varies.

The oblique loop plane inclination is determined by the initial bank angle.

Check the oblique loop for constant inclination against the turn and slip indicators, and at the entry, close to the top point of the oblique loop and before the recovery from the oblique loop, check the bank angle by reference to the natural horizon and flight director indicator.

After bringing the aircraft into the level flight, roll out the aircraft and check its attitude with respect to the selected landmark.

The climb on the ascending portion of the oblique loop in less than that on the normal loop, whereas the altitude loss on the descending portion of the maneuver is greater. If the flight altitude at the top point of the oblique loop is close to the limit value recommended for executing the wingover, perform the halfroll towards the bank angle and bring the aircraft into the horizontal flight. In case of the altitude deficiency at the recovery, roll out the aircraft and bring the latter from the dive at the limit gload (angle of attack).

CHANDELLE

79. It is allowed to perform the chandelle at the maximum and reheat power settings from the altitudes and at the airspeeds specified for entering the normal loop.

The chandelle may be performed in the some way as the oblique loop or the ascending spiral.

Accomplish the chandelle in the manner of the oblique loop as the ascending portion of this loop, with the halfroll at the top point of the maneuver.

Bring the aircraft into the chandelle in the manner of the ascending spiral by smoothly deflecting the control stick backward and towards the turn. At the initial stage, set a bank of 10 to 15°. As the aircraft flies along the spiral, continuously increase the bank angle of up to 70° by the end of 2/3 of the chandelle.

During the further flight of the aircraft, gradually decrease the angles of bank and Pitch, deflecting simultaneously the control stick forward over the diagonal, and the pedals, in the direction opposite to the turn so as to bring the aircraft into the level flight at the airspeed of at least 300 km/h, turning the aircraft through 180°.

The altitude climbed per chandelle is dependent on the entry altitude and airspeed, engine power setting, gload and the type of flying.

ZOOM

80. It is allowed to execute the zoom at the angles of up to 70°, with the engines running at the maximum and reheat power settings throughout the entire operational range of altitudes at the airspeeds not exceeding the maximum permissible ones.

At the zoom angles of 45° and more, the entry airspeed should not be less than 600 km/h at the full reheat power setting, 700 km/h, at the minimum and maximum power settings.

Enter the zoom, with the gloads being within 3.5 to 5.0 g at the angles of up to 45° and within 5.0 to 6.0 g at the angles exceeding 45°. Check the zoom angle value and absence of the bank by reference to the flight director indicator. Recover from the zoom km/h initially turning the aircraft, and, with the zoom angles exceeding 45° by executing two half rolls.

For recovering the aircraft from the zoom by making a turn, bring the aircraft into a turn by coordinated deflection of the control stick and pedals, with the simultaneous lowering of the aircraft nose to the horizon.

For recovering the aircraft from the zoom by making two halfrolls, turn the aircraft through 180° around the longitudinal axis and pulling the control stick backward, lower the aircraft nose to the horizon with the subsequent turn of the aircraft through 180° around the longitudinal axis.

UPWARD SPIN

81. It is allowed to execute the upward spin at the altitudes and airspeeds specified for performing the wingover.

Enter the upward spin at the airspeeds not less than recommended ones for recovering the aircraft from the zoom, depending on the zoom angle and engine power setting.

Make the upward spin by the smooth coordinated execution of the halfroll with the subsequent energetic lowering of the aircraft nose to the horizon.

When flying over the horizon, set up the engine power setting required for the further execution of the maneuver. Then, proceed with flying in the same way as during the execution of the wingover.

If, with the aircraft nose lowered to the horizon, the altitude proves to be less than tolerable one for the wingover, accomplish the halfroll and bring the aircraft into the level flight.

ROLL

82. It is allowed to perform the snap and slow, solo and multiple, horizontal, ascending and descending rolls.

For performing the snap roll, it is necessary to create an angle of pitch of 10 to 20° at an airspeed of at least 500 km/h (the higher the airspeed, the lesser is the angle), fix this angle and, by the coordinated application of the control stick towards the direction of rotation and the pedals to the same direction, turn the aircraft through 360° around the longitudinal axis for 6 or 8 s. When the aircraft approaches the level flight attitude, set up the control surfaces (rudders) to the recovery position and on completion of the rotation, to the neutral position.

For performing the slow (controlled) roll, it is necessary to create an angle of pitch of 10 to 200 at an airspeed of at least 600 km/h, fix the angle and, smoothly deflecting the control stick towards the direction of rotation, start executing the maneuver at the rotation rate of 10 to 12 s/turn. While executing the roll, deflecting the control stick and pedals, hold the aircraft nose against lowering; in this case, a certain portion of the maneuver is performed at the positive angles of attack, and the other portion, at the negative angles of attack, and the pilot experiences the alternating gloads.

The double multiple rolls are essentially two or more rolls performed in the close succession.

The multiple snap rolls are performed at an airspeed of at least 600 km/h with the aircraft nosed up through 20 to 25°

The multiple slow rolls are performed at an airspeed of at least 700 km/h and the technique of their execution is similar to that of the solo slow roll.

The horizontal, ascending and descending rolls according to their technique of execution are identical and differ only in the flight path slope.

SPIRAL

83. It is allowed to perform the shallow (at a bank angle of up to 45°) and steep (at the bank angles exceeding 45°), flat (at the flight path angles of UP to 30°) and tight (at the flight path angles exceeding 30°) descending and ascending spirals.

Bring the aircraft into a spiral from the steady descent (climb) by the coordinated deflection of the control stick and pedals.

While executing the spiral the aircraft is stable, showing no tendency for an increase in the angles of bank and pitch.

Pull the aircraft out of the spiral by the coordinated deflection of the control surfaces. For bringing the aircraft from the steep descending spiral to the level flight, first roll out the aircraft and then, decrease the angle of pitch with the simultaneous increase (decrease) of the engine thrust.

The altitude loss per turn of the spiral is dependent on the airspeed and altitude of the flight, angle of bank and engine power setting. The mean loss of the altitude in executing a turn of the spiral at a bank of 45° at the idle power setting and altitudes ranging from 5000 to 8000 m amounts to 2500 to 3000 m and at an airspeed of 550 km/h, and 4500 to 5000 m at an airspeed of 650 km/h.

SPIN

CAUTION. THE INTENTIONAL BRINGING OF THE AIRCRAFT INTO A SPIN IS PROHIBITED.

84. in case of an inadvertent stall of the aircraft, immediately push the control stick forward to the neutral position, with the ailerons and rudders set to the neutral position and the airspeed increased up to 450 km/h, bring the aircraft to the level flight, avoiding the overpowering of the stall limitation system.

In the case the aircraft goes into a spin:

1. If the spin direction is not distinct (the "falling leaf" mushing takes place), proceed as follows:

- set all the control surfaces to the neutral position;
- throttle back to the idle power;
- after ceasing the variation in yaw and roll and increasing the airspeed up to 450 km/h, start recovering the aircraft from the dive.

2. In case of the steady rotation in the spin, proceed as follows:

- fully deflect the pedal in the direction opposite to that of the spin;
- on a lapse of 2 or 3 s, set the control stick to the neutral position (in roll and pitch);
- throttle back to the idle power;
- after ceasing the rotation, set the control surfaces to the neutral position and after gaining an airspeed of 400 to 450 km/h, start recovering the aircraft from the dive.

WARNING. IF THE AIRCRAFT FAILS TO RECOVER FROM THE SPIN TO AN ALTITUDE OF 2000 M, THE PILOT MUST ABANDON THE AIRCRAFT.

FLIGHT FOR ATTAINING EXTREME HIGH AIRSPEED AND MAXIMUM MACH NUMBER

85. In order to obtain the maximum Mach numbers, accelerate the aircraft at the full reheat power. For the training purpose, accelerate the aircraft up to the maximum Mach numbers without the external stores or with two.

Program given below:

- climb to an altitude of 12,000 m at a true airspeed of 900 km/h With the engines running at the maximum power;
- execute the level flight at an altitude of 12,000 m, at the Mach number equal to 085 until the assigned distance from the airfield is obtained;
- turn the aircraft in the airfield direction at the Mach number equal to 0.8 085;
- accelerate the aircraft at the full reheat power in the level flight at an altitude of 12,000 m;
- cease accelerating the aircraft, switch off the afterburner after gaining the assigned airspeed or when the fuel remainder amounts to 1100 kg, with the distance to the airfield not exceeding 150 km;
- perform the flight at an altitude of 12,000 m, decreasing the airspeed to the Mach number equal to 0.85 and proceed with flying at this airspeed until the estimated descent line is reached;
- descend the aircraft from the estimated descent line at an airspeed of 500 km/h, with the engines running at the idle power, until the circling flight altitude is gained.

86. If the aircraft acceleration has been accomplished along the route or profile other than specified, cease the aircraft acceleration at a distance from the airfield exceeding 150 km, with the fuel remainder increased by 100 kg per 50 km of the additional air track.

FLIGHT TO SERVICE AND DYNAMIC CEILINGS

87. For the training purpose, execute the flight to the service ceiling on the aircraft carrying no external stores or carrying two airtoair guided missiles according to the program given below:

• climb to an altitude of 12,000 m at a true airspeed of 460 km/h. with the engines running at the maximum power;
• execute the level flight at an altitude of 12,000 m and at the Mach number equal to 085 to the assigned distance from the airfield;
• turn the aircraft in the airfield direction at the Mach number equal to 0.85;
• accelerate the aircraft, with the engines running at the full reheat power until altitude of 12,000m;
• the Mach number equals to 1.7 is gained in the level flight;
• climb at the constant Mach number equal to 1.65 until a vertical speed of 5 to 3 m/s is gained;
• cease the climb, switch off the afterburner, when the assigned altitude is reached or when the fuel remainder amounts to 1100 kg, with the distance to the airfield not exceeding 150 km; descend the aircraft to an altitude of 12,000 m, decreasing the airspeed to the Mach number equal to 0.85;
• perform the flight at the Machnumber equal to 6.65g, and at an altitude of 12,000 m until the estimated descent line is reached;
• descend the aircraft from the estimated descent line at an airspeed 500 km/h, with the engines running at the idle power, until the circling flight altitude is gained.

The service supersonic ceiling of the aircraft carrying two missiles R60, with the engines running at the full reheat power, in the standard atmospheric conditions, amounts to 17,500 m.

The service ceiling of the aircraft carrying two missiles R60 when executing the climb at the maximum power and at a 'true airspeed of 90 km/h in the standard atmospheric conditions amounts to 14,500 m.

88. For gaining the altitudes exceeding the service ceiling, use is made of the dynamic method of climbing, beginning with the entry into a zoom at the altitudes of 12,000 m to 13,000 m at the Mach number equal to or more than 1.8. Eater the zoom by increasing the angle of pitch by 10 to 15° from the initial one, with the gload ranging from 15 to 2.0. On reaching an airspeed of 600 to 650 km/h, pushing the control stick forward, recover the aircraft from the zoom at a gload of 0.5 and switch off the afterburners. Descend the aircraft to an altitude of 12,000 m, with the Mach number decreasing to 0.8 0.85.

Section Five

COMBAT EMPLOYMENT

89. According to the flight performances and capabilities of the aircraft armament complex the aircraft is capable of fulfilling the following combat missions:

- interception and attacks of the aerial targets in the free airspace and against the ground both in the day time and at night, under the VFR and IFR conditions, at the middle and short ranges;
- conduct of the close aerial combat;
- destruction of the visually visible ground (surface) targets.

90. The targets can be destroyed by use of missiles, rockets, guns and bombs employed as the aircraft armament in the various variants (Table 8).

Name of external stores

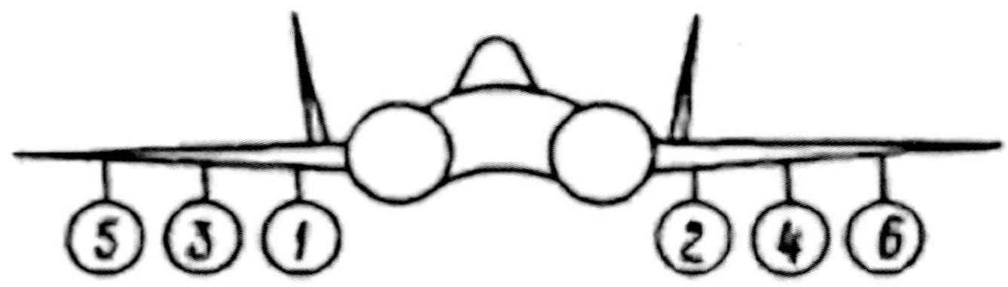

Forward view

Table 8

No.	Stations						Mass of equipped aircraft, kg	Remarks
	5	3	1	2	4	6		
1	-	-	-	-	-	-	14,375	Drop fuel tank only with ammunition allowance
2	R-60	R-60	R-27 R	R-27 R	R-60	R-60	15,360	Same
3		R-60	R-27 R	R-27 R	R-60		15,180	Same
4	R-60	R-60	R-60	R-60	R-60	R-60	14,920	Same
5	R-60	S-24 B	S-24 B	S-24 B	S-24 B	R-60	15,990	Same
6		S-24 B	S-24 B	S-24 B	S-24 B		15,810	Same
7		B-8	B-8	B-8	B-8		16,240	Same
8	R-60	ABCP-500	ABCP-500	ABCP-500	ABCP-500	R-60	16,890	Ammunition allowance is mandatory. Do not suspend drop fuel tank
9		ABCP-500	ABCR-500	ABCP-500	ABCP-500		16,710	Drop fuel tank only with ammunition allowance
10		ZB-500	ZB-500	ZB-500	ZB-500		16,140	Without drop fuel tank
11	-60	ABCP-250	ABCP-250	ABCP-250	ABCP-250	R 60	15,810	Without drop fuel tank
12	-	ABCP-250	ABCR-250	ABCP-250	ABCR-250	-	15,630	Drop fuel tank only with ammunition allowance
13	-	KMGI	KMGI	KMGI	KMGI	-	16,980	
14	-	-	KMGI	KMGI	-	-	15,710	
15	-	-	470|T	470|T	-	-	15,120	

NOTES

1. The mass of the aircraft in all armament variants is given with due regard for the ammunition allowance (AL) of the gun and the maximum weight of the ammunition modifications.

2. When the drop fuel tank is attached, the aircraft mass increases by 1300 kg.

3. The armament variants with incomplete symmetrical load of ammunition and asymmetrical suspension of the ammunition may be employed, with a misbalance being not in excess of 500 kg.

4. Furnish the aircraft with the rockets, aerial bombs, including the small load pack, of only one modification with the same fuzes.

5. When the aerial bomb calibre mounts to 250 kg and less: the ammunition .allowance is not mandatory, the drop fuel tank may be employed when the ammunition allowance is available.

6. Temporarily use variants 5, 8 and 11 without guided missiles.

7. Employment of training missiles 47OET makes it possible to train the flying personnel in employing combat missiles R27R under all conditions of its employment. In this case, the pilot's actions and the display in the cabin correspond to the actions taken during combat employment of missiles R27R. The repeated attack can be delivered upon automatic return of the system to the initial position on a lapse of 40 to 50 s from the moment of the target reset by the integrated radar/sight system after the attack accomplishment.

WEAPONS EMPLOYMENT CONDITIONS

MISSILES EMPLOYMENT CONDITIONS

91 The conditions for launching the "air-to-air" guided missiles are indicated in Table 9.

Note: Launch the missiles throughout the entire operating range of the flight conditions at the Mach numbers not exceeding 2.25 (M >, 0.6 for missiles R27R) at the steady and transient power settings of the engines from the idle power setting to the full reheat one.

Table 9

Parameters	Missiles	
	R-60	R-27R
Tolerable weather conditions and time of the day	Beyond clouds, in the day and night time, against clouds, ground and water surface, with the sun or "sun way" bearing being not less than 15°, including against ground heat-contrast targets	In any weather conditions in the day and at night
Missile launching ranges when computed by armament control system, depending on conditions for attack, km	In rear hemisphere, 0.6 to 8.0 In front hemisphere, 1.5 to 12.0	In rear hemisphere 0.6 to 17.0 In front hemisphere 2.0 to 50.0
Missile launching ranges when estimated visually (by ' to information delivered from command post), km	In rear hemisphere, 0.3 (0.7 during training flights) to 1.5 at $H_{intrcp} \leq 3000$ m; 0.3 (0.7 during training flights) to $\frac{H_{intrcp}}{2}$ at $H_{intrcp} > 3000$ m. In front hemisphere of targets flying at reheat power, 1.5 to 5 against maneuvering target; 1.5 to 12 against non-maneuvering target	
Range of flight altitudes of targets to be destroyed, m	In front hemisphere, 0 to 8000 In rear hemisphere, 0 to 20,000	In front and rear hemispheres, 30 to 23,000
Tolerable aspect angles of target	In front hemisphere, 1/4 to 4/4 at targets flying at reheat power In rear hemisphere, 0/4 to 4/4	Any
Tolerable g-loads of aircraft at moment of missile launching	-1.5 to +7.0 (up to 6 during slipping not exceeding two diameters of ball)	

92. Launch the missiles, it the necessary conditions are provided. by depressing the RAT MSS B trigger and keeping it depressed until a missile lifts off (a pair of missiles lifts off when the missiles are launched in series) or during the time of at least 3.5 to 4 s which is necessary for launching the next missile in the case the preceding missile fails to lift off. Launch the next missile furnished with an infrared homing head (a pair of missiles) 4 to 7 a after launching the first missile or after the first missile has passed by the target to avoid looking on the running engines of the previously launched missiles.

ROCKETS AND GUN EMPLOYMENT CONDITIONS

93. The conditions for launching the "airtosurface" rockets and firing the gun against the ground (surface) targets are given in Table 10.

Table 10

Parameters	Rockets		Gun
	S-8	S-24B	
Range of effective and safety firing, m	2300 to 1300	2500 to 1600	1800 to 1200
Dive angles, deg	Up to 40	Up to 40	Up to 40
Firing airspeed range, km/h	550 to 1100	550 to 1100	550 to 1100
Maximum firing altitudes, m	5000	5000	Not limited
Time of firing till ammunition allowance is completely used up, s	2	0.5	6 (in bursts with firing cut-off time interval of 1.0 s)
Aircraft engine power setting during firing	Without limitation		
Minimum altitudes of flying over exploded ammunition for recommended firing conditions, m	400	500	200

WARNING: 1. PROCEEDING FROM THE AERODYNAMIC HEATING OF THE POWDER MOTORS OF THE ROCKETS, TYPE S8, THE FLIGHT DURATION UNTIL ME LAUNCH OF THE ROCKETS AT THE AIRSPEEDS LESS THAN 900 KM/H IS NOT LIMITED AND AT THE AIRSPEEDS EXCEEDING 900 KM/H IT SHOULD NOT EXCEED 5 MINUTES FOR ROCKETS S8 IN PODS S 8MI, WITH THE AMMUNITION ALLOWANCE PARTIALLY FIRE AT THE AMBIENT AIR TEMPERATURE NEAR THE GROUND EXCEEDING 30°C (AT THE AMBIENT AIR TEMPERATURE LESS THAN 30°C WITHOUT LIMITATION).

2. IT IS PROHIBITED TO FLY THE AIRCRAFT CARRYING ROCKETS S24B AND S8 AT THE MACH NUMBERS EXCEEDING O.8.

94. The duration of the burst in firing rockets S8 is determined by the time during which the RKT MSL B trigger is kept depressed as well as by the position of the AL SINGLE 0.5 AL selector switch and when firing the gun, by the time during which the GUN firing button is kept depressed and the firing outoff time interval amounts to 1.0 s.

Perform gun firing training flights with the WEAPON TRAINING MODE selector switch turned on in the cabin power system; in this case, the gun ammunition allowance should contain not more than 75 rounds.

When firing the gun and rockets S8 against the ground (surface) targets, depending on the burst duration, aircraft airspeed and dive angle, the range for opening the fire should ensure the cessation of fire at the recommended minimum ranges. In case of the fire cessation at the shorter ranges during the recovery from the dive, the aircraft may get into the zone of splinters caused by the ammunition explosion.

Conduct the gun firing against the aerial targets within the operating ranges of the altitudes and airspeeds at gloads varying from 1.5 to 8, in bursts, with the firing cutoff time interval amounting to 1.0 s and time intervals between the bursts being within 3 During the training flights, the firing cutoff time interval in not required.

WARNING. FIRE THE GUN ONLY WHEN THE DROP FUEL TANK IS NOT SUSPENDED.

95. Employ the rockets and gun with the ASS switch turned on. When launching the rockets and firing the gun, the gas dynamic stability of the power plant may be lost in some cases for a short time, which is recovered by the antisurge system. Besides, when launching rockets, firing the gun and during the surge, owing to the increased vibration, the illumination of the REDUCE RPM LH (RH) ENG) annunciator for a short time and transmission of the VIBR LEFT (RIGHT) signal to the EKRAN annunciator panel as well as the delivery of the "LH (RH) engine vibration "voice information and reduce rpm of LH (RH) engine" command to the pilot's earphones are allowed. If the above signals do not fade out in 2 to 3 s, it is necessary to reduce the engine speed the discontinue the flight mission.

BOMBS EMPLOYMENT CONDITIONS

96. Employ the bombing armament of the aircraft in compliance with the directions of the Flight Manual and Manual on Aerial Destruction Facilities Employment Against Ground Waterborne) Objects, Part 1.

GENERAL PROCEDURES FOR EXECUTION OF FLIGHT WITH USE OF ARMAMENT

97. Before the flight involving the use of the armament, listen to the report of the aircraft technician on the aircraft readiness for the flight, the variants of the armament, the type, quantity and loading of the ammunition, the introduction of the ballistic characteristics of the armament into the integrated optoelectronic navigation/sighting system (when the bombs are to be used), the position of the TRAINING MODE selector switch, the availability of the film in the camera gun.

98. When inspecting the aircraft, proceed as follows:

- make certain that the prepared armament variant complies with the flight mission;
- check for the ground safety pins on the weapons;
- make certain that the external stores are free from the external damage, the protective caps are removed from the optical locator laser and missiles and the screw clamps are removed from the missiles and rockets, the optical locator laser fairing is clean and the weapon interlocking device access hatch is closed.

99. Before climbing into the cabin, make certain that the circuit breakers of the cabin DC power system required for execution of the flight, are turned on.

100. After climbing into the cabin, proceed as follows;

- make certain that the fire control triggers are placed in the safe positions, the MASTER switch is set to the OFF position, the ARMED SAFE selector switch is) set to the SAFE position and locked, the EMERG RELEASE button is capped, and all other controls of the armament and weapon control
- systems are turned off (the ILLUM DUMMY OFF selector switch is placed in the OFF position);
- check to see that the canopy, screens and optics of the armament control system are kept clean;
- select the armament and the armament employment mode in compliance with the armament variant with the aid of the INBD OUTBD, AL SINGLE 0.5 AL selector switches.

101. After starting the engine and disconnecting the ground power sources proceed as follows:

- turn on the ACS and WEAPON switches;
- set the controls of the armament control system to the positions corresponding to the flight mission and the armament to be employed;
- check the display on the screens of the integrated display system and, if necessary, adjust the display brightness;
- when the integrated radar/sighting system is used, set the ILLUM DUMMY OFF selector switch to the DUMMY position;
- check to see that the external store status indication system functions properly, for which purpose, set the ACS MODES selector switch to the OPTIC position.

102. Prior to taxiing the aircraft from the parking area, make certain that after giving the command to remove the chokes, the ground safety pins are removed and shown by the aircraft technician (the ground safety pine may be removed by the specialist at the check post before taxiing to the runway).

WARNING TO PRECLUDE THE MISSILE HOMING HEAD DAMAGE, TAXI, IPA UP AND LAND THE AIRCRAFT CARRYING THE MISSILES ONLY WITH THE "WEAPON" W ON" SWITCH TURNED ON.

103. After detecting, identifying the target and taking a decision for attack (after receiving the respective permission), proceed as follows:

- turn on the MASTER switch;
- check the ASS switch for turning on when the missiles, rockets and bombs are used;
- set the AIR GROUND selector switch to the GROUND position when firing the gun against the ground targets;
- specify the selection of the weapons and the armament control system function mode corresponding to the weapons to be used and to the attack conditions;
- with the drop fuel tank attached, make certain that it does not impedes the execution of the attack or the employment of the selected weapons (if necessary, jettison the drop fuel tank);
- attack the target in compliance with the directions an the employment of the selected weapons.

104. The flight mission over, turn off the MASTER switch. If the encounter with the enemy in likely to occur, make certain that the armament control system is ready for combat and set the armament control system function mode ensuring the employment of the weapons remained aboard the aircraft.

Note: Depending on the situation created in flight, the pilot may execute the emergency release or unguided launch of the various types Of the weapons suspended from the aircraft.

105. After accomplishing the flight with the use of the armament and landing the aircraft, observing the safety precautions, taxi the aircraft to the area specially allotted for the preliminary inspection of the armament and installation of the ground safety pins; this done, proceed with taxiing the aircraft to the parking area.

After taxiing the aircraft to the parking area, prior to abutting down the engines, turn off all the controls of the weapons and armament control system.

In the case the missiles (rockets) fail to lift off, the bombs fail to be released or the gun fails to be fired in flight, it is necessary, after taxiing the aircraft to the specially allotted area, to shut down the engines and act in compliance with the directions given by the armament specialists.

ACTIONS AGAINST AERIAL TARGETS

106. The armament control system ensures the employment of the missiles and gun against the aerial targets in the following modes: RIK ENCOUNTER, PURSUIT, AUTO, CLOSE COMBAT IRR CLOSE COMBAT (IRR) HELMET, OPTIC, GUN and RETICLE during the autonomous or combined operation of the integrated radar/sighting system and integrated optoelectronic navigation/sighting system. The RM (radar) mode in the basic one employed by the armament control system.

Note. The combined operation of the integrated radar/sighting and optoelectronic navigation/sighting systems is ensured by setting the COOP RETARD NO RETARD selector switch to the COOP RETARD position.

RADAR MODE

Radar Mode Peculiarities

107. The radar mode ensures:

- the detection of the aerial targets flying at the airspeeds:
- from 230 to 2500 km/h, in the front hemisphere (the ENCOUNTER (B) mode);
- from 210 to 2200 km/h, in the rear hemisphere (the PURSUIT (JI) mode);
- from 270 to 1100 km/h. in the front hemisphere (the PURSUIT mode);
- the attack of the aerial targets in the front and rear hemispheres at the altitudes ranging from 30 50 to 23,000 M.

108. The detection and lockon ranges in attacking the target in the free space and against the ground are practically the same, and are dependent on the hemisphere, where the target is attacked, the aircraft flight altitude said are equal to the following values with respect to the target having the reflecting surface 0 = 3 m2:

109. The radar ensures an attack against aerial target in the front and rear hemispheres. In attacking with the use of the aiming circle, the maximum initial elevation of the fighter over the target amounts to: 6 km in the front hemisphere, 6 km at a target altitude of over 5000 m and 4 km at a target altitude of below 5000 m in the rear hemisphere.

The maximum initial depression of the fighter with respect to the target amounts to: 10 km when attacking in the front hemisphere, 10 km at a target altitude of over 12,000 m and 6 km at a target altitude of below 12,000 m when attacking in the rear hemisphere.

At altitudes between the abovementioned values, the maximum elevation (depression) is selected by the pilot depending on the attack conditions, own and target altitudes. At flying altitude H < 1000 m the target elevation is not more than 2000 m.

110. The radar ensures the stable detection and lockon of the lowspeed targets (e.g., a helicopter) at the target airspeed not exceeding 180 km/h in the front hemisphere at a range of 23 kM and in the rear hemisphere at a range of at least 17 km.

111. When attacking the target maneuvering in altitude, it is necessary to take into account the following peculiarities:

- while scanning, the target maneuver initiation is not determined and in case of a narrow zone of search in elevation, the target leaves the zone. During the instrument direction, the stable detection of the target is not ensured on the integrated display system screen due to the high discreteness in delivering the target designation data to the antenna and the target altitude. During the manual tracking it is neces-

sary to switch on the TRACK WHILE FLYBY mode which ensures the tracking of the gated target while scanning;
• during the lockon, the stable tracking of the target is ensured, however, when the target goes to the background of the ground, when the attack is launched in the rear hemisphere, the changeover to the ground "tracking" may occur which results in the higher motion of the aiming circle, antenna position mark and aspect angle indicator on the windshield indicator screen. In this case, reset the lockon and in response to the commands delivered by the navigator from the direction control post, set required AH set' the range and detect the target;
• the target maneuver during the lockon is determined by the motion of the aiming circle up or down before the "Zoom" command is delivered and by the more energetic motion of the aiming circle after the above command has been delivered.

112. In attacking the target maneuvering in yaw, it is necessary to take into account the following peculiarities:

• while scanning, the target blip disappears when the target arrives at aspect angles 3/4 to 4/4 and when attack is launched in any hemisphere, and when the aircraft and target airspeeds are equalized, with the attack launched in the rear hemisphere. When the target maneuver in yaw is detected, switch on the AUTO mode, and after the target has turned from the headon course to the followon one, increase the flight airspeed until the stable lockon is ensured in the rear hemisphere;
• while locking on, the stable tracking of the target maneuvering from the followon course to the headon one is ensured when attacking the target, with a depression with respect to the target of more than 1000 m, in the PURSUIT mode, whereas the stable tracking of the target maneuvering from the headon course to the followon one is gained in the ENCOUNTER mode, provided the closing airspeed of at least 50 km/h is ensured.

When attacking the target with an elevation over the target or when the target maneuvers in yaw with the loss in altitude, the changeover to the ground "tracking" takes place both when the target maneuvers from the followon course to the headon one and vice versa. In this case, it is necessary to reset the lockon, switch on the AUTO mode and after detecting the target, look on the target once more. If the target course information is obtained, select the ENCOUNTER or PURSUIT mode.

113. In the duel situation of the meeting missile/rocket combat conducted by two fighters equipped with the integrated radar/sighting system having the same frequency range, the interference effect is absent in the scan mode, however, the preference for executing the attack will be given to the aircraft which will execute the autotracking lockon first.

Preparation of Integrated Radar/Sighting System for Operation

114. Before starting the engines, proceed as follows:

• set the AIR GROUND and COOP RETARD NO RETARD selector switches to the AIR and NO RETARD positions, respectively;
• set the ILLUM DUMMY OFF selector switch to the OFF position;
• turn off the COMPENS switch;
• set the AJ OFF CAJ selector switch to the AJ position;
• set the TRACK WHILE FLYBY: FHS RHS selector switch to the RHS position;
• set the RDR MODES selector switch to the ENCOUNTER or PURSUIT mode (depending • on the hemisphere in which the target is to be attacked);
• set the selector switch to the zero position;
• turn on the GUIDANCE switch (do not turn on this switch when the interceptor is guided by voice);
• set the SPAN knob to the M (mean) (C) position (set the SPAN knob to the small) (M) or G (great) (B) position when the aircraft flight is controlled from the direction control post);
• set the ACS MODES selector switch to the RIIR position;
• set the WAVES, CODE, SEPARATION selector switches located an the command radio link control panel in compliance with the selected radio data.

115. After starting the engines and disconnecting the external power source, proceed as follows:

• turn on the ACS and WEAPON switches located on the RE console;
• set the ILLUM DUMMY OFF selector switch to the DUMMY position;
• check the display on the screen of the integrated display system (adjust the brightness intensity), and in case of the interception flight, check command radio link control equipment for proper functioning by reference to the display of the FES (RHS) discrete commands and by the gradual variation in the course command.

WARNING. WHEN RDR SYMBOL ILLUMINATES ON THE INTEGRATED DISPLAY SYSTEM SCREEN, SET THE "ILLUM DUMMY OFF" SELECTOR SWITCH TO THE "OFF" POSITION, TO PRECLUDE THE PERSONNEL FROM BEING EXPOSED TO IRRADIATION IN THE PARKING AREA AND DURING TAXIING AND IN FLIGHT, SET THIS SELECTOR SWITCH TO THE "DUMMY" POSITION.

Scan Mode

116. The radar is automatically cut in for irradiation during the instrument direction in response to command "100" or "60" when the attack is launched in the front hemisphere and in response to command "36" when the attack is launched in the rear hemisphere.

When the command to cut in the radar for irradiation is not transmitted through the command radio link control panel, cut in the radar for irradiation manually.

Note: After cutting in the radar for irradiation, turn off the SCAN switch located on the radar illumination warning station control panel.

When the aircraft is directed by voice or when it approaches the search line, manually cut in the radar for irradiation.

Check the radar for irradiation by reference to the PR symbol flashup on the integrated display system screen. The view of the screen of the integrated display system in the scan mode is shown in Figs 5-2 and 5-3. If the RDR symbol fails to flash up, make several switching of irradiation.

After switching on the radar for irradiation alternately displayed in the right portion of the direct vision indicator are antenna position marks 1, 2, 3, 4, 5, 6 (marks 1, 2, 3, 4 in the manual mode of guidance) indicating the number of the line on which the space is being scanned at the given moment.

Check the directivity of the radar antenna by reference to the position of the search zone marks in elevation and azimuth on the integrated display system screen. The range scale amounts to 150 km in the ENCOUNTER mode, 100 km in the AUTO mode and 50 km in the PURSUIT mode.

117. Carry out the target detection by comparing the range delivered by the. direction control post with the target blips displayed on the integrated display system screen.

For improving the conditions for detecting the targets in directing the aircraft by voice, the gate control should be smooth in range and correspond to the information delivered from the command post.

Identify the target by depressing the INTERROG button for at least five seconds. In five seconds the identification similar to the target blip

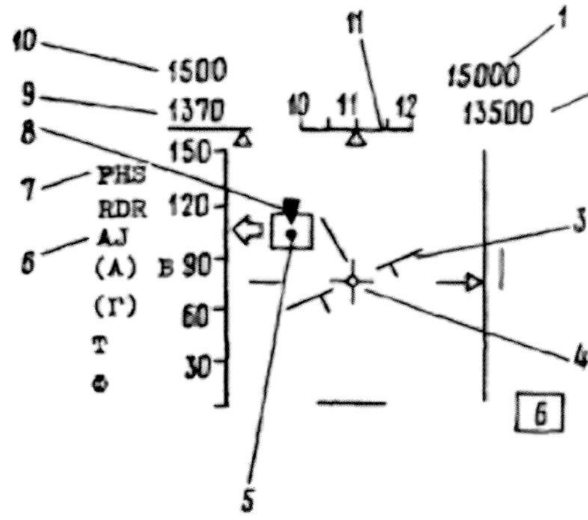

"SCAN" RADAR MODE DURING INSTRUMENT GUIDANCE

1 - target flight altitude; 2 - barometric pressure altitude; 3 - movable miniature aeroplane; 4 - control circle from ground control system; 5 - target blip; 6, 7 - "Active jamming" and "Attack hemispheres" commands; 8 - identification mark; 9 - true airspeed; 10 - assigned airspeed delivered by ground control system (to direct vision indicator); 11 - present heading scale and index; (A) B - (attack) encounter; Г - zoom; T - homing; O - reheat

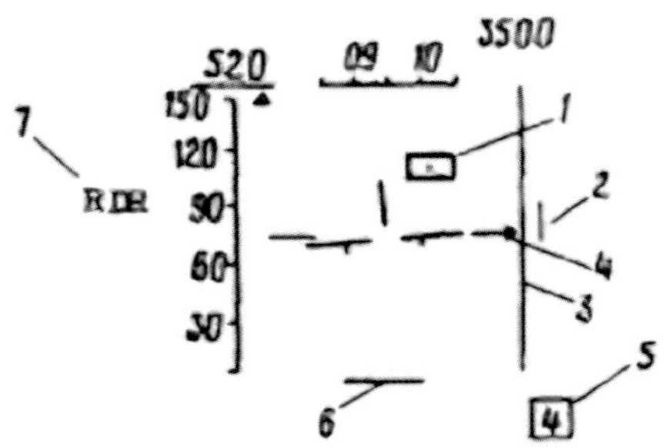

"ENCOUNTER (SCAN)" RADAR MODE DURING GUIDANCE BY VOICE AND INDEPENDENT SEARCH

1 - radar lockon gate; 2 - radar coverage zone in elevation (on direct vision indicator); 3 - windshield indicator coverage zone in elevation; 4 - horizon mark (on direct vision indicator); 5 - No. of present scan line on direct vision indicator; 6 - radar search zone in azimuth; 7 - armament control system operation character (main channel)

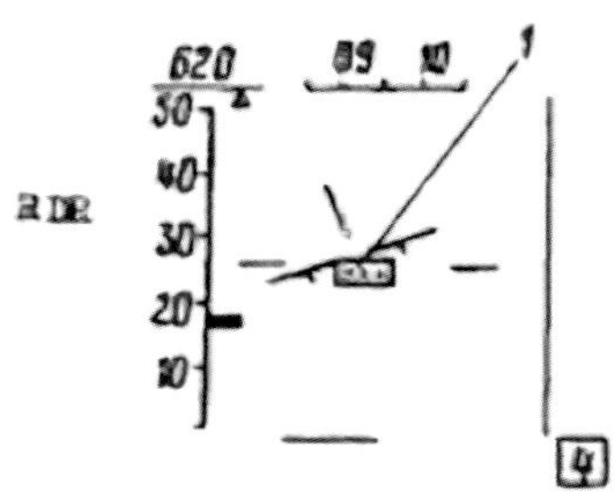

"PURSUIT (SCAN-WHILE-FLYBY)" RADAR MODE

1 - target blip

will appear above the friendly target blip. During the lockon the "Friendly" identification sign is indicated by the "C" (friendly) symbol flickering at about 2 Hz, and located at place of the heading scale with the INTERROG button depressed.

The absence of the identification mark (symbol) is indicative of the foe target.

Note: Even with the identification mark available request additional information from the guidance post that the target to be attacked is selected correctly.

After the target blip has reached the lockon range, gate it and lock an the target by depressing the WSM, LOCKON PF button. After locking on the target , symbol "A" is displayed on the screen of the integrated display system and the scan display is changed by the aiming one (Fig. 10).

118. For detecting the aerial targets, the TRACKWHILEFLYBY and AUTO modes of the radar may be used.

The TRACKWHILEFLYBY mode may be used provided the information on the aircraft direction is available, the mode is switched on by setting the RDR MODES selector switch to the ENCOUNTER, PURSUIT positions and the TRACKWHILEFLYBY FHS RHS selector switch, to the PHS Position. For tracking the target, set the AJ OFF CAJ selector switch to the OFF position. When the above selector switch is placed in the AJ position, the jammer is put into operation for the autotracking of the target in the TRACKWHILEFLYBY mode. In this case, the target is automatically gated (the most dangerous target among the other ones) and automatically locked on at the ranges close to those used for launching the missiles, which ensures the short duration and secrecy of the attack.

The target track while flyby setting is evidenced by the hop of the lockon gate onto the target on the screen of the integrated display system and by the gate stretching in range by 1.5 times. In this case, the scan zone centre will always be directed at the most dangerous target. The range mark of the automatic changeover from the track while flyby mode to the lockon one is displayed at the range scale on the integrated display system screen. After the target is tracked in the trackwhileflyby mode, maneuver in heading to place radar coverage in the middle of the integrated display system screen. In straight flight displacement of the gate and radar coverage in azimuth indicates the target maneuver direction. Maneuver the aircraft to hold the gate mid point in the centre of the integrated display system screen. For reaiming at the other target, depress the control button, gate the selected target, keeping the button depressed until the gate becomes 1.5 times greater. After this, the target is tracked in the trackwhileflyby mode. The target may be locked on manually by means of the WSM, LOCKON PF button after the gate is superimposed on the given target mark. For returning to the automatic selection of the dangerous target, depress the RESET button on the control panel.

The AUTO mode is selected by the RDR MODES (selector switch and this mode ensures the simultaneous detection of the targets flying both on the headon and followon courses as well as the gated target autotracking lockon.

Lock On Mode

119. Make sure that the integrated radar/sighting system looks on the target firmly and control the aircraft so as to align the aiming circlewith the electronic crosshairs. As soon as the aiming circle is aligned with the electronic crosshairs, the aircraft is directed to the predicted target intercept point in heading. The antenna position mark points to the target.

Before the "Zoom" command is applied, the manually selected target stepup (stepdown) separation is maintained. After the "Zoom" command is applied, the aircraft gain the target altitude. Once the target is looked on firmly. it is recommended to set AH,e a 0 which ensures smooth flying to the target altitude before the "Zoom" command is applied: The moment the "Zoom command is applied, the symbol is displayed and the aiming circle suddenly moves upward or downward. For the headup display, refer to Fig. 11.

Approach the target energetically, especially when attacking in the rear hemisphere, ensuring closure with a speed difference not less than 150 km/h at ranges greater than 15 km and with a speed difference not less than 50 km/h at ranges less than 15 km. Otherwise, breakdown of autotracking may occur.

When the integrated radar/sighting system operates in the lockon mode, the target yaw maneuver is indicated by stable change of the present aspect angle pointer position, deviation of the aiming circle from the screen centre of the integrated display system to right or to left and deviation of the antenna position mark.

When the integrated radar/sighting system functions in the ENCOUNTER or PURSUIT mode with the target maneuvering in yaw at the moment of the transition from the attack in the front hemisphere to the attack in the rear one (which is indicated by the target aspect angle index on the screen of the integrated display system and deviation of the aiming circle from the crosshairs), it in necessary to increase the airspeed of the friendly aircraft. This is required f or gaining the superiority of the friendly aircraft over the target in airspeed. If the given requirement is not observed during the operation of the

integrated radar/sighting system in the. ENCOUNTER mode, the friendly aircraft airspeed may be equalized with or become less than the airspeed of the target, resulting in the reset of the autotracking. For detecting the target after the lockon reset, it is necessary to set the RDR MODES selector switch to the PURSUIT position if it was in the ENCOUNTER _ position, to detect the target and to lock it on once more, maintaining the own airspeed at least 150 km/h greater than the target speed at a distance of more than 15 km to the target. When the distance to the target is less than 15 km, the closein or separation airspeed should be at least 50 km/h. During the repeated lockon of the target, estimate the nature of the change in the present range; if the present range has increased, estimate approximately the target airspeed (by the nature of the range change or by the data obtained from the command post) and intensively increase the airspeed.

During the lockon, when the target does not execute the yaw maneuver and when attacking the target in the rear hemisphere in the PURSUIT (P) mode, the target airspeed increase is evidenced by the retarded present range mark movement or the constant value of the present range for 10 to 12 s and the aspect angle pointer hesitation.

For ensuring stable tracking of the target, it is necessary to energetically increase the aircraft airspeed.. In this case, the lockon will continue either the aircraft lags or approaches the target.

The given peculiarities are typical for the integrated radar/sighting system operation during the lockon and in the PURSUIT AUTO modes at the background of the underlying surface (at the integrated radar/sighting system antenna tilt angles less than 5°) and with the range to the target exceeding 10 or 15 km as well as when the aircraft flies at altitudes less than 2.5 km at all ranges.

After launching missiles R27R at the ranges close to Dperm. min or against a maneuvering target at the moment the latter reaches aspect angles 4/4, index KP, "freezing" or irregular motion of the present range index to increase the reading may appear.

In this case, perform the breakaway in response to the BKWY command on the integrated display system screen or in response to the command given from the ground. After breaking the attack, reset the autotracking by depressing the RESET button.

120. When attacking the lowaltitude target (Htgt is less than 500 m) while flying along the aiming circle, the flight is executed along the variable profile. After locking on the target, arrive at an altitude of 2000 m and in response to the "Zoom" command, descend the aircraft to the flight altitude with the 500m elevation over the target.

Note: During the instrument direction, the ground automatic control system ensures the continuous support of the integrated radar/sighting system, therefore, in cage of the autotracking failure, perform the repeated search and lockon of the target without making the additional switching.

While approaching the target, the range scale in changed over from 100 km to 50 km, 25 km and 10 km, with the present range decreasing to the respective value.

WARNING: IN ALL OTHER CASES WHILE CLOSING WITH THE TARGET IN ATTACKING THE LATER BOTH IN THE REM AMD FRONT HEMISPHERES, THE ARRIVAL AT THE TARGET ALTITUDE IS ENSURED.

"Close Combat" Mode

At the ranges to the target being less than 10 km, the stable autotracking is ensured at the equalized airspeeds and on the lag, which makes it possible to launch the missiles (rockets) and conduct the maneuver aerial combat.

Aiming and Launching of Missiles

121. While aiming at the target, fly the aircraft, matching the aiming circle with the electronic crosshairs.

Check:

- the readiness for launching the selected missiles by illumination of the digital symbols on the windshield indicator and by presence of the sound signal for missiles equipped with infrared homing heads;

• the entry of the present range mark into the missile launch permissible range zone;
• the delivery of the light and voice LP (launch permit) signals.

122. Launch the missiles with the due observance of the recommendations laid down in Table 9;

• launch missiles R27R if the LP signal is applied irrespective of the position of the aiming circle with respect to the electronic crosshairs;
• launch missiles R60 if the LP signal is applied and the aiming circle centre is shifted from the centre of the electronic crosshairs by not more than one diameter of the circle.

Launch the missiles by depressing the RKT USL B trigger until the missile liftoff. The missile liftoff is indicated on the screen of the integrated display system by the fadeout of the missile readiness digital symbols and the presence symbol (line).

In case of the missile liftoff failure, the digital symbol is enclosed in a rectangle.

Notes: 1. In case of the target autotracking failure after the liftoff of missiles R27R repeat the target lockon.

2. In case of the employment of missiles R60, if the LP signal flickers (the disappearance of the present range mark), with the target being steadily locked on by the missile infrared homing head, launch the missiles by estimating the distance by eye. In this case, only the maximum range is displayed on the windshield indicator range scale, whereas the minimum range mark is set to zero.

123. After launching the missiles furnished with the radar homing head, continue target tracking, preventing the antenna * position mark from drifting beyond the integrated display system screen, until the target in destroyed or the "OTB." (BKWY) symbol is illuminated.
In response to the EM (OTB.) command, the aiming circle moves in jumps to the rightup or to the leftup and shows the recommended direction for breaking off the attack. When launching the missiles furnished with the infrared homing head, perform the breakaway just after the missile liftoff. Shift the trigger to the safety position if the attack will not be continued with the employment of the other missiles.

WARNING. IN CASE OF THE AUTOTRACKING FAILURE, AFTER LAUNCHING MISSILES R27R AT THE SHORT RANGES, ENERGETICALLY MAKE THE CORRECTIVE TURN TO THE TARGET TO ENSURE THE ILLUMINATION IN THE SCAN MODE BREAK OFF THE ATTACK AFTER DESTROYING THE TARGET.

124. After breakin off the attack, check the consumption of the ammunition allowance, reset the "Preparation" command (if this command has been applied), turn off the MASTER switch.

Set the ILLUM DUMMY OFF selector switch to the DUMMY position, and to the OFF position if the scan mode will not be used further on.

"CLOSE COMBAT" MODE

125. This mode is used in the target visual visibility conditions and ensures the look on of the non-maneuvering and energetically maneuvering target within the search zone (vertical) at the ranges from 10 km to 450 m.

For selecting the mode, it in necessary to proceed an follows:

• set the ACS MODES selector switch to the RDR position;
• set the RDR MODES selector switch to the CLOSE COMBAT position;
• set the ILLUM DUMMY OFF selector switch to the ILLUM position;
• use the SPAN knob to not the target size (the great, middle or small size)

Two vertical lines and the OPA symbol will appear an the screen of the integrated display system:

• maneuvering the aircraft, align the vertical zone with the aircraft being attacked;
• lock on the target by depressing the WSM, LOCKON PF button. The integrated display system screen displays the aiming information similar to the ENCOUNTER or PURSUIT mode, with the range scale amounting to 10 km;
• specify the aiming by aligning the aiming circle with the electronic crosshairs;
• after delivery of the LP command, launch the missiles;
• in response to the "Breakaway" command, break off the attack in the direction of the aiming circle.

126. Owing to the high airspeeds of the closure and the much time elapsed from the visual detection of the target to the delivery of the ILP command (4 to 6 s), the attack of the target in the front hemisphere calls for the special training of the pilot in executing the aiming operations and launching the missiles in case of the time deficiency (1 or 2 s).

For ensuring the better conditions for locking on the "afterburner" targets in the front hemisphere by missiles R60, it is necessary to attack the target at an aspect angle exceeding 0/4, not aligning accurately the aiming circle with the electronic crosshairs and orienting by reference to the target aspect angle symbol on the windshield indicator.

127. In conducting a close maneuver combat in the CLOSE COMBAT mode, the radar ensures the tracking of the target (after its lockingon) on the ground background at an aspect angle of 4/4.

When attacking the target accomplishing the descending maneuver in the presence of the lockon transition to the ground, reset the lockon and lock the target once more in the ascending maneuver.

IRR MODE

128. The IRR mode is used under the VFR conditions, in the complicated jammingto radar situation, in case of the radar failure and when attacking the target without own radar irradiation so as to enhance the secrecy of the target approach. The interception and attack of the aerial target in the IRR mode are executed in the rear hemisphere. Switch on the mode after the takeoff.

In the MR (GATE) mode, with the cooperation selected and the ranges measured by the integrated radar/sighting system in the quasiscan mode, the Dperm max mark (Dp,. min in equal to zero) is indicated on the screen. With the integrated radar/sighting system changed over to the main channel (and the laser range is not indicated), the D Perm max 1 and D marks are displayed, the D perm max and D per min.values may change regularly.

The target detection range in the IRR mode amounts to 25 10 km at the background of the free space and to 10 5 km at the background of the clouds. The target lockon range is less than the target detection range by 2 or 3 km.

For operating the integrated radar/sighting system in this mode, it is necessary to proceed as follows:

• set the ACS MODES selector switch to the MR position;
• use the FHS RHS selector switch to select the hemisphere to be attacked;

• use the SPAN knob to ascertain the target size (the large, middle or small size);
detect the target blip on the windshield indicator, making use of the information delivered from the command post on the target attitude with respect to the flight course;
• in the presence of the jamming and complication in discrimination of the target, remove a certain portion of the marks displayed on the windshield indicator by decreasing the sensitivity of the optical locator laser station with the aid of the IRR AMPL knob. In this case, the passive jamming symbol appears on the windshield indicator, which is indicative of the optical locator laser station desensitization; use the control knob to superimpose the gate on the target blip; depress the WSM, LOCKON PF button and keep it depressed until the target is locked on (for 2 or 3 s). The target designation to the missile infrared homing heads and preparation of the missiles are performed automatically after the optical locator laser station changes over to automatic tracking. The presentation of the mode information and aiming are similar to those as in the radar mode.
• as soon as the LP command is applied, launch the missiles by depressing the RKT MSL B trigger;
• break off the attack after launching the missiles or by the BM command (break off the attack by the BKWY command towards the aiming circle displacement).

For ensuring the preparation of the missiles for launching, if the target is locked on by the infrared homing head and no information an the target range is given, the LP command is delivered in the flickering mode. Launch the missiles in response to the LP command at the range less than the maximum permissible one determined by eye or according to the information delivered from the command post. In this case, the windshield indicator range scale reads only the maximum range, and the minimum range mark is set at zero.

CLOSE COMBAT (IRR) MODE

129. The CLOSE COMBAT (IRR) mode is used in the target visual visibility conditions when the target is attacked in the rear hemisphere at an aspect angle of up to 3/4.

The method of aiming and display are similar to those as in the CLOSE COMBAT radar mode.

For selecting the mode, set the ACS MODES selector switch to the CLOSE COMBAT (IRR) position and act as in the CLOSE COMBAT radar mode.

HELMET MODE

130. The helmet mode is used in the target visual visibility conditions at the great angles of sighting and in the deficiency of the time required for attacking the target. The helmetmounted target designation system delivers target designation information within the angles of +600 in azimuth and +6° to 14° in elevation to the radar, optical locator laser station and to the missile infrared homing heads as well as the delivery of the aiming and warning information to the helmetmounted sighting device (Fig. 5-4).

For operating the integrated radar/sighting system in the HELMET mode, proceed as follows:

• set the ACS MODES selector switch to the HELMET position;
• set the monocular of the helmetmounted sighting device to the . the operating position, adjust the brightness of the aiming mark with . the aid of . the . IRR AMPL HEL BRIGHT rheostat align the aiming mark centre on the helmetmounted sighting device with the target turning the head use the SPAN knob to ascertain the target size (the great, middle or small size);
• depress the IVSM, LOCKON PF button until the target is locked on by the radar, optical locator laser station or the missile infrared homing head and keep it depressed up to the launch (when the target is locked on by the missile infrared homing head, the audible signal is heard and the LP command starts flickering on the helmetmounted sighting device);
• release the WSM, LOCKON PF button.

Notes: 1. When the WSM LOCKON PF button in depressed, the missile infrared homing heads are target designated directly from the helmetmounted sighting device, irrespective of the target lockon either by the optical locator laser station or by the radar.

2. If the target is locked on by the radar or by the optical locator laser station, the "IRR missile lockon" command is displayed on the helmetmounted sighting device and the information representation of the target lockon mode by the radar or by the optical locator laser station and the target designation to the missile infrared homing heads are displayed on the screen of the integrated display system (with the WSM, LOCKON PF button released, target designation is delivered from the integrated radar/sighting system or from the optical locator laser station).

Launch the missiles by depressing the RKT, MSL, B trigger and aim at the target by reference to the information displayed on the screen of the integrated display system (by appearance of the LP command).

If the target is locked an by the missile infrared homing head (an audible signal is heard and the LP command is displayed on the helmetmounted sighting device), provided the aircraft is within the launch area (with the range estimated by eye), proceed as follows:

• keep the WSM, LOCKON PF button;
• track the target through the helmetmounted sighting device, launch the missiles, having estimated the range by eye.

Note. The employment of the helmetmounted target designation system calls for the thorough individual fitting of the crash helmet, since, while maneuvering the aircraft at gload n y > 6, the crash helmet may be displaced, which results in the visual lose of the information displayed on the helmetmounted sighting device.

OPTIC MODE

131. The OPTIC (optical) mode is used for aiming at the visually visible targets with the delivery of target designation to the radar, optical locator laser station and the missile infrared homing heads with the aid of the aiming mark controlled by the control knob.

For operating the integrated radar/sighting system in the OPTIC mode with the employment of the missiles furnished with the infrared homing heads, proceed as follows:

• set the ACS MODES selector switch to the OPTIC position;
• visually detect the target;
• use the FHS RHS selector switch to select the hemisphere to be attacked;
• ascertain the target size (the large, middle, or small size) with the aid of the SPAN selector switch;
• introduce the target into the field of view of the windshield indicator by executing the respective maneuvers of the aircraft;
• align the aiming mark with the target by use of the control knob;
• depress the WSM, LOCKON PF button and keep it depressed until the target is locked on by the radar, optical locator laser station or missile infrared homing head.

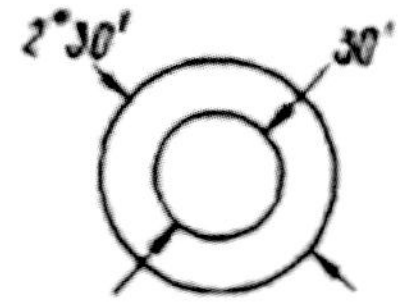	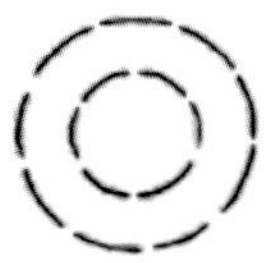	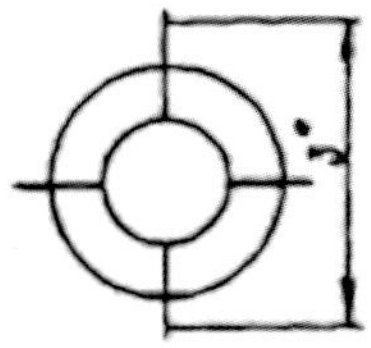	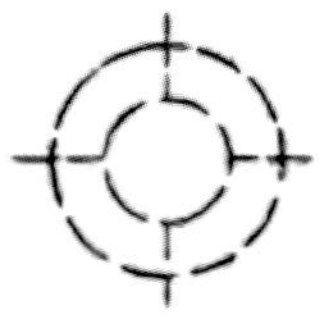	
When mode is selected	IHH MSL LOCKON command When target is locked on by radar or optical locator laser station (lamp flickers at frequency of 2 Hz) It is allowed to change over to integrated display system operation, release WSM LOCKON PP button	LP command It is allowed to depress firing button	When target is locked on by missile infrared homing head (lamp flickers at frequency of 2 Hz) It is allowed to depress firing button in determining launch permissible range visually	When helmet-mounted sighting device leaves scan zone (helmet-mounted sighting device display circuit is cut off). Displace head towards plane of aircraft symmetry and seat headrest

"HELMET-MOUNTED TARGET DESIGNATION SYSTEM" MODE
(DISPLAY ON HELMET-MOUNTED SIGHTING DEVICE)

When the target is locked on by the radar or optical locator laser station earlier than locking on the target by the missile homing head, the integrated display system screen illuminates command "All (attack) and information on the RM or IRR mode depending on the system which has locked on the target first;

- release the WS11 LOCKON PF button and proceed with further work as in the RER or IRR mode.

When the target is locked on by the missile homing head with the WSM LOCKON PF button depressed, the LP command will appear in the flickering mode and the missiles way be launched in determining the range by eye.

F Mode

132. The F mode is used for taking an aim and launching the missiles furnished with the infrared homing head in case of the failure of the abovementioned modes of the armament control system.

For taking an aim in this mode, the fixed electronic crosshairs is displayed an the windshield indicator.

For operating the integrated radar/sighting system in the F mode, proceed as follows:

- set the ACS MODES selector switch to the F position;
- use the FES RHS
- selector switch to select the hemisphere to be attacked;
- ascertain the target size (the large, middle or small size) with the aid of the SPAN selector switch;
- maneuvering the aircraft, superimpose the fixed electronic crosshairs of the windshield indicator on the target;
- as soon as the audible signal is heard (the target is locked on by the missile infrared homing head), launch the missiles having estimated the range by eye.

133. In case of absence of the electronic crosshairs on the windshield indicator when the F mode is selected, it is necessary to set the DAY NIGHT RETICLE selector switch to the RETICLE position, execute the aiming by reference to the fixed RETICLE crosshairs (Fig. 33) and launch the missiles if the audible signal is heard (the target in locked on by the missile infrared homing head), having estimated the range by eye.

GUN MODE

134. The GUN mode is used in firing the gun against the aerial targets. The mode is selected by shifting the GUN trigger to the intermediate position with the ACS MODES selector switch set in the RIM IRR CLOSE COMBAT, HELMET or OPTIC position.

It is allowed to conduct the gun fire in the AUTO mode (the selector switch is placed in the AL position) and in the CUTOFF mode (the selector switch is placed in the SINGLE 05 AL.

WARNING: 1. USE THE "AUTO" MODE ONLY IN THE COMBAT CONDITIONS.

2. IF THE GUN CONTINUES FIRING AFTER RELEASING THE FIRING BUTTON IF THE "AUTO" MODE OR IF THE BURST LENGTH EXCEEDS 2 S IN THE "CUTOFF" MODE, IMMEDIATELY SET THE "MASTER" SELECTOR SWITCH TO THE "OFF" POSITION AND DO NOT TURN ON THIS SELECTOR SWITCH AGAIN DURING THE TRAINING FLIGHTS.

It is allowed to use the mode by the "asynchronous fire" or "target predicted lead track" method.

135. The "asynchronous fire" method is realized in case of the target automatic lockon by the radar or optical locator laser station.

Prior to launching an attack, it is necessary to set the target span in meters for ensuring the continuation of the attack by the "target predicted lead track" method in case of the failure of the target lockon by the radar or optical locator laser station.

136. When attacking the visually visible target, proceed as follows:

- at a range of 1200 m and less (after the change in the display on the integrated display system screen), perform aiming by aligning the movable crosshairs with the target, with the subsequent alignment of this crosshairs with the aiming circle.

Note. At range D _ 1 km, the range scale is changed over.

- fire the gun by depressing the GUN trigger;
- break off the attack in response to the BKWY command.

137. When attacking the visually visible target (in the clouds, at night), proceed an follows:

- change the information representation on the screen of the integrated display system by depressing the control button for a short time.
- approach the target at the range less than 1200 m;
- align the aiming circle with the fixed central crosshairs by making the respective maneuvers (the coarse aiming);
- align the fine aiming mark (point) with the aiming circle (the fine aiming); fire the gun by depressing the GUN trigger; break off the attack in response to the BKWY command.

Notes: 1. The aimed firing in ensured at the ranges within 1200 to 2000 m. The effective firing range is 800 to 200 m.

2. When keeping the aiming circle on the electronic crosshairs, the fine aiming mark (point) should enter inside the aiming circle.

138. The "target predicted lead track" method is realized against the visually visible target in absence of the target automatic lockon by the radar or optical locator laser station.

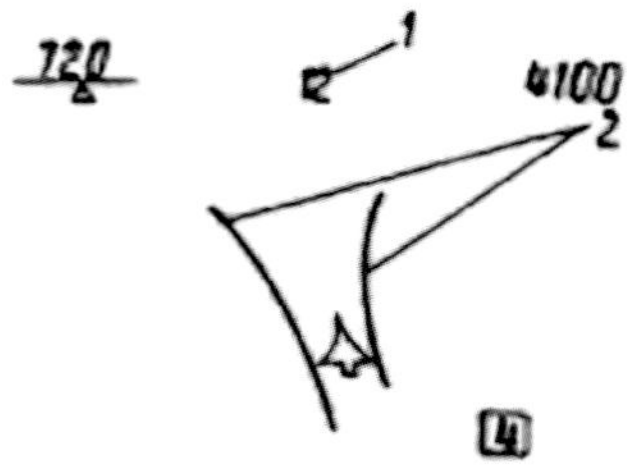

"GUN" MODE. "TARGET PREDICTED LEAD TRACK" METHOD

1 - target span; 2 - aiming track

Before launching an attack, set the target base in meters. When attacking the target, proceed as follows: maneuvering the aircraft, inscribe the target into the predicted lead track contour; fire the gun by depressing the GUN trigger; break off the attack at the safe range estimated by eye.

Note. The most accurate firing is effected in decreasing the speed of the target movement along the screen of the integrated display system at the moment the target is inscribed into the predicted lead track contour (due to the angular rate of the own aircraft turn) and the aircraft airspeed gets equalized with the target airspeed.

RETICLE MODE

139. The RETICLE mode is used when it is impossible to use the other aiming methods.

When firing the gun with the use of the fixed RETICLE for conducting the accompanying barrage fire, after the visual detection and identification of the target, set the DAY NIGHT RETICLE selector switch to the RETICLE position and shift the GUN trigger to the intermediate position.

Occupy the initial position for launching the attack. On approaching a range of about 400 m, estimate the true airspeed of the target, its aspect angle and, according to these data, select the firing zone (determine the maximum and minimum leads), verify the laying, decrease the angular rate of turn of the own aircraft and at the moment the target enters the

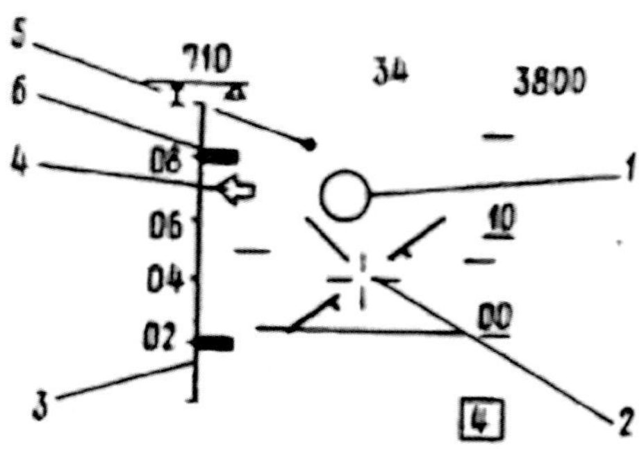

"GUN" MODE. "ASYNCHRONOUS FIRING" METHOD AT VISUALLY INVISIBLE AERIAL TARGET AT D < 1200 M

1 - coarse aiming circle; 2 - fixed crosshairs; 3 - $D_{perm.min}$ mark; 4 - present range mark in permissible firing range; 5 - fine aiming mark; 6 - $D_{perm.max}$ mark

firing zone (at the point corresponding to the maximum lead) open the fire and cease it at the moment the target leaves the firing zone (after passing the point an the sight RETICLE, corresponding to the minimum lead).

If the target has not been hit, verify the conditions for the attack (the aspect angle and airspeed of the target) and launch the attack again as described above, having increased the airspeed of the own aircraft (if it is possible).

The maximum total angular correction (lead 4t max depends on the airspeed, aspect angle of the target and, when firing the gun, it is equal (in ails) to the half of the product of the target flight true airspeed (in the tens of kilometers) by the numerator of the target aspect angle visible by the pilot in the eighth fractions of the unit.

The minimum total angular correction for the second burst should be two times less than the maximum one. For example, at the target aspect angle of 1/8 and target true airspeeds of 600, 800 and 1000 km/h, the maximum total corrections for the fire are equal to 30, 40, 50 mils, respectively, and the minimum ones, to 15, 20, 25 mils, respectively.

The greatest probability of the target destruction in conducting the accompanying barrage fire is ensured when firing in the bursts of 1 to 15 s in duration at a distance of 200 to 400 m.

PECULIARITIES OF LAUNCHING MISSILE R60 IN FRONT HEMISPHERE OF TARGETS

140. Missiles R60 are launched in the front hemisphere against the following targets flying at the reheat power settings and at the aspect angles from 1/4 to 4/4:

- against the nonmaneuvering targets flying at the altitudes of up to 8000 m, with the own aircraft true airspeed not exceeding 1100 km/h;
- against the targets maneuvering with the gload being up to 50, at the altitudes of 3000 to 5000 m and with the gload being up to 8.0 at the altitudes below 3000 m, with the own aircraft true airspeed not exceeding 850 km/h.

Attack the targets in the front hemisphere in the RDR, HELMET, OPTIC or F mode of operation of the armament control system.

The maximum range for launching the missiles against the maneuvering targets amounts to 5 km and against the nonmaneuvering targets, up to 12 km, with the maximum ranges at which the "afterburner" targets are locked on being within 5 to 12 km.

Launch the missiles at the ranges of at least 1500 m with the subsequent energetic breakaway for ensuring the flight at the range of at least 30° with respect to the target.

Note. For ensuring the effective and safe attack against the target, perform aiming at the aspect angle more than 1/4 with the use of the target designation modes.

VECTORED FLIGHT

141. During the instrument guidance, proceed as follows:

- establish the communication with the guidance control post;
- after receiving the "Fulfil commands" command, fly the aircraft in response to the guidance commands in the manual mode of the aircraft control;
- execute the directional control by aligning the directional circle (the smaller circle) with the electronic crosshairs on the screen of the integrated display system, checking the coincidence of the selected course and present heading on the combined course indicator;
- execute the climb in compliance with the interception flight program. Select the reference altitude for attacking the target independently after receiving the "Vertical" command, making use of the information on the target altitude, delivered to the integrated display system over the command radio link;
- when the "F" (reheat) command is illuminated, switch on the engine afterburner and execute the reheat program, accelerating the aircraft to the assigned airspeed value delivered to the integrated display system over the command radio link (during the instrument guidance, the true selected and own airspeeds of the aircraft are displayed on the screen of the integrated display system).

In case of the failure of the target designation commands delivered over the command radio link, change over to the manual control mode, having turned off the GUIDANCE switch and set the elevation (depression) in compliance with the target flight altitude.

When the aircraft is guided by voice, fly the aircraft annually, fulfilling the commands on the course, airspeed and altitude delivered from the guidance control post. According to the received information on the target flight altitude, not the selected elevation (depression) for the attack with the aid of the AR selector switch. The view of the integrated display system screen at the stage of the (ground guidance in shown in Figs 7 and 8.

INTERCEPTION FLIGHT PROGRAM

142. When intercepting the aerial target, depending on the airspeed and altitude of the target flight and the range of the target detection by the ground radars, the various power settings (the reheat, nonreheat and combined power settings) may be used.

The flight reheat power settings (flying at the reheat power setting from takeoff to the end of attack) are used at the minimum available tine required for executing the combat mission and for attacking the high altitude target.

The flight program is delivered from the guidance control post by voice or is determined by the pilot according to the data received from the ground automatic control system over the command radio link.

143. The reheat program comprises the following legs:

- the takeoff, climb to an altitude of 1000 m with acceleration up to M = 0.85;
- the climb to an altitude of 10,000 m at M = 0.85;
- the acceleration at the constant altitude of 10,000 m up to the indicated airspeed of 1250 to 1300 km/h;
- the acceleration with the climb at the constant indicated airspeed of 1250 to 1300 km/h up to the selected true airspeed or M = 2.2;
- the climb at the selected true airspeed or 11 = 2.2 till the "Zoom" command is given;
- the precision zoom after delivering the command from the integrated radar/sighting system (the irregular deflection of the aiming mark upward) or from the guidance control post by voice the closure with the target to the missile launching range and the missile launch. In response to the "Breakaway" command, the disengagement of the afterburner and return of the aircraft to the landing airfield.

144. The combined program of the flight consists of the following legs:

- the takeoff at the maximum power setting, the climb to an altitude of 1000 m with the acceleration of the aircraft to M 0.85;
- the climb to an altitude of 12,000 m at the maximum power setting at M 0.85;
- the flight to the target at an altitude of 12,000 m and M = 0.85;
- on illumination of the "F" (reheat) command or in response to the command and delivered from the guidance control post by voice, switch on the afterburner and accelerate the aircraft at an altitude of 12,000 m to an indicated airspeed of 1250 to 13W km/h (for improv-

ing the characteristics, it is allowed to accelerate the aircraft with a certain decent but not lower than to an altitude of 11,000 m);
• the climb at the constant indicated airspeed of 1250 to 1300 km/h until the assigned true airspeed or M 2.2 is obtained;
• the climb at the assigned true airspeed (Mach number) till the "Zoom" command in given;
• the precision zoom after delivering the command from the integrated radar/sighting system (the irregular deflection of the aiming mark upward) or from the guidance control post by voice;
• the closure with the target to the missile launching range and the missile launching;
in response to the "Breakaway" command, the disengagement of the afterburner and return of the aircraft to the landing airfield.

Notes:

1. The indicated programs of the flight are realized at the ambient air temperatures close to the standard ones.

2. When intercepting the highaltitude targets in the front hemisphere at the ambient air temperature exceeding the standard one by 10°C and more, go off the program at M = 1.8 to 1.9.

3. When the highaltitude targets are attacked in the rear hemisphere at the ambient air temperature exceeding the standard one by 10°C and more, the maximum airspeed of the targets being intercepted is decreased by 100 km/h.

4. If the altitude from which the attack is launched in below 10,000 m, go off the program to the target attack altitude and accelerate the aircraft to the terminal speed at this altitude.

145. When flying the aircraft at the altitude above 10,000 m and intercepting the targets flying at the altitude below 10,000 m, descend to the assigned altitude with the simultaneous setting of the assigned airspeed, but not in excess of 1000 km/h. At the altitudes below 3000 m in the IFR conditions and at night, maintain the airspeed on the flight path by increasing the engine power.

When descending in the day time under the VFR conditions, it in allowed to increase the rates of descent observing the flight safety precautions.

146. At the great distances from the airfield, after accomplishing the attack against the target, proceed with flying towards the airfield at an altitude of 12,000 m and M = 0.8 to 0.85

147. After attacking the highaltitude target, execute the descent to an altitude of 12,000 m, with the airspeed decreasing to M = 0.8 - 0.85.

148. After attacking the target flying at the altitude below 12,000 m, climb to an altitude of 12,000 m for returning to the landing airfield at M = 0.85 and the maximum power setting.

149. When approaching the landing airfield, execute the descent from an altitude of 12,000 m to the circling flight altitude at an airspeed of 500 km/h, with the engines running at the idle power setting.

ACTIONS AGAINST GROUND TARGETS

150. When acting against the ground targets, the armament control system in employed in the OPTIC, PITCHUP, and RETICLE modes.

EMPLOYMENT OF ROCKETS AND GUN

151. The rockets, types S24B, S8 and the gun may be employed against ground targets, with the armament control system operating in the OPTIC and RETICLE modes.

OPTIC Mode

152. In the OPTIC model the aiming procedure may be executed both with and without the preliminary fix of the target. The preliminary fix in performed provided the time sufficient for attacking the target is available. For aiming by this method proceed as follows:

• check to see that the ACS MODES selector switch in placed in the OPTIC position, the AIR GROUND selector switch is placed in the GROUND position (only for the gun), the AL SINGLE 0.5 AL selector switch is placed in the position corresponding to the flight mission assigned (only for the rockets);
• turn on the MASTER switch;
• go into a dive;
• shift t the GUN or MET MSL B trigger to the preliminary position;
• make certain by reference to illumination of the "A" (attack) symbol that, with the dive angles exceeding 10° and the target present range being lose than Dperm.max + 500 m, the laser range finder in automatically triggered If the "A" symbol is not available, trigger the laser range finder manually by depressing the control button;
• depress the WSM LOCKON PF button and keep it depressed; in this case, the aiming mark will assume the position for fixing the target;
• align the aiming mark with the target and release the WSM LOCKON PF button. At the moment the button is released, the target is preliminarily fixed;
• align the aiming mark with the target once more and on reaching the permissible firing ranges, depress the GUN or RKT MSL B trigger (the D pres index is located within the effective fire zone, and the audible signal is heard);
• after ceasing the fire, break off the attack.

Notes: 1. When attacking the ground targets, the minimum time between the release of the WSM. LOCKON PF button and the depression of the GUN. trigger should be at least 2 s, the maximum time, not more than 10 s, the most accurate fire is effected within the interval of 4 s.

2. It is allowed to refax the target by repeatedly depressing the WSM, LOCKON PF button while attacking the target.

3. If, after depressing the control button, the "A" symbol is not illuminated, it is allowed to continue attacking the target. In this case, the target range is measured by the less accurate elevation method.

4. On reaching the minimum permissible firing range, the BKWY flickering symbol appears on the screen of the integrated display system.

153. Execute the aiming procedure in the OPTIC mode without the preliminary fix in case of the deficiency of the time (altitude) required for attacking the target. When aiming without the preliminary fix of the target, it is necessary after going into a dive triggering the laser range finder, to proceed as follows:

• determine by eye the corrections for the wind velocity and the target speed;
• bring off the aiming mark with respect to the target by the correction value and, on reaching the permissible firing ranges, depress the GUN

or RKT. MSL B trigger (the D pres index is within the effective fire zone and the audible signal is heard in the earphones);

154. The RETICLE mode in used in case of the failure of the OPTIC mode. For performing the operations in the RETICLE mode, proceed an follows:

- not the RETICLE DAY NIGHT selector switch to the RETICLE position, check to see that the AL SINGLE 0.5 AL selector switch is placed in the position in compliance with the flight mission assigned (for the rockets) and the AIR GROUND selector switch is placed in the GROUND position (for the gun);
- turn on the MASTER switch;
- go into a dive;
- shift the GUN or MET MSL B trigger to the preliminary position;
- determine by eye the target range, correction for the wind velocity and the target moving speed with the due allowance for the ballistic characteristics of the weapons employed (Table 13);
- bring off the crosshairs centre with respect to the target for the correction value and an reaching the recommended firing ranges, depress the GUN or MET MSL B trigger;
- after ceasing the fire, break off the attack.

Table 11

Type of weapon	Firing range, m	Airspeed during firing, km/h	Angle of dive, deg	Angular corrections in plane of aircraft symmetry, mils
S-24	1800	800	20	60
S-8	1600	800	20	43
Gun	1500	800	20	14

EMPLOYMENT OF AERIAL BOMBS

155. The aerial bombs may be employed with the armament control system operating in the OPTIC, PITCHUP and RETICLE modes.

OPTIC Mode

156. In the OPTIC mode, the aiming procedure any be executed both with and without the preliminary fix of the target. The preliminary fix is performed provided the time sufficient for attacking the target is available.

157. The aiming with the preliminary fix is ensured during the bombing at the dive angles of up to 40°:

- check to see that the ACS MODES selector switch is placed in the OPTIC, AL SINGLE 0.5 AL position in compliance with the mission assigned;
- turn on the MASTER switch;
- go into a dive on the combat course, with the dive angle of up to 40°;
- shift the RKT KSL B trigger to the preliminary position;
- make certain by reference to illumination of the "A" (attack) symbol, that, with the dive angles exceeding 10° and the target present range being about 3500 m, the laser range finder is triggered. When the "A" symbol is not displayed, trigger the laser range finder manually by depressing the control button;
- depress the WSM, IMKCU PF button located on the throttle control lever and keep this button depressed; in this case, the aiming mark will assume the position required for fixing the target;
- align the aiming mark with the target and release the WSM, LOCKON PF button. At the moment the button is released, the target is preliminarily fixed;
- align the aiming mark with the target once more and depress the RKT MSL B trigger.

The aerial bombs are dropped as soon as the RKT MSL B trigger is depressed, provided the angle of sighting and angle of bomb drop are equal.

If the bomb drop angle exceeds the sighting one, then, after depressing the RKT ML B trigger, the presentation displayed on the windshield indicator is changed the assigned gload circle and the present gload vector appear, the scale range is replaced by the scale showing the time remaining till the bomb drop. Further on, keeping the firing button depressed till the bomb drop, it is necessary to keep the end of the present gload vector aligned with the assigned gload circle centre. Just before dropping, an audible signal in heard in the earphones.

158. The aiming without the preliminary fix is used during the bombing from the level flight or bombing from a dive at the angles of up to 40° in case of the deficiency of the time (altitude) required for accomplishing the attack. The correction for the wind velocity and the target moving speed is made by eye. For executing the aiming procedure by said method, proceed as follows:

- check to see that the ACS MODES selector switch is placed in the OPTIC position;
- intercept the bombing course in the level flight (go into a dive on the combat course, with the angle of dive of up to 40°);
- shift the RKT MSL B trigger to the preliminary position;
- trigger the laser range finder (in case of bombing from the level flight) 3 or 5 s before superimposing the aiming mark on the target (carry out the check by reference to illumination of the "A" symbol on the screen of the integrated display system;
- determine by eye the corrections for the longitudinal and lateral drifts of the aerial bomb by wind for the time of its fall and the target moving Speed;
- bring off the aiming mark centre with respect to the target for the correction value, depress the RKT ML B trigger and keep it depressed until the bombs are dropped. The aerial bombs are dropped at the moment the RKT MSL B trigger is depressed provided the angle of sighting and the angle of bomb drop are equal. If the bomb drop angle exceeds the sighting one, then, after depressing the RKT KSL B trigger, the presentation displayed on the windshield indicator is changed the present gload vector and the assigned gload circle appear thereon, the range scale is replaced by the scale showing the time remaining till the bomb drop;
- proceed with level flying or diving until the bombs are dropped, aligning the end of the present gload vector with the assigned gload circle centre. 1.5 to 3 s before zeroing of the remaining time an audible signal is heard in the earphones and on expiration of this time, the aerial bombs are dropped. Check the aerial bomb drop by refer-

ence to the fadeout of • the aerial bomb station status indexes an the windshield indicator;
• after dropping the aerial bombs, energetically maneuver the aircraft to break off the attack.

PITCHUP Mode

159. The PITCHUP mode ensures the tons bombing at the angles of 110 to 130°.

In this mode, the aiming procedure in executed in the following sequence:

• check to gee that the ACS MODES selector switch is placed in the PITCH position;
• intercept the bombing course;
• 3 to 5 s before superimposing the aiming mark on the target, trigger the laser range finder by depressing the control button;
• on aligning the aiming mark with the target, depress the RKT MSL B trigger and keep it depressed until the aerial bomb drop. At the moment the trigger is depressed, the range scale in replaced by the scale showing the time remaining till the beginning of going into the pitchup;
• proceed with the level flight, aligning the assigned gload command circle with the end of the present gload vector;
• I assigned gload command circle;
• 2 present gload vector 1.5 to 3 s before going Into the pitchup, an audible signal is heard in the earphones;
• At the moment of zeroing of the remaining time, the audible signal is not heard in the earphones, and the "G" (zoom) symbol appears and the assigned gload command circle moves in jumps upward.
• In case oil overflying the target with lateral deviation, movable mark may deflect to the left (right) too. Displacement of the movable mark should not be followed up.
• The scale shows the time remaining till the aerial bomb drop;
• bring the aircraft into a halfloop, aligning the end of the present gload vector with the command circle centre. In this case, the gload will be within 4 to 5 g; at the pitch angles of 90 to 95°, the assigned gload vector (command circle) may deflect to the right, then to the left with the subsequent return to the initial position. The command circle deflection should not be followed up by the aircraft piloting;
• 1.5 to 3 s before the drop an audible signal is heard in the earphones;
• check the drop of the aerial bombs by reference to the fadeout of the station status symbols on the screen of the integrated display system.

Note. It is allowed to release the RKT MSL B trigger after fixing the target, with the subsequent depression of the trigger while executing the halfloop.

RETICLE Mode

160. In case of the OPTIC mode failure, it is allowed to perform the bombing from the level flight and dive, taking an aim with the use of the fixed RETICLE by reference to the computed angle and observing the flight conditions during the aiming procedure.

161. For carrying out the bombing with the aid of the fixed RETICLE, it is necessary to set the DAY NIGHT RETICLE selector switch to the RETICLE position and depress the MANUAL PREPARE lightbutton.

When bombing in the conditions when the computed correction is within the vertical line of the RETICLE, it is necessary to align the computed point with the target in the assigned conditions and press the RKT MS1 B trigger.

While bombing in the conditions when the computed correction is but of the limits of the vertical line of the RETICLE, it is necessary, aligning the target with the outer lead circle to vertical line intersection point, to read out the time according to the computed data and depress the RKT USL B trigger.

EMPLOYMENT OF MISSILES R60 AGAINST ST GROUND HEAT CONTRAST TARGETS

162. Launch the missiles against the ground heatcontrast targets singly under the get visual visibility conditions, with the use of the OPTIC, go and RETICLE modes of the armament control system, at the dive angles of 15 to 45° and ranges of 1. 5 to 3 km.

Note. When attacking the ground targets, the proximity optical fuze is not cut off.

Launch an attack from the direction ensuring the maximum heat contrast of the target.

Take an aim by superimposing the circle on the target with the aid of the control button or by maneuvering the aircraft.

After locking on the target by the missiles and delivering the LP command. launch the missiles at the recommended ranges and, energetically maneuvering the aircraft, break off the attack to fly over the burst area at an altitude of at least 500 m.

MISSILE EMERGENCY LAUNCH AND EXTERNAL STORE EMERGENCY RELEASE

163. It is impossible to fly or land the aircraft carrying the weapons and if the necessity arises to release them from the tactical considerations, it is necessary to execute the emergency launch or release of the weapons.

164. The emergency (unguided) launch of missiles R60 is usually executed throughout the entire operational range of altitudes and airspeeds of the aircraft flight at the gloads not exceeding 2.0 g and the sideslip not exceeding half the diameter of the ball. whereas missiles R27R are launched in the name conditions, but from the level flight (n_y = 1. 0).

Perform the emergency launch of rockets S24B together with launcher, the empty, partially or fully equipped pods, type B8, in the level flight, without the banking and sideslipping, at the altitudes above 300 m and airspeeds of 500 to 800 km/h, with the Mach are not exceeding 0.8.

Perform the aerial bomb "armed" and "safe" releases under the conditions required for the bomb combat employment. Determine the minimum safe bombing altitudes depending on the fuzing of the aerial bombs in compliance with the Combat Employment Manual.

Drop the belly drop fuel tank, filled to capacity or empty, when flying level, without banking and slipping, at the airspeeds ranging from 450 to 950 km/h, at Mach numbers not exceeding 0.85. Drop the belly drop fuel tank with partially used fuel when flying without banking and slipping at the airspeeds ranging from 600 to 700 km/h. at Mach numbers not exceeding 0.85 and n = 1.5 to 2.0.

CAUTION. DROP OF THE EMPTY BELLY DROP FUEL TANK LEADS TO THE AIRCRAFT OVERBALANCE WHICH INCREASES AS THE AIRSPEED RISES. AT THE AIRSPEED OF 900 KM/H, THE AIRCRAFT OVERBALANCE AT THE MOMENT OF THE TANK DROP LEADS TO REACHING nmw 1.5 AND ALTITUDE LOSS DOWN TO 150 M.

Perform the emergency drop of the empty, partially or fully equipped small load packs in the level flight without slipping and banking , at H 4 5000 m. within the airspeed range of 500 to 900 km/h, at H > 5000 m with the airspeeds not exceeding 1100 km/h (M < 1.5).

The minimum altitude of dropping the packs is 100 m.

Releasing each pair of the small. load packs at the airspeeds exceeding 900 km/h decreases n to 0.5. The aircraft overbalance is easily counteracted there is no tendency to overswing, but don't fail to keep the EMERG RELEASE button depressed to drop all four small load packs. Depress the button once more if not all small load packs are dropped (the button may be released inadvertently because of a negative gload).

165. Before launching or releasing the external stores, select the recommended flying regime.
Launch the missiles (rocket by depressing the STBY LAUNCH button and keep it depressed until the missile (rocket) liftoff.

For releasing the aerial bombs, select the release mode with the aid of the ARMED SAFE selector switch, depress the EMERG RELEASE button and 'keep this button depressed till the aerial bombs are released.

After launching or releasing the external stores, check their liftoff (release) visually and on the screen of the integrated display system, set the ARMED SAFE selector switch to the SAFE position (or check to see that this selector switch is placed in the said position).

SAFETY PRECAUTIONS TO BE OBSERVED DURING EMPLOYMENT 0F ARMAMENT

166. After launching missiles R27R, R60 and firing the gun, check the engines for the proper power setting.

Perform the "armed" release of the aerial bomb in the flight conditions ensuring the recovery from the splinter effective zone or flying over the burst area at the altitude of at least 700 m.

If it is impossible to release the external stores in flight, take a decision for landing the aircraft depending on the limitations of the aircraft landing mass with the given ammunition allowance.

If missile R27R or R60 fails to lift off with the launching conditions an ensured and the firing button depressed, continue accomplishing the mission with the remaining ammunition allowance aboard the aircraft.

WARNING. IF MISSILE R27R FAILS TO LIFT OFF IN RESPONSE TO DEPRESSION OF THE FIRING BUTTON, IT IS PROHIBITED TO PERFORM THE UNGUIDED LAUNCH OF THE MISSILE WITH THE USE OF THE "STBY LAUNCH" BUTTON.

When it is necessary to release all the external stores in the combined variants of the guided missiles and rockets or the aerial bombs, depress the EMERG RELEASE button to release the external stores or depress the STBY LAUNCH buttons to perform the unguided launch in the sequence dictated by the presence of the necessary flying conditions.

In case of the failure of the aerial bombs release over the firing and emergency circuits when the bombs are "hungup", make certain that the bombs are reliably suspended by executing a number of maneuvers within the range of the operating gloads and make a decision for leading the aircraft; in this case, check to see that the ARMED SAFE selector switch is placed in the SAFE position, and if the guided missiles are not available aboard the aircraft, additionally turn off the WEAPON switch. After the landing roll, stop the aircraft an the runway for inspection of the ammunition and for unscrewing the fuzes therefrom.

WARNING. Never taxi the aircraft carrying the fuzed aerial bombs which have failed to be released over the firing and emergency circuits.

In case of the demolition of missiles R27R or R60 after their launch on the flight path, fly the aircraft so as to pass through the centre of the missile demolition cloud.

In case of the unsteady operation of the aircraft power plant after the employment of the missiles (rockets) or gun and if the engine operation parameters reach the inadmissible values, act in compliance with the recommendations land down in Section "Emergencies in Flight".

WARNING. NEVER TAXI THE AIRCRAFT CARRYING THE FUZED AERIAL BOMBS WHICH HAVE FAILED TO BE RELEASED OVER THE FIRING AND EMERGENCY CIRCUITS.

In case of the demolition Of missiles R27R or R60 after their launch On the flight Path, fly the aircraft so as to pass through the centre Of the missile demolition cloud.

In case of the unsteacy operation of the aircraft Power Plant after the employment of the missiles (rockets) or gun and if the engine operation parameters reach the inadmissible values, act in compliance with the recommendations laid down in Section "Emergencies in Flight".

Section Six

EMERGENCIES IN FLIGHT

167. The recommendations on the actions in emergencies shall be obligatorily fulfilled. It is supposed that in all other complicated situations, the pilot acts depending on the arised situation.

On detecting some failure of the aircraft equipment, calmly check to see whether a mistake has been made in handling the valves and selector switches, estimate the situation and take a competent decision. Report the failure and the decision taken to the flight control officer. Further on, act depending on the created situation with due regard for the directions given by the flight control officer of command post.

In any emergency situation, when the pilot may need the help from the command post, turn on the DISTRESS switch of the identification system.

On illumination of the All (emergency reception) lamp on the control panel of the communication radio set, turn on the ER switch, listen to the information transmitted over the emergency receiver and immediately deliver this information to the flight control officer on the operating communication channel with indication of the own coordinates.

If the pilot's life is endangered in flight, the pilot must immediately abandon the aircraft.

ENGINES AND AIRCRAFT ACCESSORIES GEAR BOX FIRE

168. The symptoms:

- the critical mode warning lightbutton flickers;
- the LH ENG (RH ENG) FIRE signal illuminates;
- the "LH (RH) engine fire" ("Aircraft accessories gear box fire") voice information is heard in the pilot's earphones.

The additional symptoms:

- smoke or flame observed by the pilot through the rear view mirror, when turning the aircraft and also frow. the ground or by the other aircraft;
- the appearance of smoke in the cabin.

169. The actions to be taken in the parking area (while taxiing):

- shut down the engines at once (brake the aircraft wheels);
- close the fuel shutoff valve on the engine caught by the fire;
- set the EXTING selector switch to the required position;
- after stopping the aircraft, leave the aircraft.

170. The actions to be taken during the takeoff run (till the unstick):

- shut down the engines, for which purpose set the throttle control levers to the STOP position;
- take all measures to stop the aircraft;
- close the shutoff valves;
- set the EXTING selector switch to the required position;
- after the aircraft has come to a halt, immediately leave the aircraft.

If there is a danger of collision with the obstacles or in case of another immediate danger to the pilot's life, abandon the aircraft at an airspeed exceeding 75 km/h, if the airspeed is below 75 km/h, jettison the canopy and after running off the runway, evade the obstacle (retract the lending gear, if required) and deenergize the aircraft.

171. The actions to be taken after unsticking:

(a) in case of the aircraft accessories gear box fire:

- proceed with the takeoff;
- set the EXTING selector switch (aircraft accessories gear box);
- climb to the required altitude In the direction safe for the pilot ejection and fall of the aircraft;
- make certain that the aircraft accessories gear box is still on fire, shut down the engines (if possible) and abandon the aircraft;
- if the fire in the aircraft accessories gear box has been extinguished (which is indicated by the fadeout of the signal and absence of the additional symptoms of the fire), immediately land the aircraft;

(b) in case of the fire in the engine compartment:

- proceed with the takeoff;
- set the EXTING selector switch to the LH (RH) ENG position; without changing the engine power, climb to the required altitude and gain the necessary airspeed in the direction safe for the pilot ejection and fall of the aircraft;
- shut down the engine caught by fire;
- close the shutoff valve of the engine caught by fire;
- make certain that the engine is still on fire, shut down the second engine (if possible) and abandon the aircraft;
- if the fire has been extinguished, which in indicated by the fadeout of the LH (RH) ENG FIRE signal and absence of the additional symptoms of the fire, immediately land the aircraft.

172. The actions to be taken in flight:

(a) in case of the fire in the aircraft accessories gear box:

- shut down the engine afterburner;
- set the EXTING selector switch;
- turn the aircraft into the direction safe for the pilot ejection and fall of the aircraft;
- having made certain that the aircraft accessories gear box is still on fire, shut down the engines (if possible) and abandon the aircraft;
- if the fire has been extinguished (which is indicated by the fadeout of the ICA FIRE signal and absence of the additional symptoms of the fire), immediately land the aircraft;

In case of the fire in the engine compartment:

- shut down the engine afterburner;
- set the throttle control lever to the STOP position and close the shutoff valve of the engine caught by fire;
- disconnect the automatic flight control system by depressing the AFCS MODE OFF red button on the control stick for a short time (the DAMPER mode is preserved);
- set the EXTING selector switch to the *LH (RH) ENG* position;
- having made certain that the engine compartment is still on fire, shut down the ines (if possible) and abandon the aircraft;
- if the fire has been extinguished (which in indicated by the fadeout of the 1E (RE) EKG FIRE signal and absence of the additional symptoms of the fire), immediately land the aircraft.

FAILURE OF ENGINE

173. The symptoms (with the throttle control lever position remaining invariable):

- change in the engine(s) sound;
- the decrease in the longitudinal acceleration (flight airspeed);
- the onset of the turning and rolling moments;
- the fadeout of the LH (RH) ENG REHEAT annunciator;
- the sharp drop of the engine speed.

174. The actions to be taken at the takeoff run with the concrete runway length equal 1800 m and over:

(a) at a speed below 200 km/h or takeoff mass exceeding 16,500 kg:

- discontinue the takeoff;
- take all the measures to stop the aircraft movements;
- shut down the failed engines if necessary, shut down the second engine.

In case of risk of collision with obstacles or direct threat to the pilotla life at a speed exceeding 75 km/h, eject if a speed is below 75 km/h, jettison the canopy for emergency; having rolled out from the runway, turn away from the obstacle (if necessary, retract the landing gear) and deenergize the aircraft.

175. The actions to be taken after unaticking:

at an airspeed exceeding 200 km/h and takeoff mass below 16,500 kg;
- when taking off at the reheat power setting, place the throttle control levers of both ngines to the MAXIMUM position;
- continue executing the takeoff with one engine running.

WARNING: THE ENGINE FAILURE IMMEDIATELY AFTER UNSTICKING CAN CAUSE THE ALTITUDE LOSS UP TO TOUCHING THE RUNAWAY WITH THE WHEELS FOLLOWED BY UNSTICKING AFTER SHORTTIME RUN;

- hold the aircraft free from banking and yawing retract the landing gear after stable unsticking before reaching an airspeed of 350 km/h. climb at a rate of 1 to 2 m/s, then increase the rate of climb up to 3 to 5 m/s;
- at an altitude of at least 100 m retract the flaps;
- set the throttle control lever of the failed engine to the STOP position If the engine speed is decreased to less than 45%;
- if necessary, release the external stores when flying over a safe area and execute landing.

Note. If the runway length is below 1800 m, continue executing takeoff only in case of one engine failures at speeds ensuring the aircraft unsticking within the runway limits specified for the actual takeoff mass according to Fig. 56a. As for other cases, take the actions as per Para. 174a.

176. The actions to be taken in flight:

- counteract the turning and banking momental
- immediately shut down the engine afterburners
- disengage the AFCS by depressing the AFCS MODE OPP red button on the control stick for a short time (the DAMPER mode is maintained)
- set the throttle control lever of the failed engine to the STOP position if the engine speed in decreased to less than 45%.
- proceed to the nearest airfield and execute lending with one engine running.

When flying with one engine running, maintain an airspeed of 400 to 600 km/h, counteract the turning moment by manipulating the rudders, while avoiding the rudder sharp deflection.

In case of wind from the failed engine side during landing, check to see that the cronswind component should not exceed 10 m/s.

POWER PLANT SURGE

177. The symptoms:

- the Presence of the sound in the area of the air intakes or engines an well as the longitudinal jerks caused by the sharp increase in the engine thrust.
- the absence of the engine speed rise with the increase in the temperature during acceleration
- the engine speed drop with the possible rise of the exhaust gas temperature, with the throttle control lever position remaining invariable
- the extension of the air intake ramps
- the illumination of the LH (RH) ENG START (3AnYCK AEB. MPABJ) annunciatorl
- the possible flickering of the light button accompanied by the illumination of the REDUCE RPM LH (RH) ENGINE annunciator and the OVERHEAT LEFT (RIGHT) signal on the EKRAN annunciator panel delivery of the "LH (RH) engine overheat" voice information and the "Reduce RPM of LH (RH) engine" command.

178. The actions to be taken;

- counteract the turning and roll moments
- shut down the afterburners of both engines
- Set the throttle control lever of the surged engine to the STOP position if the surge has not been eliminated by the antisurge system for 5 or 8 s
- check the ramp position.

Note: If, prior to relighting the engine the ramp assumes the 100 % position by reference to the ramp position indicator, it in necessary to decrease the airspeed, shift the ramp manually to the 0% position against the ramp position indicator by depressing the RMP EMG RETR selector switchs.

- relight the engine;
- check to see that the engine has gained the idle speed;
- approach the airfield and land the aircraft.

INADVERTENT SHUTDOWN OF BOTH ENGINES IN FLIGHT

179. The symptoms:

- the energetic deceleration of the aircraft;
- the sharp drop of the engine speed and exhaust gas temperature.

180. The actions to be taken:

- disconnect the automatic flight control system by depressing the APCS MODE OFF red button located on the control stick for more than 3s till the DAMPER lightbutton extinguishes;
- set the throttle control lowers to the STOP position if the engine speed ha. dropped to less than 45%;
- relight the engines at a time (it is allowed to relight the engines in succession, beginning with the left engine);
- approach the airfield and land the aircraft.

WARNING. IF NEITHER OF WE ENGINES HAS BEEN REMGHTPED TO AN ALTITUDE OF 2000 M, ABANDON THE AIRCRAFT.

AIR RELIGHT

181. The reliable air relight is ensured at the rotor rundown and windmilling speed being 12% and more.

The engines are relighted at the following airspeeds:

- 400 to 1000 km/h at the altitudes below 12,000 m;
- 550 km/h to M = 1.8 at the altitudes ranging from 12,000 to 17,000 m the air relight is ensured from the speed of at least 50%.

Notes:

1. For improving the air relight reliability, it is necessary to maintain the airspeed close to the upper limit, which results in increasing the windmilling speed.

2. The amount of oxygen aboard the aircraft makes it possible to perform five air relights with due allowance for five startings of the turbostarter on the ground.

The basic type of the air relight is the antiblowout one (the automatic or manual) air relight in case of the HP rotor speed drop to 40%. The air relight is executed as follows:

(a) the automatic air relight is executed if the HP rotor speed drops below 50 % and the throttle control lever assumes the position from the idle to the full reheat power by automatically cutting'in the air relight system; in this case, the LH (RH) ENG START annunciator located on the right console illuminates, the engine speed increases, and the exhaust gas temperature rises to the values specified by the throttle control lever position;

(b) the semiautomatic air relight in executed by setting the throttle control lever to the STOP position for 2 or 3 s and then, to any position (from the idle to the maximum power); in this case, the LH (RH) ENG START annunciator illuminates an the right console;

(c) the manual air relight in executed by setting the throttle control lever to the STOP position, turning on the LH (RH) ENG AIR RELIGHT switch and shifting the throttle control lever to the idle power position. After the engine has gained the idle speed, turn off the AIR RELIGHT switch, but not later than 100 s after the moment of its turningon. The annual air relight is used in case of the failure of the automatic and (or) semiautomatic air relights; in this case the LH (RH) ENG AIR RELIGHT annunciator does not illuminate.

The time required for relighting the engine in flight does not exceed 70 s.

When any method of the engine relighting is used, after the engine has gained the power determined by the position of the throttle control lever, check the engine operation by smoothly shifting the throttle control lever. If the engine is beyond the control, set the throttle control lever to the STOP position and repeat the air relight at the high airspeed.

FAILURE OF ENGINE OIL SYSTEM

183. The symptoms:

- the KSTs lightbutton flickers;
- the LH (RH) ENG OIL PRES annunciator illuminates;
the OIL PRESSURE LEFT (RIGHT) or OIL TEMP LEFT (RIGHT) signal illuminates on the EKRAN annunciator panel;
- the "Low oil pressure in LH (RH) engine" or "LH (RH) engine oil temperature exceeds tolerable value", or "Chip in oil of LH (RH) engine", voice information and the "Reduce RPM of LH (RH) engine" command are delivered.

184. The actions to be taken:

- shut down the engine afterburner;
- decrease the engine power till the failure signal is not applied;
- shut down the engine if the signal continues illuminating;
- approach the airfield and land the aircraft with one engine running.

FAILURE OF ENGINE SPEED AND EXHAUST GAS

TEMPERATURE CONTROL SYSTEM

185. The symptoms:

- the KSTs lightbutton flickers;
- the illumination of the LEFT (RIGHT) ENG STBY SYS signal on the EKRAN annunciator panel;
- the delivery of the "LH (RH) engine from the standby system" and the "Watch temperature and RPM" command.

186. The actions to be taken:

- discontinue the flight mission;
- prevent the throttle control lever from sharp movements;
- watch the exhaust gas temperature and engine speed, avoiding increasing the engine steed in excess of the maximum tolerable values;
- set the engine power required for returning the aircraft to the airfield;
- approach the airfield and land the aircraft.

INADVERTENT SHUTDOWN OF AFTERBURNER

187. The symptoms:

- the fadeout of the LH (RH) ENG REHEAT annunciator located on right console;
- the energetic deceleration of the aircraft (the change in the longitudinal gload);
- the appearance of the rolling and turning moments.

188. The actions to be taken:

- counteract the rolling and turning moments;
- set the throttle control levers to the MAXIMUM position;
- Set the REHEAT-EMERG OFF selector switch to the EMERG OFF position;
- do not switch on the afterburner for the second time;approach the airfield and land the aircraft.

FUEL PRESSURE DROP DOWNSTREAM OF SERVICE TANK BOOSTER PUMP

189. The symptoms:

- the KSTs lightbutton flickers;
- the illumination of the "BOOSTER SWITCH OFF" on the EKRAN annunciator Panel;
- The delivery of the "No fuel feed. Shut Down Afterburner."

190. The actions to be taken;

- shut down the Afterburner and decrease the engine power;
- perform the flight at an altitude not above 5000 m;
- avoid the negative and near zero gloads;
- approach the airfield and land the aircraft.

FAILURE OF ENGINE AUTOMATIC FUEL CONTROL SYSTEM

EXCESSIVE VIBRATION, ENGINE OVERHEAT

191. The symptoms:

- the KSTs lightbutton flickers;
- the illumination of the REDUCE RPM LH (RH) ENG Annunciator;
- the illumination of the OVERHEAT LEFT (RIGHT) and/or LEFT (RIGHT) OVERSPEED and/or FUEL PRES LEFT (RIGHT) signals on the EKRAN Annunciator panel;
- the delivery of the "LH (RH) engine overheat" and /or LH (RH) engine vibration" and/or "LH (RH) engine speed exceeds tolerable" voice information and "Reduce "LH (RH) engine speed" command.

192. The actions to be taken:

- switch off the engine afterburner;
- reduce the speed of the failed engine;
- if the signal is still alive on the EKRAN annunciator panel, shut down the engine;
- approach the airfield and land the aircraft with one engine running.

FAILURE OF AIRCRAFT ACCESSORIES GEAR BOX

193. The Symptoms:

- the KSTs lightbutton flickers;
- the illumination of the OIL annunciator;
- the illumination of the OIL PRES ACCRY GBX and/or ACFT ACCRY GBX VIBR signals On the EKRAN annunciator panel;
- the delivery of the "Low oil pressure in aircraft accessories gear box Aircraft accessories gear box vibration exceeds tolerable one voice information and/or "Discontinue flight mission" command.

194. The actions taken during the takeoff run:

- discontinue the takeoff;
- shut down the engines.

195. The actions to be taken after unsticking and in the case it is impossible to safely discontinue the takeoff:

- after the aircraft unsticks, do not retract the landing gear and flaps;
- execute the landing approach along the shortest route;
- decrease the engine power;
- watch the pressure in the hydraulic systems;
- if the pressure is available, land the aircraft and shut down the engines during the landing roll;
- if the pressure drops in both hydraulic systems, provide the conditions for abandoning the aircraft and abandon the aircraft.

196. The actions to be taken in flight:

- switch off the afterburner and decrease the engine power;
- discontinue the flight mission and turn to the nearest airfield;
- watch the pressure in the hydraulic systems;
- if the pressure is available, execute the descent, approach and land the aircraft;
- shut down the engines during the landing roll;
- if the pressure drops in both hydraulic systems, provide the conditions for abandoning the aircraft and abandon the aircraft.

FAILURE OF AIR INTAKE AUTOMATIC CONTROL SYSTEM

197. The symptoms:

- the KSTs lightbutton flickers;
- the illumination of the UPPER INLET signal on the EKRAN annunciator panel after retracting the landing gear;
- the delivery of the "Air intake upper inlet" voice information and the "Do not exceed M = 0.81, command;
- the indicator shutters (one or both) assume the TO/LAND position.

198. The actions to be taken:

- do not exceed M = 0.8;
- execute the circling flight and land the aircraft.

199. The symptoms:

- the KSTs lightbutton flickers;
- the illumination of the LEFT (RIGHT) AIR INTK signal on the EKRAN annunciator panel;
- the delivery of the "Failure of LH (RH) engine air intake automatic control system" voice information;
- the ramps have been looked in the failure position;
- during acceleration of the aircraft, with the engines running at the similar speed, one of the ramps fail to move from the 0% mark as read off the ramp position indicator by the moment of gaining M 1.6; during acceleration of the eirefaft, with the engines running at the similar speed, difference in position of the LH and RH ramps exceeds 15% as read off the ramp position indicator.

200. The actions to be taken:

- discontinue the flight mission;
- shift the throttle control levers to the IDLE position;
- decelerate the aircraft to the airspeed of 800 to 900 km/h;
- at the altitude of below 10,000 m, get the ramp to the 0% position against the ramp position indicator by depressing the pressure selector switch labelled RAMP EMERG RETR;
- in case of appearance of the abnormal sound, increase the engine speed;
- return to the airfield at the subsonic airspeed.

Note: When the pressure selector switch labelled RAMP EMERG RETR LH (RH) is released, tolerable in ramp return extension by not more than 8%.

FAILURE OF DC GENERATOR

201. The symptoms:

- the KSTs lightbutton flickers;
- the illumination of the DC GEN signal on the EKRAN annunciator panel;
- the delivery of the "DC generator failure. Flight time 15 minutes" voice information and the respective command;
- the voltmeter reads the voltage drop to 22 V instead of the usual reading of 27 to 28 V;
- the appearance of flags D, LOC, and DCA, fadeout of the indicator lamps on the control board, the course pointers assume the arbitrary position.

202. The actions to be taken:

- disconnect the automatic flight control system by depressing the AFCS MODE OFF red button on the control stick for more than 3 s till the fadeout of the DAMPER lightbutton or with the aid of the AFCS switch;

• switch off the engine afterburner;
• land the aircraft within the minimum possible time on the nearest airfield set the CHANNELS selector switch to the MANUAL position, the present heading readings will be restored, the LOC flag will disappear.

Notes:
1. In case of the DC generator failure, the armament control system is automatically disconnected from the aircraft electrical system, and the NAVIG LANDING mode is not selected.

2. When performing the flight at night or in the IFR conditions, set the NAV LTS selector switch to the FLASH position, and the ITS selector switch to the ITS: LAND position on the runway landing heading after flying over the inner homing radio station.

The time of the safe flight with the failed DC generator amounts to 15 35 minutes, with the consumers automatically cut off, after which the voltage should be within 21 to 22 V against the voltmeter.

203. The symptoms (except for those mentioned above):

• the voltage drop in the mains below 20 V;
• the decrease in the lamp brightness;
• the failure of the radio communication and other systems and equipment;
• the air intake ramps have locked in the position, at which the voltage has dropped, and are not emergency controlled.

204. The actions to be taken:

• change over to the visual piloting and visual orientation;
• calculate the remaining fuel proceeding from the readings of the fuel flowmeter at the moment of deenergizing and the fuel flow per minute at the given power setting;
• accomplish the landing within the minimum possible time on the nearest airfield, flying the aircraft against combined instrument DA200, the altimeter, airspeed indicator and compass KE13;
• perform emergency extension of the landing gear.

CAUTION: 1. IT SHOULD BE BORNE IN MIND THAT THE FLAPS AND AIR BRAKES CANNOT BE EXTENDED, THE BRAKE CHUTE MAY NOT BE DEPLOYED AND THE WHEEL ANTISKID SYSTEM MAY FAIL TO ACTUATE, THEREFORE, THE PRESSURE IN THE BRAKES SHOULD BE INCREASED SMOOTHLY.

2. IF THE VOLTAGE IS LESS THAN 18 V, THE ROLL, PITCH AND YAW ARE NOT INDICATED.

FAILURE OF AC GENERATOR DRIVE

205. The symptoms:

• the KSTs lightbutton flickers;
• the illumination of the DISCON GEN DRIVE or AC GEN signal on the EKRAN annunciator panel;
• the automatic cutout of the roll and pitch trimming mechanisms.

206. The actions to be taken:

• disconnect the automatic flight control system by depressing the AFCS MODE OFF red button located on the control stick for more than *3* s until the *DAMPER* lightbutton extinguishes or with the aid of the AFCS switch located on the right console;
• depress backward (turn on) the GEN DRIVE EMERG OFF selector switch located on the left console for not more than 25 s (to preclude the electric valve failure);
• check inverter PTO for cutin;
• proceed with landing the aircraft.

CAUTION: NEVER TURN ON THE AC GEN SWITCH FOR THE SECOND TIME.

FAILURE OF TWO GENERATORS

207. The symptoms:

• the KSTs lightbutton flickers;
• the illumination of the DC GEN signal on the EKRAN annunciator panel;
• the illumination of the AC GEN signal in the EKRAN priority;
• the delivery of the "DC generator failure. Flight time 15 minutes" voice information and the respective command;
• the automatic cutout of the control system of the roll and pitch trimming mechanisms and the weapon control system;
• the voltmeter reads a voltage of 21 to 22 V.

208. The actions to be taken:

• disconnect the automatic flight control system by depressing the AFCS MODE OFF red button located on the control stick for more than *3* s until the DAMPER lightbutton extinguishes or with the aid of the AFCS switch located on the right console;
• switch off the engine afterburner;
• discontinue the flight mission;
• land the aircraft within the minimum possible time on the nearest airfield.

Note. The flight time with the failed generators amounts to 10 to 25 minutes.

FAILURE OF AUTOTUTIC FLIGHT CONTROL SYSTEM

209. The symptom: the dashes and drift of the aircraft in roll and pitch, the inadvertent change in pitch or yaw.

210. The actions to be taken:

• counteract the dash, drift or change in roll, pitch or yaw;
• disconnect the automatic flight control system by depressing the AFCS MODE OFF rod button located on the control stick for more than 3 s until the DAMPER lightbutton extinguishes or with the aid of the AFCS switch located on the right console;
• trim out the aircraft control forces;
• discontinue the flight mission.

211. The symptom: appearance of longitudinal or lateral selfsustained oscillations with a constant or increasing amplitude.

212. The actions to be taken:

• fix the control stick without counteracting the arised oscillations;
• disconnect the automatic flight control system by depressing the AFCS MODE OFF red button located on the control stick for more than 3 s until the DAMPER lightbutton extinguishes or with the aid of the AFCS switch on the right console;
• after cessation of oscillations trim out the control stick forces;
• discontinue the flight mission.

212a. The symptoms:

- the KSTs lightbutton flickers;
- the illumination of the LSA FAILURE signal on the right console;
- dashes of the aircraft in pitch are possible.

212b. The actions to be taken:

- never exceed the 14 degrees angle of attack.

CAUTION: 1. In CASE OF THE LSA FAILURE THE EXTENSION OF THE WING LEADING EDGES IS FOLLOWED BY THE NEGATIVE PITCHING MOMENT. WHEN FLYING THE AIRCRAFT WITH THE CUTOFF AFCS AT ALTITUDES FROM 0 TO 4000 M AND MACH NUMBER EQUALLING 0.8 TO 1.05, ANY ATTEMPT TO COUNTERACT THE ARISED LONGITUDINAL OSCILLATIONS IN PITCH BY PULLING AND PUSHING THE CONTROL STICK MAY CAUSE THE LONGITUDINAL OVERSWINGING OF THE AIRCRAFT. IN CASE OF THE OVERSWINGING WITH A CONSTANT AND INCREASING AMPLITUDE, NEVER COUNTERACT THE LONGITUDINAL OVERSWINGING BY MANIPULATING THE CONTROL STICK. TO ELIMINATE OVERSWINGING, BRING THE AIRCRAFT INTO A CLIMB AND DECREASE THE AIRSPEED TO 800 KM/H. CONTINUE THE AIRCRAFT FLIGHT AT TILE AIRSPEED LESS THAN 800 KM/H, WITH THE CONTROL STICK FORCES TRIMMED OUT AND THE WAIST AND SHOULDER BELTS LOCKED IN THE TIGHT POSITION.

213. The symptoms: the failure to execute the yaw and glide slope maneuver or increase the roll in excess of 35° in the AP mode during the landing approach. The readings of the command bars on the flight director indicators and the integrated display system command circle do not correspond to those of the deviation bars on the flight director and combined course indicators.

214. The actions to be taken:

- disconnect the automatic flight control system by depressing the AFCS MODE OFF red button located on the control stick for a short time (the DAMPER mode is preserved);
- execute the landing approach in the manual control mode by reference to the deviation bars on the combined course indicator, checking the landing approach against the automatic direction finder.

FAILURE OF AUTOMATIC TRANSMISSION RATIO CONTROLLER

215. The symptoms:

- the KSTs lightbutton flickers;
- the illumination of the ARI signal on the EKRAN annunciator panel;
- the delivery of the "ARI failure". Before landing set EASY" voice information and the respective command;
- the disproportionately easy control of the aircraft in pitch, not corresponding to the flying regime.

216. The actions to be taken;

- immediately bring the aircraft into a climb, preventing the "sawtooth" manipulations with the control stick (to preclude the aircraft overswinging);
- disconnect the automatic flight control system by depressing the AFCS MODE OFF red button located on the control stick for more than 3 s until the DAMPER lightbutton extinguishes;
- decrease the airspeed to 500 km1h;
- bring the aircraft into the level flight;
- report to the flight control officer;
- change over the automatic transmission ratio controller to the EASY position (until the API TAKEOFF LAND annunciator lights up) by setting the selector switch to the EASY position;
- approach the airfield and land the aircraft.

217. The symptoms:

- the KSTs lightbutton flickers;
- the illumination of the API signal an the EKRAN annunciator panel;
- the delivery of the "AFI failure. Before landing set EASY" voice Information and the respective command;
- the disproportionately heavy control of the aircraft in pitch, not corresponding to the flying regime.

218. The actions to be taken:

- disconnect the automatic flight control System by depressing the AFCS M0DE. OFF red button located on the control stick for more than 3 s until the DAMPER lightbutton extinguishes;
- decrease the airspeed to 500 km/h;
- change over the automatic transmission ratio controller to the EASY position (until the API TAKEOFF LAND anntinciator illuminates) by setting the selector switch to the EASY position;
- approach the airfield and land the aircraft.

CAUTION: IF THE AUTOMATIC TRANSMISSION RATIO CONTROLLER FAILS TO BE SET TO THE "EASY" POSITION OR IS PARTIALLY SET TO THIS POSITION THE "API TAKEOFF LAND" ~ANNUNCIATOR FAILS TO ILLUMINATE, DURING THE LANDING APPROACH IT IS NECESSARY TO INCREASE THE GLIDING SPEED BY 40 TO 50 KM/H.

INADVERTENT EXTENSION OF HYDRAULIC CYLINDER ROD OF STALL LIMITATION SYSTEM

219. The symptoms:

- the illumination of the COC STOP signal on the annunciator panel after extending the landing gear;
- the control stick stop is perceptible when deflecting the control stick backward.

220. The actions to be taken:

- fully shift the stabilizer trimming mechanism for the pitchup (backward);
- fly the aircraft during the landing approach and landing, overriding the stop of the hydraulic cylinder of the stall limitation system, bearing in mind that with the decrease in the airspeed, the required forces will increase;
- after the touchdown, hold the control stick in the backward position until the aircraft nose lowers, preventing sharp lowering of the nose wheels on the runway.

FAILURE OF STABILIZER TRIMMING MECHANISM

221. The symptoms:

- the inadvertent onset of the push or tractive forces on the control stick, with the flying regime being invariable;
- the push or tractive forces cannot be eliminated from the control stick with the aid of the trimming mechanism slider.

222. The actions to be taken:

• disconnect the automatic flight control system by depressing the AFCS MODE OFF red button located on the control stick for a short time (the DAMPER mode is preserved);
• eliminate the push or tractive forces with the aid of the stabilizer trimming mechanism slider;
• discontinue the flight mission.
• If the forces cannot be eliminated completely, it is necessary, for decreasing the tractive forces, to execute the flight at an airspeed of 600 to 700 km/h (without exceeding M = 0.8) and for decreasing the push forces, perform the flight at an airspeed of 500 km/h;
• land the aircraft in the usual order.

With the decrease in the airspeed, the tractive forces will increase. If necessary, fly the aircraft with the use of both hands.

FAILURE OF BOTH HYDRAULIC SYSTEMS

223. The symptoms:

• the KSTs lightbutton flickers;
• the illumination of the TWO H/STS FAIL and EMERG PUMP UNIT;
• the illumination of the BOOST RTD SYS signal on the EKRAN annunciator panel and the HAIN En SYS (OBMAR MUO) signal in the priority on the EKRAN annumciator panel;
• the delivery of the "Booster hydraulic system failure. Airspeed not in excess of 600 km/h and the Main hydraulic system failure. Perform emergency extension" voice information and the respective commands;
• the drop of the pressure in both hydraulic systems, the pointers are within the red sectors of the combined hydrogas system indicator.

224. The actions to be taken:

• disconnect the automatic flight control system by depressing the AFCS MODE OFF red button located on the control stick for more than 3 s until the DAMPER lightbutton extinguishes or with the aid of the AFCS switch located on the right console;
• if the pressure in the booster hydraulic system has not increased (all the indicator lamps illuminate) provide conditions for catapulting without sharp manipulations by controls and abandon the aircraft;
• if the pressure in the booster hydraulic system has increased, stop fulfilling the flight mission and, without decreasing the engine speed to loss than 85%, arrive at the airfield area (the bailout area) and abandon the aircraft having first provided the safe conditions for this purpose.

FAILURE OF BOOSTER HYDRAULIC STSTEM

225. The symptoms:

• the KSTs lightbutton flickers;
• the illumination of the BOOST HYD SYS signal on the LICRAS annunciator panel;
the delivery of the "Booster hydraulic system failure". Airspeed not in excess of 600 km/h voice information and the respective command;
• the drop of the pressure in the booster hydraulic system, the pointer is Within the red zone of the combined hydrogas system indicator.

226. The actions to be taken:

• disconnect the automatic flight control system by depressing the AFCS MODE OFF red button located on the control stick for more than 3 s until the DAMPER lightbutton extinguishes or with the aid of the AFCS switch located on the right console;
• do not use the air brakes and do not execute the excessive maneuvers;
• land the aircraft with the flaps retracted, on the nearest airfield within the minimum possible time;
• perform the emergency extension of the landing gear.

FAILURE OF MAIN HYDRAULIC SYSTEM

227. The Symptoms:

• the KSTs lightbutton flickers;
• the illumination of the MAIN HYD SYS signal on the EKRAN Annunciator panel;
• the delivery of the "Main hydraulic system failure. Perform lending gear emergency extension" voice information and the respective command;
• the main hydraulic system pressure indicator pointer is within the red zone of the combined hydrogas system indicator.

228. The actions to be taken:

• disconnect the automatic flight control system by depressing the AFCS MODE OFF red button located on the control stick for more than 3 s until the DAMPER lightbutton extinguishes or with the aid of the AFCS switch located on the right console;
• perform the air intake ramp emergency retraction.

Note. In case of main hydraulic system failure, simultaneous retraction of ramps of two air intakes is effected by depression of any of two selector switches labelled RAMP EMERG RETR LH (RH);

• land the aircraft with the flaps retracted on the nearest airfield within the minimu possible time;
• perform the landing gear emergency extension.

DEPRESSURIZATION OF CABIN AT ALTITUDES ABOVE 12,000 M

229. The symptoms:

• the decrease in the cabin pressure differential;
• the appearance of the fog in the cabin for a short time;
• the pressure is created in the tensioner (chambers) of the highaltitude pressure suit.

230. The actions to be taken:

• descend at the maximum possible vertical speed to the cabin "altitude" below 12,000 m (in this case, the pressure differential will not be more than 0.05 kgf/cm2);
• check to see that the CABIN SUPPLY valve knob is fully shifted forward;
• thoroughly check functioning of the oxygen set and the oxygen consumption;
• descend to an altitude of 4000 m, with the amount of oxygen being 15 % and less.

231. The actions to be taken in case of the canopy glass breakdown or separation of the hinged canopy:

- immediately decrease the airspeed;
- lower the crash helmet light filter;
- descend to the cabin "altitude" below 12,000 m;
- proceed to the nearest airfield.

Note. When the cabin "altitude" exceeds 7000 m, the decompressive disorders (the highaltitude pains in the muscles, etc.) may appear. For relieving the hightaltitude pains, it is necessary to descend to the cabin "altitude" below 7000 m.

FAILURE OF CABIN AIR TEMPERATURE AUTOMATIC CONTROL SYSTEM

232. The symptom: the sharp rise or drop of the cabin air temperature irrespective of the conditions and regime of the flight.

233. The actions to be taken:

- set the CABIN HEAT selector switch to the COLD position for a short time, if it is hot in the cabin or to the HOT position, if it is cold in the cabin;
- set the CABIN HEAT selector switch to the neutral position;
- repeat the actions every two or three minutes, if necessary;
- while descending, increase the engine speed for the faster heating of the cabin;
- descend to an altitude below 11,000 m and disconnect the cabin pressurization line by setting the CABIN SUPPLY valve knob to the CLSD position (fully backward), if the high temperature is maintained in the cabin.

If the extremely unsatisfactory temperature is still maintained, decrease the airspeed, depressurize the cabin and act according to the situation, up to the cabin jettison.

WARNING. NEVER OPEN OR PARTIALLY OPEN THE CANOPY IN FLIGHT.

FOGGING OR FROSTING OF CANOPY GLASS PANELS

234. The actions:

- check to see that the CABIN SUPPLY valve knob is placed in the OPEN position (it should be fully shifted forward);
- check the canopy air tightness by reference to the presence of the excessive pressure;
- set the BLOWING valve knob to the position for the canopy blowing;
- set the CABIN HEAT selector switch to the HOT position for a short time (repeat, if required);
- set the CABIN HEAT selector switch to the AUTO position after defrosting or defogging the canopy glass panels. If the fogging cannot be removed by speed at the descent, increase the engine speed and extend the air brakes if necessary.

SMOKE IN CABIN

235. The actions to be taken:

- tightly tighten the oxygen mask;
- change over for breathing with pure oxygen, for which purpose set the additional feed knob to the 100% 02 Position;
- shut off the cabin pressurization line by setting the CABIN SUPPLY valve knob to the CLSD (backward) position. If the excessive pressure is created in the oxygen mask as measured against the combined lifesupport system indicator and the cabin "altitude" amounts to 12,000 m, descend at the maximum permissible vertical speed to the altitude below 11,000 m. While descending, the pressure should decrease to zero against the scale of the combined lifesupport system indicator;
- If the smoke still penetrates into the cabin, decrease the airspeed, depressurize the cabin and act according to the situation, till the canopy jettison.

WARNING. NEVER OPEN OR PARTIALLY OPEN THE CANOPY IN FLIGHT.

FOG IN CABIN

236. The actions to be taken:

- set the CABIN HEAT selector switch to the HOT position for a short time, then to the neutral position; repeat the actions, if required;
- set the CABIN HEAT selector switch to the AM position after defogging.

FAILURE OF OXYGEN EQUIPMENT

237. The symptoms:

- the oxygen is not fed to the mask when the cabin "altitude" exceeds 2000 m (the oxygen flow indicator pointer does not respond to the deep inhalation and exhalation);
- the sharp decrease in the oxygen amount as measured against the combined liresupport system indicator;
- when the cabin is depressurized at the altitudes above 12,000 m, no oxygen is fed to the high altitude pressure suit tensioner chambers

238. The actions to be taken:

- feed the oxygen from the parachute oxygen set with the aid of the EMERG OXYGEN knob;
- descend at the maximum possible vertical speed to an altitude below 4000 m;
- take off the oxygen mask at the altitude below 4000 m and at the airspeed not exceeding 700 km/h.

FAILURE OF RADIO COMMUNICATION

239. The symptoms:

- the sidetone monitoring is not available, but the transmission is heard in the line;
- the sidetone monitoring is available, but the transmission is not heard in the line.

240. The actions to be taken:

- check the patch wire bundle for reliable coupling with the crash helmet plug connector;
- check the positions of the control and adjustment elements of the radio set;
- switch off the noise suppressor;
- check the radio set for proper functioning on the other communication channels.

If the cause of failure is not revealed and the radio communication is not reestablished, proceed as follows:

• fly to the landing airfield, without interruption of the radio communication in the transmission mode;
• turn on the DISTRESS switch;
• periodically identify yourself by depressing the IDENT button, having first made assured that the aircraft transponder function selector switch (CO) is placed in the emergency position;
• report the changeover to the reception with the use of the automatic direction finter to the flight control officer;
• set the PCEH ADF and ANT COMP, selector switches to the ADF and ANT positions, respectively;
• turn on the RC switch;
• periodically set the ANT COMP selector switch to the COMP position when the flight is executed by reference to the automatic direction finder for determining the direction of the flight to the homing radio station;
• before the landing approach, set the CO switch to the GCA position.

If the transmissions are heard in the line, but the sidetone monitoring is not available, perform the commands delivered from the ground control posts.

FAILURE OF AIR DATA COMPUTER SYSTEM

241. The symptoms:

• the illumination of, the ADCS signal on the EKRAN annunciator panel;
• the delivery of the "Air data computer system failure" voice information;
• the pointers of the true airspeed indicator, Machnumber indicator and altimeter hesitate or are set to the extreme positions;
• the considerable discrepancies between the readings of the Machnumber indicator, altimeter, airspeed indicator and rateofclimb indicator and the flight profile and the engine power setting.

242. The actions to be taken:

• disconnect the automatic flight control system by depressing the AFCS M0DE OFF red button located on the control stick for a short time;
• proceed with landing the aircraft;
• maintain the airspeed to be not more than 600 km/h against the airspeed indicator;
• check the altitude against the combined lifesupport system indicator (the cabin "altitude" at the altitudes above 2000 m is approximately equal to the half of the flight altitude, at the altitudes below 2000 m it is approximately equal to the flight altitude), and at the altitudes below 1000 m, the altitude is also checked by reference to the radio altimeter;
• approach the airfield and land the aircraft.

FAILURE OF NAVIGATION SYSTEM

243. The symptoms:

• the preset course is not followed up for the programmed point;
• the wrong range readings (the range counter is closed by the shutter);
• when the aircraft makes the turns, the course scale on the combined course indicator moves irregularly or is motionless (the LOC flag appears).

244. The actions to be taken:

• take a decision to return to the airfield or to proceed with the flight;
• set the CHANNELS selector switch to the MANUAL position;
• set up the channels of the navigation and landing radio beacons of the lending airfield;
• check the followup of the course, azimuth and range on the combined course indicator;
• approach the selected airfield by reference to the shortrange radio navigation system or automatic direction finder;
• execute the landing approach as in case of the landing approach to the unprogrammed airfield.

245. The symptoms:

• the fadeout of the CORR lamp on the control panel.
• the radio beacon relative bearing (azimuth) reading on the combined course indicator from the shortrange radio navigation system is wrong (determined by comparing the reading with the radio station relative bearing from the automatic direction finder).

246. The actions to be taken: proceed with the further flight and approach the airfield by reference to the automatic direction finder and course selector.

247. The symptom: after accomplishing the intermediate approach, the localizer and glide slope flags (or one of them) do not disappear and the respective deviation bars are motionless on the combined course indicator.

248. The actions to be taken:

• select the LANDING mode annually with the aid of the LANDING switch;
• execute the landing approach by reference to the automatic direction finder and course selector, if the LANDING mode fails to be selected.

249. The symptoms:

• on the glide slope, during the automatic landing approach, the aircraft has begun going into the level flight the LEVELLING mode of the automatic flight control system has been selected; in this case, the LEVELLING lightbutton located on the control stick lights up;
• the localizer or glide slope flag appears on the combined course indicator.

250. The actions to be taken:

• switch off the LEVELLING mode by depressing the AFCS MODE OFF red button located on the control stick for a short time; in this case, the DAMPER mode will be preserved;
• make a decision to continue the approach maneuver or to go around;
• execute the approach in the manual control mode by reference to the automatic direction finder and course selector.

FAILURE OF INFORMATION DIRECTIONAL/VERTICAL GYRO SYSTEM E KBK

251. The symptoms:

• the illumination of the CAGE lightbutton located on the face side of the flight director indicator;
• the roll and pitch readings on the flight director indicator (possibly, the yaw reading on the combined course indicator) differ from the aircraft actual attitude;

• appearance of the LOC flag on the combined course indicator;
• the illumination of MAIN DIR VERT GYRO signal on the EKRAN annunciator panel.

252. The actions to be taken:

• switch off the automatic modes of the automatic flight control system by depressing the AFCS MODE OFF red button located on the control stick for a short time; in this case, the DAMPER mode is preserved;
• set the EKB (inertial directional/vertical gyro system) selector switch to the STBY position; in this case, the CAGE lightbutton should extinguish and the LOC flag on the combined course indicator should disappear;
• check for the correct roll, pitch and yaw readings by comparing them with those of the flight and navigation instruments and with the interrogated airfield bearing;
• estimate the possibility of the flight mission continuation.

253. The symptoms:

• the illumination of the CAGE lightbutton after.setting the EKB selector switch to the STBY position (the LOC flag has not disappeared);
• the roll and pitch readings on the flight director indicator (possibly, the yaw reading on the combined course indicator) differ from the aircraft actual attitude.

254. The actions to be taken:

• bring the aircraft into the level flight by reference to combined instrument DA200, compass KE13, altimeter and airspeed indicator;
• set up the constant airspeed of.the flight;
• repeat the'start of the main inertial vertical gyro system by depressing the CAGE lightbutton;
• perform the steady level flight for 2 to 3 min (until the failure signal extinguishes on the EKRAN annunciator panel) during which set the CHANNELS selector switch to the MANUAL position and depress the M HDG SLAV button for 10 to 15 s;
• check for the fadeout of the indication and restoration of the correct roll, pitch and yaw readings by comparing them with those of the flight and navigation instruments and with the interrogated airfield bearing;
• estimate the possibility of the flight mission continuation taking into account the fact that the navigation system will operate with great errors, especially when executing the maneuvers.

If the indication is still illuminated and the correct readings of the angles have not been restored:

• fly the aircraft by reference to combined instrument DA200 in combination with the readings of the altimeter, airspeed indicator and compass KE13;
• fly to the landing airfield;
• maintain the direction of the flight by reference to the KE13 compass, bearings and automatic direction finder.

FAILURE OF PITOT STATIC TUBE

255. The symptoms:

• the readings of the airspeed and Mach number indicators do not correspond to the flying regime and engine power setting;
• the readings of the altimeter and rateofclimb indicator do not correspond to the flight director indicator readings and the flying regime;
• the malfunction of the air intake ramp automatic control system, engine power control system, automatic transmission ratio controller, automatic flight control system and navigation system.

256. The actions to be taken:

• check the HEAT, W/SHIELD switch for being cut in;
• turn on the HEAT: W/SHIELD switch if it has been cut out (the serviceability of the pressure instruments should be restored on a lapse of two or three minutes of the flight);
• set the pitotstatic tube selector switch to the STBY position.
• If the readings have been restored, continue fulfilling the flight mission and on a lapse of two or three minutes, check the pitotstatip tube system for proper functioning with the IMA selector switch placed in the MAIN position;
• If the serviceability of the pressure instruments cannot be restored from the standby and main pitot static tube systems, proceed as follows: switch off the automatic flight control system by depressing the AFCS MODE OFF red button located on the control stick for more than 3 s until the fadeout of the DAMPER lightbutton or with the aid of the APCS switch located on the right console;
• discontinue the flight mission;
• switch off the afterburner, decelerate the aircraft to the subsonic airspeed and proceed with landing the aircraft;
• if necessary, perform the emergency air intake ramp retraction.

While intercepting the airfield area and executing the landing approach, check:

• the flight airspeed by reference to the angleofattack indicator, flight director indicator and tachometer indicator;
• the flight altitude by reference to the combined lifesupport system indicator (the cabin "altitude" at the altitudes above 2000 m is approximately equal to the half of the flight altitude, at the altitudes below 2000 m it is approximately equal to the flight altitude), and at the altitudes below 1000, check the flight altitude also by reference to the radio altimeter.

FAILURE OF LANDING GEAR EXTENSION INDICATION SYSTEM

257. The symptom: the fadeout of one (two or three) greenlamp indicating the landing gear extended position and the red lamp indicating the landing gear intermediate position.

258. The action to be taken:

• check the indication system for serviceability and the presence of the normal pressure in the main hydraulic system.
• If the indication system is unserviceable (the lamps fail to light up during the check) and the pressure is normal in the hydraulic system, proceed as follows:

  • remain the landing gear valve in the landing gear extension position;
  • report the matter to the flight control officer and fly over the start area at an altitude of 100 m for making the visual check of the landing gear extension from the ground;
  • after receiving the landing gear extension confirmation,land the aircraft.

ONE OF MAIN IANDING GEAR LEGS EXTENSION FAILURE

259. The symptoms:

- one of the green lamps indicating the main landing gear extended position is dead;
- the red lamp indicating the landing gear intermediate position burns continuously

260. The actions to be taken:

With the normal pressure in the main hydraulic system, proceed as follows:

- make two or three attempts to extend the landing gear once more; while doing so, create the alternating gload at the airspeed not exceeding 700 km/h (in this case, the landing gear control valve should be shifted from the landing gear extension position to the retraction one not earlier than on a lapse of 15 s);
- make several energetic wing rocks, with the landing gear control valve placed in the landing gear extension position and the airspeed not exceeding 700 km/h.

If the abovementioned actions do not succeed in extension of the main landing gear, Proceed as follows:

- retract the landing gear;
- make certain that all the landing gear legs are uplocked;
- perform the emergency extension of the nose landing gear;
- check the extension of the nose landIn gear by reference to the indication system
and restoring of the pressure in the main hydraulic system;
- land the aircraft on the unpaved runway, resting on the front landing gear and fuselage.

In this case, the landing approach should be executed with the fuel remainder being minimum one. Before landing the aircraft, jettison the hinged canopy. Avoid flaring out and holding off. Before the touchdown, shut down the engines. After touching down, deploy the brake chute and switch off the storage battery.

NOSE LANDING GEAR EXTENSION FAILURE

261. The symptoms:

- the green lamp indicating the nose landing gear extended position fails to come on;
- the red lamp indicating the nose landing gear intermediate position burns contlnUously or flickers.

262. The actions to be taken:

- check to see that the pressure is normal in the main hydraulic system;
- make two or three attempts to extend the nose landing gear once more, while doing so, create the alternating gload at the airspeee. not exceeding 700 km/h (in this case, the landing gear control valve should be shifted from the landing gear extension position to the retraction one not earlier than on a lapse of 15 s);
- if the above actions do not succeed in extension of the nose landing gear, it is necessary to retract the landing gear, perform the emnergency nose landing gear extension and check the latter by reference to the indication system;
- perform the emergency extension of the main landing gear if the nose landing gear has extended, but has failed to be held by the downlock, create the alternating gload;
- if, after the above actions, the nose landing gear fails to be downlocked, the landing may be executed by touching the concrete runway with the main landing gear.

When taking a decision for this type of landing, proceed as follows:

- drop the warloads and drop fuel tanks in a safe area;
- lock the seat belts in the tightened position;
- execute the landing approach and landing proceeding according to the established order, with the minimum fuel remainder;
- after touchdown, shut down the engines by setting the throttle control lever to the STOP position and hold the control stick fully pulled until the aircraft stops;
- as soon as the nose fuselage lowers to a position close to the horizontal one (at a speed of 200 to 220 km/h), deploy the drag chute.

Note: Deployment of the drag chute in the landing position leads to abrupt lowering of the nose fuselagel the horizontal stabilizer effectiveness is not sufficient for counteracting this lowering;

- apply the brakes only after touching the concrete runway with the nose fuselage with due account of its remaining length;
- as soon as the aircraft stops, turn off the BAT GND SUP switch and abandon the aircraft.

EMERGENCY EXTENSION OF LANDING GEAR

263. The actions to be taken (irrespective of the position of the landing gear control valve):

- set up an airspeed of 500 km/hi;
- Pull backward the landing gear emergency extension handle as far as it will go; the nose and gear should extend;
- check the extension of the nose landing gear by reference to illumination of the green signal on the landing gear and flaps position indicator panel;
- turn the landing gear emergency extension handle to the right (clockwise) and pull the handle backward all the way home the main landing gear should extend;
- leave the landing gear emergency extension handle in the extended position.

If any of the landing gear legs has failed to extend during the emergency extension, abandon the aircraft.

BREAKDOWN OF MAIN LANDING GEAR WHEEL TIRE

264. The symptoms:

- the heavy buffeting of the aircraft;
- the onset of the turning moment and banking of the aircraft towards the destroyed tire.

265. The actions to be taken during the ta ' keoff run till the nose wheel liftoff:

- hold the aircraft against the turning;
- discontinue the takeoff;
- deploy the brake chute;
- shut down the engines, if necessary;
- apply the normal braking.

266. The actions to be taken dnring the takeoff run after lifting the nose wheel and when it is impossible to safely discontinue the takeoff:

- retain the aircraft against turning;
- continue the takeoff;
- after unsticking, brake the wheels;
- do not retract the landing gear if the flight mission is not a combat one;
- disengage the nose wheel brake before executing the landing approach;
- land the aircraft with the minimum fuel remainder aboard.

267. Actions to be taken during the landing roll:

- retain the aircraft against turning;
- deploy the brake chute; if necessary, shut down the engines;
- apply the normal braking.

FORCED OFF FIELD LANDING OF AIRCRAFT WITH ENGINES RUNNING

268. A decision to execute the forced offfield landing is made by the pilot. It is allowed to land the aircraft on the grounds known to the pilot or on the grounds selected, inspected and estimated from the air.

When making a decision to land the aircraft on the known or selected ground, it is necessary to know or determine the possibilities of landing the aircraft with the landing gear extended or on the nose landing gear and fuselage.

After making a decision to perform the offfield landing, proceed as follows:

- report the decision made and the place of forced landing to the flight control officer or to the command post;
- jettison the drop fuel tank and external stores (perform the emergency launch of the missiles) in the safe area;
- extend the landing gear and flaps;
- jettison the canopy at the altitude below 5000 m and at airspeed of 400 to 700 km/h (prior to jettisoning the canopy bead yourself towards the instrument board); lock the shoulder straps in the restrained position;
- before the touchdown, shut down the engines;
- after the touchdown, deploy the brake chute;
- switch off the storage battery;
- when landing the aircraft, with the landing gear extended, decelerate the aircraft, taking into account the solidity of the soil, the ground length and the presence of obsticles.

EMERGENCY ESCAPE FROM THE AIRCRAFT IN FLIGHT

269. A decision on the emergency escape is made by the pilot individually or by the order of the flight control officer or by the command delivered from the command post.

270. Before escaping from the aircraft, if the situation permits, proceed as follows:

- turn on the DISTRESS switch to apply the respective signal;
- fly towards the friendly territory if the flight has been performed near the state frontier or towards the coastal line if the flight has been performed over the water surface; report the approximate bailout area over the radio;
- climb to an altitude of 2000 to 3000 m when flying at the low altitude or descend to an altitude of 3000 to 4000 m when flying at the high altitude;
- decrease the airspeed to 400 600 km/h during the climb or the level flight;
- manually lower the crash helmet light filter, withdraw the legs from the pedals and place them in front of the seat pan;
- abandon the aircraft, if possible, beyond the clouds.

WARNING: IN THE URGENT CASES, JUST AS DURING THE TAKE-OFF RUN AND LANDING ROLL AT AN AIRSPEED OF 75 KM/H AND MORE, BAIL OUT IMMEDIATELY.

271. For bailing out, proceed as follows:

- take hold of the bifurcated ejection handle with both hands, with the palms each other, if one of the hands is injured, take hold of this handle with the other hand;
- tightly press the shoulder against the seat back, the head, against the headrest, the elbows, against the body;
- energetically pull the ejection handle upward till the seat starts moving, applying a force of 20 to 25 kgf thereto.

If the canopy has failed to be jettisoned by means of the ejection handle, but the foot and shoulder restraint mechanism and the arm guard have actuated, proceed as follows:

- sink the ejection handle to the initial position;
- jettison the canopy, for which purpose sharply pull the canopy jettison handle with the right hand.

WARNING. NEVER USE THE CANOPY AUTONOMOUS JETTISON HANDLE AND THE EJECTION HANDLE AT THE SAME TIME.

- hold the ejection handle with the left hand to preclude the fall of the handle from the hands under action of the air flow;
- this done, abandon the aircraft in the usual manner.

272. After deploying the life parachute canopy, proceed as follows:

- make certain that the parachute canopy has normally deployed;
- raise the crash helmet light filter;
- remove the oxygen mask at altitude below 4000 m:,
- make certain that the survival kit has been normally released, the life raft and radio beacon have been released and inflated.

273. Before the touchdown, proceed as follows:

- determine the direction of the drift and the approxizate area of the touchdown;
- turn downwind at an altitude of 150 to 100 m;
- join the legs together, then slightly bend them in knees and bring them forward, with the feet being parallel to the ground;
- touch down on the full feet, slightly springing and not tending to hold on the legs;
- after touching down, disconnect the parachute canopy by releasing quickdisconnect locks.

Notes:
1. When abandoning the aircraft at the low altitude, it is necessary to prepare the legs for the touchdown and turn downwind. Take the other actions before the touchdown versus the available time.

2. In case of dragging after the touchdown, it is necessary to collapse the parachute canopy, for which purpose it is necessary to turn on the side and release one of quick disconnect locks with the aid of the free hand.

274. When alighting on the water surface, if there is a sufficient altitude margin, proceed as follows:

• make certain that the parachute has normally deployed and determine the direction to the coast;
• fit the main sling of the harness system under the thighs;
• inflate floats ASP 74 by pulling the inflation plugs;
• disconnect the hoses of the oxygen mask and the highaltitude partial pressure suit from the parachute harness system, for which purpose withdraw the pin of the emergency connector of the oxygen set and disconnect the latter;
• disconnect the ventilation and antig suit hoses from the upper block of quickdisconnect assembly;
• detach the oxygen mask tightness compensator connector;
• detach the crash helmet communication and power supply wire bundle from quickdisconnect assembly;
• disconnect the intermediate strap of the oxygen mask corrugated hose and withdraw the oxygen mask together with the connector block of the oxygen set;
• close the air relief valve on the VMSK marine survival suit;
• taking hold of the right shoulder strap with the left hand, release the central lock of the harness system with the right hand and bring the leg straps from the halfrings;
• hold yourself in the harness system, joining both hands together and lowering them down;
at an altitude of 7 to 10 m, alternately transfer the hands to the free ends of the parachute harness system;
• after touching the water surface, straighten out the body and legs, unclasp the hand fingers and slip out of the parachute harness system;
• turn downwind (with your back turned to the wave edge) and, if required, boostinflate the life belt floats or VMSK marine survival suit;
• tear off the calibrated seam of the survival kit line having cut off the orange end from the main line and use the life raft and survival kit.

For delivering the signals to indicate your present position, make use of the signalling means available in the survival kit.

275. When alighting on the water surface with the parachute harness system undetached and the survival kit unreleased, proceed as follows:

• raise the crash helmet light filter and take off the oxygen mask.

WARNING: 1. KEEP YOUR BREATH FOR A WHILE BEFORE TAKING OFF M. OXYGEN MASK AT THE MOMENT OF TOUCHMG THE WATER TO PRECLUDE THE SUCTION OF THE WATER THROUGE THE OXYGEN HOSES.

2. IF THE PILOT IS DRAGGED BY THE PARACHUTE OVER THE WATER SURFACE, GET RID OF THE PARACHUTE CANOPY BY RELEASING QUICKDISCONNECT LOCKS.

• inflate the life belt floats, for which purpose energetically pull the inflation plugs of the floats;
• close the VMSK marine survival suit air relief valves;
turn downwind (with your back turned to the wave edge) and, if required, boostinflate the life belt floats or VMSK marine survival suit;
• detach the hoses of the oxygen mask and highaltitude partial pressure suit from the parachute harness system, for which purpose withdraw the pin of the oxygen set emergency connector and detach the oxygen set;
• disconnect the hoses of the antiS and ventilated suits from the upper block of the quickdisconnect assembly and detach the communication connector;
• sharply pull the line to open the flaps of the survival kit pack;
• pull up the life raft by the line and sharply pull it to engage the life raft inflation system;
• tear off the calibrated seam of the survival kit line having cut off the orange end from the main line;
• if the life belt is used, for getting rid of the parachute harness system, open the main sling connector lock, then the central lock of the parachute harness system, bring the leg straps from the halfrings, throw the right shoulder strap of the harness system over the back, slightly bleed the left float of the life belt and pass it together with the hand under the left shoulder strap and free yourself from the parachute harness system;
• use the life raft and survival kit. If it is impossible to get rid of the parachute harness system and get in the life raft, it is necessary to take hold of it.

276. If one of the floats of the emergency escape system fails to be inflated from the gas inflation system, proceed as follows:

• shift the noninflated float to the breast;
• make certain that the noninflated float envelope flaps are open (if necessary, open them, for which purnose withdraw the pin from the loop);
• inflate the float with air by blowing the latter with the mouth through the boost inflation pipe; catch the valve stem with the teeth, open the valve, having sunk the valve stem; after exhalation, release the valve stem.

EMERGENCY ESCAPE FROM AIRCRAFT ON GROUND

277. For the quick escape from the aircraft on the ground, without ejection, the pilot should do the following:

• shut down the engines;
• depress the side triggers of the ejection handle and hinge it forward through 90°;
• get rid of the parachute harness system, having first released the lock on the main strap;
• pull the oxygen set connector pin;
• take off the oxygen mask (if possible);
• disconnect the snap hook of the survival kit line from the ring on the pilot's outfit.
• use the service handle to open the canopy. If the canopy cannot be open, jettison it with the aid of the autonomous jettison handle, having preliminarily bent towards the instrument board;
• sharply stand up, disconnect the couplings of the antig device, cabin ventilation system and electric wire bundle of the upper block of the quickdisconnect assembly;
• abandon the aircraft and go away to a safe distance.

278. If the time is available, remove the survival kit from the aircraft, for which purpose proceed as follows:

• turn in the cabin with the face towards the seat;
• take hold of the orange ribbons provided an the rear cover of the survival kit and disconnect the latter from the harness system attachment straps,for which purpose sharply pull them up;
• turn the survival kit until it assumes the vertical position with respect to the trunnions; in this case, the quick disconnect assembly will be detached and the locking wire on one of the trunnions will be broken;
• remove the survival kit from the seat pan by energetically moving the kit up and immediately throw it from the cabin since on a lapse of 4 s the pack will be automatically disconnected and opened;
• abandon the aircraft and, taking the survival kit, go away to a safe distance.

AUTONOMOUS CANOPY JETTISON

279. Prior to jettisoning the canopy, proceed as follows:

- tighten the shoulder and waist straps;
- lower the crashhelmet light filter and tighten up the oxygen mask;
- reduce the speed down to 400-700 km/h,

280. Use the CANOPY JETTISON handle to jettison the canopy as follows:

- taken hold of the CANOPY JETTISON handle with the right hand and deflect it horizontally, then pull the handle briskly as far as it will go;
- the canopy jettisoned, return the CANOPY JETTISON handle to the initial position.

Section Seven

FLIGHT PERFORMANCE AND AERODYNAMIC FEATURES

AERODYNAMIC FEATURES OF AIRCRAFT

The aerodynamic configuration, stability and controllability characteristics, flight-performance of the aircraft are optimized for performing the primary mission conduct of the aerial combat.

The aircraft design features the following aerodynamic peculiarities:

- the aerodynamic configuration of the integral layout with the loadcarrying airframe;
the highlift profiles of the wing;
- the wing twist;
- the front extensions of the centre section of the highly swept wing;
- the leading edge flaps of the wing with smooth contour;
- the twofin vertical tail;
- the differentially deflectable stabilizer for the roll control together with the ailerons;
- the employment of the automatic flight control system for improving the stability and controllability characteristics.

LONGITUDINAL STABILITY AND CONTROLLABILITY

The aircraft with the automatic flight control system switched on and with all variants of the external stores suspended, is considered to be stable with respect to the gload throughout the entire operating range of centers of gravity and flying regimes.

The margin of the aircraft gload longitudinal stability with the Mach number varying, is dependent on the position of the aircraft aerodynamic centre, the centre of gravity and presence of the external stores.

The dependence of the aircraft longitudinal moment coefficient on the angle of attack within the Mach numbers ranging from 0.5 to 1.0 is nonlinear one, i.e.., the position of the aircraft aerodynamic, centre also depends on the aircraft angle of attack. The minimum stability margin corresponds to the small and negative values of the angle of attack. With the increase in the Mach number from 0.5 to 0.9, the angleofattack boundary corresponding to the respective increase in the stability margin shifts from 75 to 450 With the angle of attack exceeding 7.5° at M = 0.5 and 4.50 at M = 085 to 0.95, the longitudinal stability margin increases and is maintained invariable until the angle of attack reaches 14° and, then it decreases again. At the angles of attack exceeding 19°, the longitudinal stability margin with respect to the gload increases and is maintained invariable until the angle of attack reaches 26°.

The minimum stability margin with respect to the gload of the aircraft carrying six missiles, with the centre of gravity being in the extremely rear position (the fuel remainder amounts to 3000 3200 kg) in the flight configuration at the subsonic airspeed is 2 % of the mean aerodynamic chord. This calls for the greater attention on the part of the pilot when flying the aircraft at the small and negative angles of attack and at the high subsonic airspeeds. The launch of the missiles increases the stability margin by 4 % for the account of the shift of the centre of gravity forward by 1 to 1.5 % of the mean aerodynamic chord and the shift of the aircraft aerodynamic centre backward by 2.5 % of the mean aerodynamic chord.

When the Mach number exceeds 0.95, owing to the substantial shift of the aircraft aerodynamic centre backward, the gload stability margin considerably increases.

When flying the aircraft with the wing leading edges retracted (failed), at the angles of attack of 12 or 13° the aircraft begins to undergo the intensive aerodynamic buffeting which is easily ascertained by the pilot and is a reliable warning of the failure of the wing leading edges to extend. At the angles of attack exceeding 13 or 14°, the aircraft gload stability decreases till the aircraft becomes neutral. Besides, at the angles of attack exceeding 17 or 18°, the aircraft tends to the sharp inadvertent change in the initial angle of roll ("turningout"). In this case, when the wing leading edges fail to extend, the stall limitation system is reset to an angle of 15°.

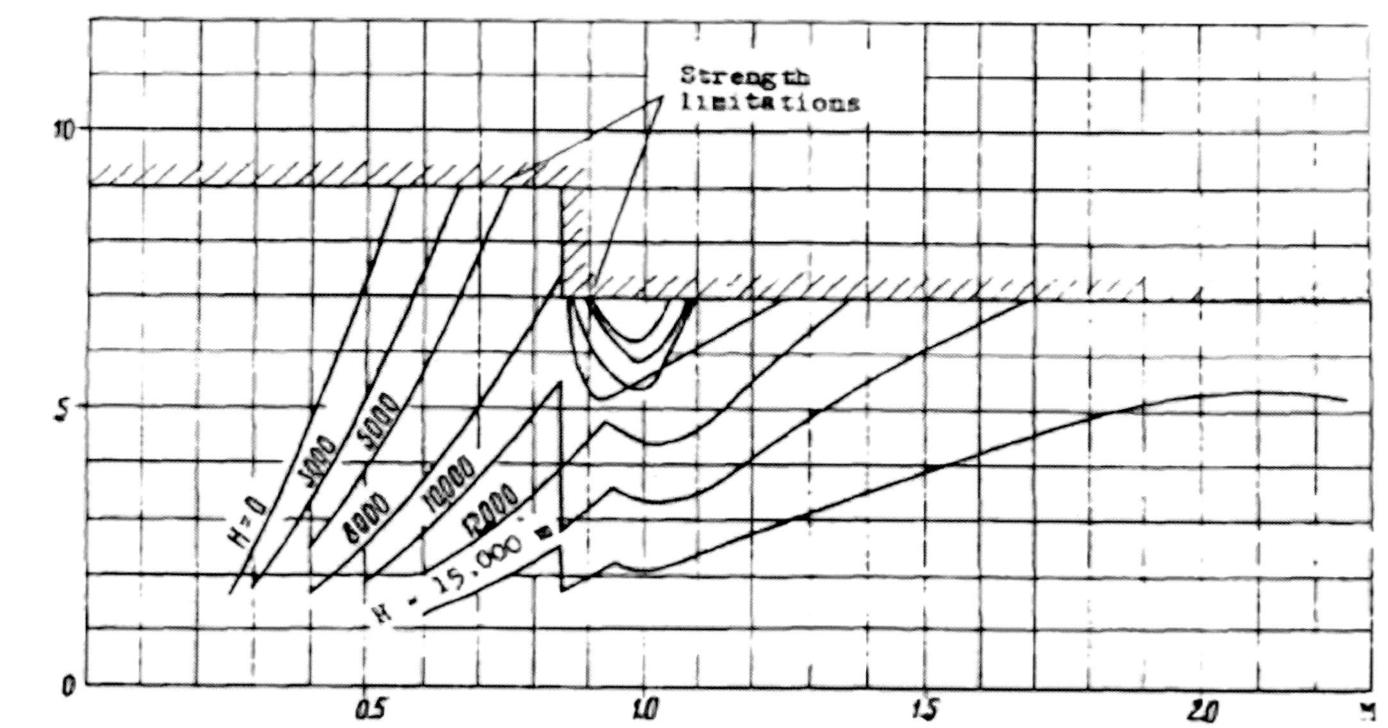

MAXIMUM G-LOAD OF AIRCRAFT WITHOUT EXTERNAL STORES (X = 25.5 % MAC) OR WITH MISSILES

In the takeoff/landing configuration with all variants of the external stores suspended, the aircraft is stable with respect to the gload up to the angle of attack of 15° at the great angles of attack, the gload stability decreases till the air craft becomes neutral.

When the automatic flight control system is switched off (faulty), the aircraft becomes neutral within the angles of attack of 14 to 19° at the Mach numbers less than 0.7, and at the Mach numbers more than or equal to 0.7. the aircraft is unstable with respect to the gload. The satisfactory characteristics of the longitudinal stability and controllability within the given range of the angles of attack are ensured by the longitudinal stability augmentor which functions only when the automatic flight control system is put into operation. In this connection, when the automatic flight control system is switched off (fails), the tolerable angle of attack is set to be 13°.

For improving the longitudinal controllability characteristics, the longitudinal channel of the control system is provided with the nonlinear mechanism varying the control sticktostabilizer gear ratio versus the control stick travel and also with the automatic transmission ratio controller and longitudinal stability augmentor which, within the angles of attack of minus 2 to +5° and 16 to 26° increases the amount of the control stick displacement per unit of gload, which results in the more smooth control of the aircraft within the indicated range of the angles.

During vigorous maneuvers at low altitudes (H < 3000 m) with gload close to the maximum permissible value, within the airspeed range of 700 to 900 km/h in the process of decelerating the aircraft the automatic transmission ratio controller deflects the stabilizer for pitchingup. In this case, the pitch position of the control stick remains unchanged. Due to above actions the gload is spontaneously increased by 0.5 to 1.0 and may inadvertently exceeds the maximum permissible value.

To exclude inadvertent increase of the maximum permissible gload, it is recommended to enter the aircraft carrying no external stores or carrying missiles into a maneuver within the above range of altitudes and speeds with a gload not exceeding 8. Counteract the increase of a gload over 8 by pushing the control stick forward.

The use of the wing leading edges automatically deflected in response to the angle-ofattack transmitter signal, alongside with the improvement of the lateral stability and controllability characteristics, practically makes it possible to eliminate the stall aerodynamic buffeting within the operating range of the angles of attack.

At the supersonic airspeeds, the attainment of the great angles of attack and, consequently, the available maximum gload value are limited by the full deflection of the control stick backward.

During vigorous maneuvers at altitudes less than 2000 m. and at M = 1.1 to 1.3 the maximum operational gload may be exceeded due to small gradient of forces on the control stick for creating the unit of acceleration. Therefore. it in recommended to create gload more than 5 under above conditions for the time not less than 5 s.

At .9 1. 5 and also at M> 9000 m and M< 0.8 the aircraft is unstable with respect to the airspeed, and at 9000 m and M > 0.8 the aircraft becomes neutral with respect to the airspeed. These features do not substantially influence on the piloting technique.

In the transonic range of airspeeds, during the deceleration, the aircraft behavior depends on the initial gload and the flight altitudes.

during the deceleration in the straight flight from M = 1.2 to 1.1 to M = 0.8 to 0.7 at H< 9000 m, the trim change results in the sharp reduction of the gload, till the negative gload is reached, which calls for the timely pull of the control stick, and at H> 9000 m, the gload practically does not exceed the rated value, when decelerating the aircraft during the execution of the maneuver at > 3 g within the same range of the airspeeds, the gload rises by 1 or 2 g as compared with the rated value, in thin case, the gload rise should be counteracted by the timely push of the control stick.

The automatic extension of the wing leading edges leads to the onset of the diving moment which is equivalent to the additional deflection of the stabilizer through 2.5° which is partially compensated by the longitudinal stability augmentor.

The change in the engine power from the idle power to the maximum one leads to the origination of the pitching moment requiring the forward deflection of the control stick approximately through a length of 15 mm at an airspeed of 600 km/h and through a length of up to 40 mm at an airspeed of 300 to 350 km/h

The extension of the air brakes is accompanied by the origination of the longitudinal moment f or pitchup at M<1. 0 and for pitchdown at M > 1. 0. In thin case, the maximum gload rise, with the control stick looked in the required position, amounts to about I g.

The simultaneous extension of the flaps and wing leading edges practically do not trim the aircraft fully.

In the flight configuration, the aircraft is stable with respect to the bomb gload up to the angles of attack of 19° and with respect to the rocket gload, up to the angles of attack of 17° . However, at the angles of attack exceeding 15°, the gload stability margin decreases, especially it decreases at M> 0.8.

When accelerating the aircraft in the level flight, with the rockets or 2 (4) bombs within the Mach numbers of 0.9 to 1.1, the pitchup moment sharply increases and it is trimmed by pushing the control stick beyond the neutral position. During the further acceleration, the control stick returns to the position close to the neutral one and the further acceleration is performed without complications. When accelerating the aircraft in the level flight, with 4 bombs (pods B8) attached at the altitude below 1500 m, the aircraft flying at M 0.95 to 1.05 is not trimmed and goes into a climb with the control stick fully pushed forward.

The nature of the airspeed stability change with the bombs and rockets attached is just the same as in the case the aircraft carries no external stores. However, at M > 0.85, the pitchup moment rises more intensively than in the case the aircraft carrying no external stores. Owing to this, at the moment of release of the 250 to 500kg bomb (at the moment of the rocket launch) in the SALVO and 0.5 AL (modes at the Mach numbers ranging from 0.85 to 1.1, the pitchdown moments will develop and create the negative gloads of up to 2 g at the low and middle altitudes, resulting in the lose of up to the 200m altitude during the counteraction of the pitchdown moment, depending on the ammunition caliber. In this case, the ABCP aerial bomb release (rocket launch) should be executed at the regimen free from these phenomena.

When the aircraft flies without the external stores and when it carries the missiles in the DAMPER mode in flights at Mach numbers of 0.95 to 1.3, the longitudinal channel can possibly set up spontaneous oscillations with a frequency of 1.0 to 1.5 Hz. The oscillation frequency in gload within Mach numbers reaching from 0.98 to 1.05 reaches 1 unit at altitudes below 1000 m. The oscillation amplitude decreases as the altitude increases. The flight becomes free from this phenomenon at Mach numbers less than 0.95 or more than 1.3.

Flights executed at altitudes below 4000 m and Mach numbers of 0.8 to 1.05 (at the airspeeds exceeding 800 km/h) are accompanied by increased response of the aircraft to the control stick deflection in the longitudinal channel. The natural frequency of the aircraft oscillations in the longitudinal channel at these regimes is close to the maximum permissible frequency of the pilot's control actions. Owing to this fact, the pilot's attempt to counteract disturbances caused by the engine thrust variation, air brake extension (retraction), atmospheric turbulence, aircraft overbalance when performing deceleration in level flight can result in the aircraft inadvertent overswinging in the longitudinal channel. The aircraft overswinging is promoted by loose attachment of the pilot by the shoulder and waist straps, and by poor trimming of the control system in the longitudinal channel due to decrease of the stick force gradient for creating the gload unit in the process of the load feel mechanism operation on the characteristic flat branch. It should be borne in mind that the precise sighting is hampered at these regimes and it is not advisable to execute the formation flights at distances less than 300 m due to overswinging which can possibly arise at these regimens.

To rule out inadvertent overswinging of the aircraft, fly the aircraft at altitudes below 4000 m and at Mach numbers of 0.8 to 1.5 by smoothly deflecting the control stick. It the aircraft overswinging occurs, do not counteract the oscillations, smoothly bring the aircraft into climbing and lock the control stick.

Disengagement of the DAMPER mode involves the most probable overswinging. Therefore, it is recommended to execute the flight with the aircraft flight control system failed at altitudes below 4000 m and airspeed less than 800 km/h. If the aircraft flight control system fails at supersonic airspeeds, the airspeed transonic range should be passed with a gload exceeding 1 unit and with the waist and shoulder straps locked in the tightened position.

LATERAL STABILITY AND CONTROLLABILITY

In the yaw and lateral control channels, the aircraft is stable in the entire range of the operating altitudes, airspeeds and gloads. The suspended missiles practically do not influence the directional stability margin.

The efficiency of the lateral control element is characterized by a roll rate whose maximum value within 140 to 160 deg/s is reached at the altitudes of 1000 to 5000 m at the airspeeds of 800 to 900 km/h. With the airspeed increasing, the efficiency of the lateral control elements changes for the worse and at an airspeed of 1500 km/h near the ground it is equal to 40 deg/s.

At supersonic airspeeds and altitudes below 5000 m, when deflecting the control stick in roll, the restraining forces may develop due to the excess of the aerodynamic hinge moment of the ailerons over the available booster moment. The restraining force disappears with the decrease in the airspeed.

The efficiency of the rudders for counteracting the developed asymmetry moments is quite sufficient in the entire available range of the angles of attack. At a subsonic airspeed, the efficiency of the rudder is dependent on the angles of attack; with the increase in the angles of attack up to 25°, the rudder efficiency drops three times as compared with the angles of attack of 10 to 12°. With the increase in the flight altitude, the aircraft damping properties decrease to the rudder deflecting, the aircraft response delay increases.

With the disturbances acting thereon, the aircraft response delay increases, and the developing oscillations are not damped any longer, the roll, yaw and pitch rates develop with a certain delay and with the subsequent rises relative to the sustained value and the aircraft slowly gains the sustained rotation. During the shorttime effect on the aircraft, the aircraft angular rates do not reach the maximum values. For reducing the tendency of the oscillatory motion at the high altitudes and great Mach numbers, use is made of the dampers employed in the automatic flight control system.

For Increasing the directional stability of the aircraft and for improving the lateral controllability characteristics at the great angles of attack and at the subsonic airspeeds ($M < 0.8$), use is made of the automatically deflectable wing leading edges. The edges automatically deflect through an angle of 20° when the aircraft reaches an angle of attack of 9° with the due regard for the angle of attack increase rate signal.

VARIOUS MACH NUMBERS

The set deflection of both ailerons up through 5 ° decreases the yaw moments developing during the deflection of the lateral control elements at the great angles of attack or at the high Indicated airspeeds, which makes it possible to improve the lateral controllability. The decrease In the aileron deflection angles at the angles of attack exceeding 18° is also intended for this purpose.

At the subsonic airspeeds, with or without the external stores suspended, with the increase in the angle of attack, the lateral controllability changes for the worse and at the angles of attack of 20° the roll rate with the control stick deflected in roll through the half of its travel, amounts to 40 50 deg/s, and when the control stick is fully deflected in roll, the roll rate amounts to 25 30 deg/s.

When flying along the stall limitation system operation boundary, the available roll rate amounts to 20 deg/s.

At a nose, 20° the aircraft stalls at angles of attack 30 to 32°.

The employment of the stall limitation system warns on the aircraft stall during functioning of the wing leading edges in the automatic mode and in case of the failure of the

wing leading edges.

The lateral controllability is maintained till the angles of attack of up to 30°. However, beginning from the angles of attack of 22 to 23° the aircraft rotation in the roll control is of the oscillatory nature at the mean angular rate of 10-20 deg/s. At Mach numbers less then 0.85 and with the increase of the angle of attack from 18 to 21° during correction of the control sticktoaileron transfer ratio the control stick drop to 30 mm is perceived when it is fully deflected in roll. With the angle of attack decreased from 21° to 18° and the control stick deflected in roll, the control stick is shocked. The mentioned phenomenon is not a handicap to the aircraft control.

At the subsonic speeds and at the angles of attack of 20 to 24° the deflection of the control pedals through the distance exceeding 2/3 of the travel causes the increase roll rate up to 20 deg/s on a lapse of 1 or 2 s and the gain of the steady value of 40 deg/s on a lapse of 4 to 6 s from the moment the pedals are deflected.

At Mach number exceeding 0.85 and the angles of attack of 15° and more with the wing leading edges retracted, the aircraft has the sudden roll response to the control stick deflection.

Besides that, interaction between the longitudinal motion and lateral motion appears to a great extent at the mentioned Mach number.

At the angles of attack exceeding 10° and with the control stick fully deflected In roll the roll rate increases greatly (in excess of 90 deg/s) which causes overrun of the angle of attack and operation of the stall limitation system (SOS) with the control stick being motionless in pitch. In the process of overrun even after operation of the limiter the aircraft may reach the angle of attack exceeding 15° temporarily; in this case, the aircraft abruptly

changes the direction of rotation to the side opposite to the control stick deflection.

Upon decrease of the angle of attack, the roll controllability is restored. To avoid the flight in such regime, it is necessary to check the control stick deflection in roll more closely since the lateral control at Mach numbers exceeding 0.85 is very effective.

At Mach number less than 0.85 selfadjustment of the angle of attack by 34° is possible with the active control in roll at the angles of attack less than 19°.

During flight with the vertical speed of 120 to 140 m/s and increase of the angle of attack at the Mach number ranging from 0.85 to 0.92 the wing leading edges are extended and the stall limitation. system is not set to the angle of attack of 15°. Therefore in the above mentioned Mach number range the stall limitation system does not prevent the inadvertent excess of the angle of attack.

PECULIARITY OF STABILITY AND CONTROLLABILITY DURING FLIGHTS WITH EXTERNAL STORES SUSPENDED AND FAILURES OF ONE ENGINE

The efficiency of the aileron practically does not depend on the external stores suspended from the wing.

When the external stores are suspended symmetrically, the roll and yaw moments are mutually trimmed and the presence of the external stores is mainly indicated by the change in the aircraft mass, moments of inertia, drag and by the inconsiderable shift of the aircraft aerodynamic center.

When the drop fuel tank is suspended from the fuselage, the gload stability margin decreases, especially at the angles of attack exceeding 15° and Mach number exceeding 0.7. Within the range of the angles of attack below 5° as well as at the angles of attack of 14 to 21° the aircraft flight control system set to operate in the DAMPER mode when carrying the drop fuel tank and six guided missiles, as well as when the remaining fuel exceeds 2500 kg the aircraft has a small longitudinal margin which calls for a particular attention on the part of the pilot in flying the aircraft. During horizontal flight at the altitudes less than 2000 m and within the airspeed range of 800 to 1000 km/h the decrease of the longitudinal stability margin may cause the aircraft overswinging.

In the roll control of the aircraft with suspended drop fuel tank at Mach number from 0.75 to 0.85 and the angles of attack exceeding 15° , at the initial stage on a lapse Of 1.5 or 2 s the aircraft roll up to 30° in the direction, opposite to the control stick fully deflected with the subsequent response to the control stick position occurs. When the control stick is deflected through 1/2 to 2/3 of the full travel the mentioned phenomenon does not exist, the aircraft rotates with rate fax equal to 20°/s.

When the external stores are suspended asymmetrically, the values of the ailerons and rudder deflections due to the trim for the flight are actually dependent on the flying regime.

The satisfactory lateral controllability of the aircraft at M< 0.85, with the missiles suspended asymmetrically is maintained till an angle of attack of 17° is reached. At the angles of attack exceeding 15° the sustained angular rate is gained on a lapse of 3 or 4 s. At the angles of attack of 20 to 22°, the available angular rate amounts to 10 15 deg/s; in this case, the aircraft rotates with halts, changing the direction of its rotation. At the angles of attack exceeding 22°, the roll controllability with the aid of the control stick is absent. In the lateral control channel, the aircraft carrying the external stores on one side is trimmed by deflecting the control stick in roll through up to 1/3 of the control tick full travel at the angles of attack not exceeding 12° when missile R27R is attached together with one or two missiles R60, and at the angles to attack not exceeding 14° when two missiles R60 or one missile R27R is suspended. With the increase in the angle of attack the required deflections of the control stick due to the trim increase. At the angles of attack exceeding 22°, the aircraft is not trimmed with the control stick fully deflected in roll. Therefore, the maximum permissible angle of attack, with the missiles suspended asymmetrically, is set to be 13°.

In case of the failure of one engine, the turning and rolling moments towards the failed engine originate. The lesser the airspeed and the higher the engine speed at the moment of the engine failure, the greater the slipping and roll originate during the engine failure. At the supersonic airspeeds and Mach numbers less than 0.55, the complete elimination of the slipping requires more than 1/4 of the pedal travel, i.e., it is necessary to override the "rigid" spring.

With four small load packs suspended, the satisfactory lateral controllability is maintained at the Mach numbers below 0.7 till the angle of attack of 19° at which the available roll rate amounts to 30 to 40 deg/s. At Mach numbers of 0.7 to 0.75 and above 0.75 the lateral controllability characteristics are satisfactory to the angles of attack of 17 and 15°, respectively.

The loss of lateral controllability and appearance of the roll response of the aircraft happen at the Mach numbers below 0.7 when a > 24°. When the Mach numbers increases, the angle of attack in the loss of the lateral controllability decreases and at M = 0.8 it amounts to 15 to 16°.

At the altitudes of 5000 m and below and Much numbers exceeding 0.7 it is impossible to reach the angles of attack with the unsatisfactory lateral stability since the aircraft is limited first in value max. Therefore, at the abovementioned altitudes the aircraft may be piloted to the angles of attack of 19° . At the altitudes exceeding 5000 m and Mach numbers exceeding 0.7, never exceed the angle of attack of 15° to avoid operations in the conditions of the lateral controllability loss.

When the load packs are released in the level flight at airspeeds of 650 to 900 km/h the aircraft is not practically overtrimmed. When controlled in roll at the angles of attack exceeding 10°, the aircraft shows the tendency for spontaneously increase of its angle of attack.

When one equipped small load pack is asymmetrically suspended in suspension station 3 or 4 (the other stations are not used), the control stick deflection in roll to counteract the asymmetry does not exceed 1/3 of its full travel to the angle of attack of 19° in all flight regimes.

Near the ground with the full deflection of the pedals and the deflection of the control stick in roll through 1/5 of its full travel, the aircraft with one engine running at the maximum power, in the takeofflanding and en route configurations with or without external stores (including asymmetrically arranged stores) is trimmed at the airspeed of 300 km/h, the full reheat power at the airspeed of 400 km/h. With the increase of the airspeed or flight altitude the trim deflections of the control stick and pedals are decreased.

In the en route configuration without external stores and with asymmetrically arranged guided missiles when one engine of the aircraft is running, at the angles of attack less than 19° the roll angular rate amounts to 30 deg/s. At Mach numbers exceeding 0.55 and the angles of attack of 20° and more. with the control stick deflected in roll, the halts or changes in the direction of rotation occur. Due to this reason, with one engine running the maximum permissible angle of attack is estimated as 19°.

AIRCRAFT TAKEOFF AND LANDING CHARACTERISTICS

According to the takeoff and landing characteristics, the aircraft may be operated from the concrete runways of the third class.

With the aircraft mass preset, the takeoff and landing speeds depend on the highlift devices (the flaps, wing leading edges) and the angle of attack of the aircraft at the moment of unstick or touchdown. The forwarddown vision angle of the pilot amounts to 14°, the maximum angle of attack with respect to the tail fuselage touchdown, with the shock strut uncompressed, amounts to 15°, and to 9.5°, when the shock strut is completely compressed.

Due to great thrusttoweight ratio during the takeoff at the full reheat power, the time required for accelerating the aircraft from the nose wheel liftoff speed to the unstick one amounts to 1 s. This calls for the severe requirements for the piloting technique during the takeoff at the reheat power settings.

Since the air intake inlets are changed over during the takeoff at an airspeed of 200 km/h, for precluding the influence of the longitudinal retrimming on the aircraft control, the nosewheel unstick speed in set to between 230 to 250 km/h. The control stick should be retained In the neutral position until the said speed is gained.

The turning moment caused by the effect of the crosswind component of up to 15 m/s is counteracted by deflecting the pedals through 1/2 of the full travel of the pedals.

In the takeoff/landing configuration, within the operating range of angles of attack, the aircraft is stable with respect to the airspeed and gload. After unsticking, the stabilizer trim deflection amounts to 11 or 12°.

The retraction of the landing gear, flaps and wing leading edges practically does not cause the substantial retrimming of the aircraft, exerting a certain affect on the piloting technique.

It in allowed to land the aircraft carrying the external stores suspended asymmetrically if it is impossible to jettison the (one bomb of 250 to 500 kg, one or two rockets S24B, one or two pods B8).

When landing the aircraft carrying the external stores on one side (one aerial bomb of 250 to 500 kg, one or two rockets S24B, one or two pods B8), the amount of travel of the control stick in roll and the pedals by the flareout Initiation and touchdown moment in the half of the control stick travel and up to 1/3 of the pedal travel.

Proceeding from the trimming conditions, the landing of the aircraft carrying two aerial bombs ABCP500 suspended from stations 1, 3 or 2, 4 in not ensured.

During the landing roll, after landing the aircraft with the nose wheel lifted, at an airspeed of 200 km/h an insignificant pitchup moment practically not affecting the aircraft control originates at the moment the air intake inlets are changed over from the axial inlets to the upper ones.

Section Eight

OPERATION OF SYSTEMS AND EQUIPMENT

LIST OF UNITS, SYSTEMS AND EQUIPMENT

The aircraft is equipped with:

- two engines RD33 with remote aircraft accessories gear box and automatic control equipment;
- armament control system SYV29 comprising integrated radar/sighting system RLPK29 and optoelectronic navigation/sighting system OZPRNK29;
- automatic direction finder ARK19;
- radio altimeter A037;
- marker receiver A611;
- communication radio set R862;
- aircraft intercom system SPY9;
- ground guidance command reception equipment 35022004;
- IFF transponder SRO2 and IFF interrogatorresponder SRZ15;
- voice information reporting system P591B;
- ATS transponder SO69;
- radar illumination warning station SPO-15LM;
- the air conditioning, cooling and pressurizing system;
- the hydraulic system comprising two autonomous systems main and hydraulic systems;
- the air system comprising main and auxiliary systems;
- the fuel system with inert gas system;
- the pitotstatic tubes: main pitotstatic tube PVD-18G-5M and standby pitot static tube PBD7;
- stall limitation system SOS3;
- the taxiing and landing lights;
- cabin light indication system VSS1;
- navigation light system ANO-7;
- oxygen equipment set KNO5 or KKO15;
- fire extinguishing system IS-5MG;
- emergency pump unit NS58; .
- the integrated information and measuring system with recorder TESTER
- recording and indicating system EKRAN;
- automatic flight control system SAE-451-03;
- automatic transmission ratio controller ARY292;
- brake chute PTK29;
- ejection seat K36DM, Series 2;
- chaff dispensing system SVP.
- The aircraft is armed with built-in gun GSh301. Besides, provision is made on the aircraft for employment of the "airtoair" guided missiles, types R27 and R60, rockets, types S24B and S8, arranged in the pods, type E8, the aerial bombs of 250 and 500 kg in weight and a drop fuel tank of 1500 litre capacity.

POWER PLANT

The aircraft power plant comprises two turbojet engines, two air intakes and the aircraft accessories gear box.

ENGINE

Twoshaft bypass engine RD33 with two spool compressor annular combustion chamber, singlestage turbines of the LP and HP rotors, common afterburner and variablearea supersonic allmode nozzle.

Each engine has the bench thrust: at the FULL REHEAT power setting 8300 kgf, at the MINIMUM REHEAT power setting 5600 kgf, at the MAXIMUM power setting 5040 kgf, at the IDLE power setting not more than 180 kgf.

The engine includes:

- the low pressure rotor (LPR) with axialflow fourstage fan and singlestage axial-flow turbine with cooled nozzle guide vanes and rotor blades;
- the highpressure rotor (HPR) with axialflow ninestage compressor with variable inlet and guide vanes of the first two stages and the axial flow singlestage turbine with cooled nozzle guide vanes and rotor blades;
- the annular straight flow combustion chamber with fuel nozzles combined into the common manifold;
- the afterburner with preliminary mixing of flows upstream of the frontal device and afterburner ignition flame path;
- the supersonic jet nozzle with smooth control of the outlet and throat sections;
- the engine accessories gear box coupled with the highpressure rotor through the torsion shaft;
- the engine automatic adjustment and control system;
- the engine starting system;
- the oil system;
- the air bleed system;
- the parameter monitoring system.

Engine Automatic Adjustment and Control System

The engine automatic adjustment and control system is used for feeding the fuel to and automatically controlling the engine running at the sustained and transient power set tings, throughout the entire range of airspeeds and altitudes of the aircraft flight. The system is a hydromechanical one comprising the limit regulator unit ensuring the control of the hydromechanical units in the engine parameter control and limitation modes during the antisurge and afterburner ignition.

- The system comprises:
- the lowpressure fuel system;
- the main manifold fuel system;
- the antisurge protection system;
- the afterburner fuel system;
- the variablearea jet nozzle control system

All the hydromechnical units of the fuel system are mounted on the engine, whereas the limit, regulator and limit command units are mounted on the aircraft.

The fuel system is controlled by the throttle control lever.

From the aircraft fuel system the fuel is fed to the booster pump of the lowpressure fuel system, which delivers the fuel to the fuel filter. The filtered fuel is simultaneously delivered to fuel flow control unit, afterburner fuel pump, and highpressure fuel flow control unit.

In the fuel flow control unit the fuel pressure is considerably increased, metered and fed through the fuel distributor to manifolds 1 and 2 of the main combustion chamber manifold burners.

Besides, the fuel flow control unit feeds the fuel to the compressor guide vanes control system, afterburner ignition system's automatic control units, main and afterburner fuel manifolds.

The afterburner fuel pump increases the fuel pressure and delivers the fuel to nozzle/reheat controller for metering it and, then to afterburner fuel flow distributor and to the afterburner burner manifolds.

The highpressure pump feeds the fuel to nozzle/reheat controller and to outer flap controller controlling the feed of fuel to the hydraulic cylinders controlling the throat and outlet areas of the jet nozzle.

The input information transmitters deliver the electric signals on the engine parameters to the limit regulator and limit command units. The limit regulator unit shapes the control electric commands and delivers them to the electromagnetic actuators of the hydromechanical units. The limit command unit compares the input information with the limit parameter values and if the parameters exceed the rated values, the unit delivers the respective information to the pilot.

The lowpressure fuel system is used for increasing the pressure of fuel delivered from the aircraft fuel system, filtering the fuel and delivering it to the units of the other fuel systems of the engine running at all power settings.

The system comprises the booster pump, the fuel filter, and the fuel drain valve.

The main manifold fuel system is used for ensuring the start, shutdown and operation of the engine running from the IDLE power setting to the maximum one. The system comprises fuel flow control unit , engine speed drop, fuel shutoff valve, fuel flow distributor, temperature sensor, fuel manifolds 1 and 2 with burners, compressor guide vane control hydraulic cylinders.

The main manifold fuel system ensures the execution of the following functions:

- the automatic start of the engines an the ground and in the air;
- the automatic maintaining of the highpressure rotor speed set by the throttle control lever and corrected with respect to engine inlet air temperature at the throttle power settings and also at the maximum power setting in case of the failure of limit regulator unit ;
- the limitation of the minimum fuel consumption;
- the limitation of the critical power settings;
- the assurance of the transient power settings;
- the control of the compressor guide vanes;
- the delivery of the electric signals versus the throttle control lever position;
- the delivery of the controlling fuel pressure to the afterburner fuel manifold automatic control equipment;
- the feed of fuel to the antiicing system valve;
- the improvement of the gas dynamic stability of the engine during the employment of the weapons and elimination of the surge;
- the maintaining of the required pressure value downstream of the pumping unit in the windmilling mode;
- distribution of the fuel between fuel manifolds I and II of the burners;
- the engine shutdown.

The engine is started by the auto start control unit of fuel flow control unit by metering the fuel by the ratio of the fuel pressure at the outlet of fuel flow control unit and the difference between the air pressure downstream of the compressor and the ambient air pressure. For ensuring the fuel metering when starting the engine in the air, the auto start control unit is provided with the altitude mixture control.

The engine parameters are maintained at the sustained throttle power settings by metering the amount of the fuel fed to the combustion chamber and by varying the jet nozzle throat area. In this case, the fuel flow control unit ensures the feed of the fuel in

the mount required for maintaining the amount determined by the position of the throttle of the engine control lever and corrected versus the air temperature at the inlet.

The no variation does not exceed +1.2 % at the throttle power settings and 115 % at the idle power setting.

The change in the engine power from the idle power to the maximum one is effected by readjusting the control, shifting the throttle control lever within the range of the idle stop to the maximum one.

The limitation of the minimum consumption of the fuel fed to the combustion chamber with the engine inlet air pressure correction is performed to ensure the stable ignition of fuel in the combustion chamber.

The sustained parameters are maintained at the maximum and reheat power settings by the joint operation of the limit regulator unit, fuel flow control unit and nozzle/reheat controller that ensures the higher accuracy in maintaining the power as compared with the throttle and idle power settings. The maximum and reheat power settings are maintained by the limit regulator unit ensuring the preset parameters of the engine versus the air temperature at the inlet of the engine.

The Antisurge Protection.

For precluding the simultaneous operation of the speed limiter channel and the hydromechanical controller of the fuel flow control unit, the limit regulator unit delivers the signal for readjusting the hydromechanical controller by 3 % above preset program controlled by the limit regulator unit.

If preset value gets exceeded by 40 to 60°C for 0.2 s, the limit regulator unit shapes the "High temperature" signal in response to which the "Engine overheat. Reduce LH (RH) engine speed" information is delivered to the pilot and the REDUCE RPM (RH) ENGINE red annunciator illuminates on the cabin starboard and at the same time the "High temperature" signal is recorded by the TESTER recorder system and the OVERHEAT LEFT (RIGHT) signal is displayed on the EKRAN annunciator panel.

In case of the failure of any of the limitation channels of the limit regulator unit, detected by the builtin test system, the unit is changed over to the hydromechanical controller whose adjustment is degraded by approximately 6 or 7%.

During operation of the antisurge protection channel from the surge transducer, in case of the transmission of the "High temperature" command from the temperature limitation channel, the abovementioned commands are prolonged for the time the "High temperature" command is available +0.5 s, but not more than 8 s. If the duration of the "High temperature" command exceeds 2.4 s, the fuel feed is cyclically shut off: 2.4 s the shutoff, 1.2 s the feed.

The antisurge protection channel is selected by the antisurge system switch located in the cabin. The selection of the surge elimination system is interlocked at the altitudes below 3000 m at the mach number less than 1.15.

When the FB signal in applied to the limit regulator unit (when launching the missiles (rockets) and firing the gun), the unit delivers the "Antiblowout start" command with a duration of 8 s to the engine auto start control unit as well an to the fan control channel and to the air intake control unit. In this case, the H< 3000 m and M< 1.15 interlock is cancelled, thus enabling the antisurge protection system to operate in response to the surge transducer and, irrespective of the latter, in response to the "High temperature" command (when it is applied) in all flight conditions.

The afterburner fuel system is used for metering the fuel and feeding it to the afterburner fuel manifold burners.

The afterburner fuel system comprises the afterburner fuel pump, nozzle/reheat controller, afterburner fuel flow distributor, two air reducers, afterburner fuel flow control unit.

The system ensures: ignition of the afterburner, operation of the afterburner at the minimum reheat power, metering of the fuel fed to the afterburner at the various stages of afterburning, reheat acceleration and engine speed drop and afterburner selection interlock. The system is controlled by the throttle control lever in response to the hydraulic commands delivered from the temperature sensor, the fuel flow control unit and in response to the electric signals delivered from the limit regulator unit.

When shifting the throttle control lever to the reheat power range, the afterburner fuel pump increases the fuel pressure and delivers the fuel to the inlet of the nozzle/reheat controller and afterburner fuel flow distributor. At the same time, the movement of the throttle control lever gives access of the fuel to manifold I of the burners of the afterburner fuel flow distributor and to the pressure switch. The pressure switch actuates and generates the signal to the limit regulator unit to ensure the fuel flow rate and to change the nozzle throat area, corresponding to the minimum reheat power setting and also to cut in the afterburner fuel flow control unit. The afterburner fuel flow control unit ensures the feed of the fuel in the certain portions and at the required pressure to the spray and swirltype burners which ignite the afterburner. The fuel injected by the spray burner into the main combustion chamber gets ignited and the torchlike flame is transferred by the gas flow downstream of the lowpressure turbine, over the socalled "flame path" where it ignites the fuel injected by the swirltype burner. The generated flame ignites the fuel fed to the afterburner through fuel manifold I located in the afterburner flame holder. To limit the flame effect on the turbine, mounted downstream of the afterburner flame holder are two ionizinggap fire detectors which, when the flame spreads over the flame holder, generate a signal to switch off the afterburner ignition system until the ignition cycle is fully completed. At the same time, the signal on the afterburner selection is applied to the annunciator panel arranged in the cabin and to the TESTER system.

On receiving the signal on the actual ignition of the afterburner, the limit regulator unit cancels the fuel flow rate limitation signal, which causes the increase in the fuel flow fed to the afterburner chamber and increase in the nozzle throat area in compliance with the position of the throttle control lever.

The fuel fed to the afterburner burner fuel manifolds is metered by the nozzle/reheat controller according to the assigned programs.

In case of the sharp shift of the throttle control lever in the range of reheat power settings, the fuel flow rate rise will be determined by the flow restrictor throughput.

When the throttle control lever is sharply shifted from the idle stop to the full reheat one, the afterburner fuel will be ignited only after gaining 72 to 76 % of the reduced speed of the highpressure rotor.

In case of the afterburner flameout (the throttle control lever is placed within the reheat power sector), 05 s later on, the fuel feed is automatically decreased to the amount corresponding to the minimum reheat power.

In case of the emergency disengagement of the afterburner with the aid of the REHEAT EMERG OFF switch, the limit regulator unit delivers the command irrespective of the position of the throttle control lever, to reduce the afterburner fuel flow rate. With the decrease in the pressure, the fuel feed is shut off first in manifolds II and III and, then in manifold 1, which results in disengagement of the afterburner.

Variablearea jetnozzle control system. The nozzle throat area control is required for maintaining speed in compliance with the assigned program of control.

The control of outlet nozzle area is required for maintaining the preset pressure differential between the bleed point on the nozzle flap and the atmosphere, which makes it possible to ensure the complete expansion of the gases in the jet nozzle. The outlet nozzle area is varied by the nozzle outer flaps controller operating together with the nozzle/reheat controller.

The nozzle pressure differential is corrected during the atmospheric pressure variation.

The nozzle f lap position is changed by the jet nozzle control system which is fed with fuel from the highpressure pump. The extension or retraction of the hydraulic cylinder rods of this system varies the diameter of the hydraulic drive rim, which causes the change in the jet nozzle throat or outlet area.

Operation of the control system at the cruising power settings. For ensuring the stable operation of the system at the cruising power settings, the limit regulator unit via the nozzle/reheat controller maintains the throat area in compliance with the control program.

The hydromechanical controller sets up the compressor speed corresponding to the given position of the throttle control lever and temperature by varying the flow rate of the fuel fed to the main combustion chamber. The limit regulator unit, having received the information of speed and temperature, determines speed and forms value according to this speed. On receiving also the information on speed, the limit regulator unit determines present value and compares it with the assigned one. If the difference is available between present and assigned values the limit regulator unit generates a signal to change nozzle throat area so that the present speed of the fan corresponds to the assigned value.

If the present fan speed is less than the assigned value, the nozzle throat area increases, and vice versa.

Starting System

The engines are started on the ground with the aid of the ground power unit or the aircraft storage batteries through the aircraft accessories gear box by turbine engine power unit GTDZ.

The engine starting system ensures:

- the autonomous start of the engines on the ground in any sequence;
- the sequential start of both engines;
- the cold crank of the engines and turbine enginepower unit GTDZ;
- the automatic, semiautomatic or manual air relight of the engines;
- the antiblowout air relight (when employing the weapons).
- The engine starting system comprises:
- the engine auto start control unit (APD);
- the cranking system (turbine enginepower unit GTDZ;
- the ignition system of the main combustion chamber;
- the engine oxygen supply system.

The autonomous start of the engines is effected in the leftright engine sequence.

After depressing the GND START button, turbine enginepower unit GTDZ is started, engine ignition units AZP are cut in (the LH (RH) START annunciator illuminates, the friction clutch of the engine being started is coupled with turbine
enginepower unit GTDZ and the highpressure rotor acceleration occurs. When the rotor speed gains 35 %, the engine ignition units are cut off (the LH (RH) START annunciator goes out)
and when the rotor speed gains 50% turbine engine power unit GTDZ and engine auto start control unit APD are cut out.

The further increase in the highpressure rotor speed occurs without accompanying of turbine enginepower unit GTDZ.

When the engines are started on the ground, the oxygen is fed from the engine oxygen supply system only to turbine enginepower unit GTDZ.

If the highpressure rotor speed does not reach 50 % for 50 a or when the throttle control lever is placed in the STOP position, the starting cycle is discontinued.

The sequential start of both engines occurs automatically when the START TWO LHRE selector switch is placed in the START TWO position, the throttle control lever on both engines is placed in the idle power position and when the GND START button is depressed.

Turbine enginepower unit GTDZ through the aircraft accessories gear box first puts the right engine into operation, ensuring its start and accompanying up to speed of 50%.

When the minimum 50% command is delivered from the preliminary command unit of the right engine, turbine enginepower unit GTDZ is switched off for 10 s to ensure the shockless coupling of the shaft of turbine enginepower unit GTDZ with the shaft of the left engine. On expiration of 10 s, the left engine is started and is accompanied up to 50%.

In case of the discontinuation of the engine start because of the fuel ignition failure or the speed increase failure, it is allowed to crank the abortive engine, with the other engine running.

For cranking any engine (turbostarter) on the ground, set the throttle control lever to the STOP position and the PWR SYS START TS CRANK ENG CRANK switch to the EKG CRANK position.

After depressing the GND START button, the highpressure compressor rotor of the engine (turbostarter) accelerates in the sequence described above, but without feed of the fuel to the engine (turbostarter) combustion chamber, with the engine (turbostarter) ignition unit switched off.

In all kinds of the air relight at the altitude above 6000 m, the electromagnet of the start control unit of the fuel flow control unit is cut in additionally to correct the flow of fuel fed to the combustion chamber, said unit being cut off when > 75%-85% if preliminary command unit BPK, Series 2 is used).

When relighting the engine, the ignition unit is cut in (the START LH (RH) annunciator illuminates) and the oxygen supply valve electromagnet gets energized for 20 s during the automatic and semiautomatic relights, for the time the AIR RELIGHT switch is turned on during the manual relight and for 8 s during the antiblowout relight.

Air Bleed System

The engine compressor air bleed system is used for ensuring the normal operation of the engine support seals, warning on the fan inlet icing, bleeding the air for conditioning the cabin, cooling the blades and jet nozzle vanes of the high and lowpressure turbines. The blades and vanes are cooled off at all power settings.

AIR INTAKES

The air intakes are used for the effective deceleration of the air flow coming to the engine inlet during the entire range of the tolerable flight regimes.

The aircraft is provided with the flat variable supersonic air intakes with horizontal braking surfaces located under the wing extension.

At the supersonic airspeeds, the deceleration of the air flow to a supersonic airspeed is effected in the system of four oblique shocks developed at the breaks of the braking surfaces and the terminal normal shock located near the air intake lip.

For precluding the flow of the compressed air into the external flow, the air intake is provided with the side slits.

The number of the oblique shocks, the angles of their inclination as well as the position of the terminal normal shock depend on the Mach number, angle of attack of the aircraft and angle of turn of the air intake ramp variable panel.

The law of the ramp movable panel control is selected proceeding from the conditions of coordination of the rate of air flowing through the engine and air intake, and assurance of the sufficient margins of the power plant stability.

The angle of the first stage is constant, whereas the angle of the second, third and fourth stages vary as integral one from 0 to 100% of the complete travel by reference to the ramp position indicator.

For improving the characteristics of the air flow in the air intakes, the fourth stage of the ramp and the throat of the air intake are provided with the perforation and slit for sucking the boundary layer. The air is discharged from the sucking duct to the wing extension upper surface.

The air intake duct has the rectangular inlet and round outlet sections.

For eliminating the effect of the boundary layer growing on the lower surface of the wing extension on the air intake characteristics, a drain slit of 60 to 70 mm in height is made between the extension and the air intake lip, in which the boundary layer is drained to the sides with the aid of the shaped drain ramp surfaces. A certain portion of the boundary layer is taken for the aircraft system requirements.

To protect the engine against penetration of the soil particles through the air intake during the start, takeoff run and landing roll, provision is made for the upper duct.

It is formed in the wing extension, when the front variable panel of the air intake ramp is completely extended shutting off the main duct.

The necessary amount of air is taken from the upper surface of the wing extension through the spring loaded doors which open under action of the rarefaction in the duct. The rear movable panel of the air intake ramp is retracted, ensuring the maximum bore of the duct.

Air Intake Automatic Control System

The air intake ramps are controlled by two air intake automatic control systems. Each of these systems controls their own ramp at the takeoff, landing and during the flight. When in flight the air intake automatic control system functions according to three programs versus the altitude and mach number. The front doors of the air intake ramps assume the retracted position till the engines are started on the ground. The rear panel of the air intake ramp is coupled with the body and is disconnected from the front panel thus ensuring the inspection of the air intake ducts and the first stages of the compressors. The upper inlet doors are fixed in the closed position.

While starting the engine, with the throttle control lever set to the idle power stop and the operating pressure available in the main hydraulic system of the aircraft, the air intake upper inlet doors get unlocked and then, the door of the air intake ramp front panel, facing the engine being started, moves towards the takeoff/landing position, closing the axial inlet and thus protecting the engine against penetration of the foreign objects. The air passes into the engine through the air intake upper inlet.

After starting both engines the serviceability of the units and assemblies of the air intake automatic control system is automatically checked, and the check results are displayed on the LH (RH) INLET CHECK annunciator located on the cabin right side at the moment the engine speed increases up to 80-90% for a short time (1 or 2 s).

At the takeoff, the air passes into the engines through the upper inlets of the air intakes. When airspeed of >200km/h is gained, the ramp front panel doors are retracted. If, because of any trouble, the front panel door of any of the air intake ramps fails to be retracted, then after retraction of the LH landing gear, the UPPER INLET signal appears on the EKRAN annunciator panel and the "Air intake upper inlet. Do not accelerate over Mach number 0.8" voice information is heard in the earphones.

After opening the axial inlet and retracting the landing gear, the third program of the air intake ramp control, with the ramp travel of up to 35%, is selected, which is realized at the altitudes of up 3000 m at M <1.5 and at the altitudes above 3000 m at M < 1.15.

During operation of the air intake according to the third program of control, in case of the automatic control equipment failure as a result of which the ramp may be extended in excess of 40%, the air intake automatic control system locks the ramp in the 55% position, the LEFT (RIGHT) AIR INTK signal is displayed on the EKRAN annunciator panel and the "Failure of automatic control equipment of left (right) air intake" voice information is delivered into the earphones.

The second program of the air intake ramp control, limiting the air intake ramp travel up to 60%, is selected at the altitudes above or equal to 3000 m at M 1.15 to 1.5.

The first program of control within the ramp travel from 0 to 100 % is executed at Mach numbers exceeding 1.5. The air intake ramp displacement may be checked by the pilot against air intake ramp position indicator EPK. Since the air intake ramps are controlled versus the reduced speed of the engines, then, with the difference in the speeds of the engines, the respective difference will be in the positions assumed by the ramps and the indicator abutters on ramp position indicator EPK.

The system ensures the emergency retraction of the ramp and makes it possible to execute the forced retraction of the ramp owing to the aerodynamic forces when there is no pressure in the hydraulic systems.

During the automatic control of the air intakes, the air intake automatic control systems execute the correction of the control programs for the ramp extension in response to the antisurge protection system actuation signal by 10 % of the ramp travel. When the air intake ramp automatic control system employs the second and third programs, the extension of the air intake ramp in response to the antisurge protection system correction signal does not exceed the value of the horizontal limitation section of the respective program.

When extending the landing gear, both air intake ramps assume the 0 % position by reference to ramp position indicator EPK.

After touching down, during the landing roll at V < 200 km/h, the air intake axial inlet is closed by the ramp front panel doors. On ramp position indicator EPK, the shutter should set at the TO/LAND mark.

When shifting the throttle control lever to the STOP position, the ramp front panels move to the retracted position O % against ramp position indicator EPK) and the upper inlet doors are fixed in the closed position.

Apart from the main control, the air intake automatic control system employs the standby control of the air intake ramps, which is selected in case of the failure of the automatic control equipment.

The air intake automatic control system provides for the manual emergency retraction of the ramps to the 0 % position, effected by the pilot. The emergency r traction mode interlocks the system and after using said mode. the operation of the system on the main channel is impossible.

TAKEOFF RUN AND LANDING ROLL MODE

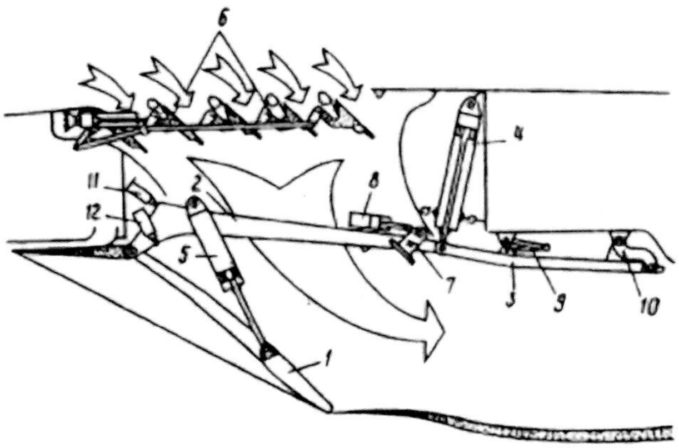

FLIGHT REGIME

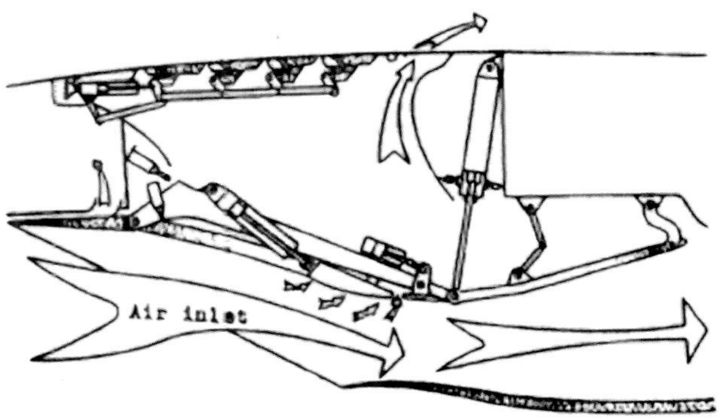

Arrangement of ramp panels in various flight regimes

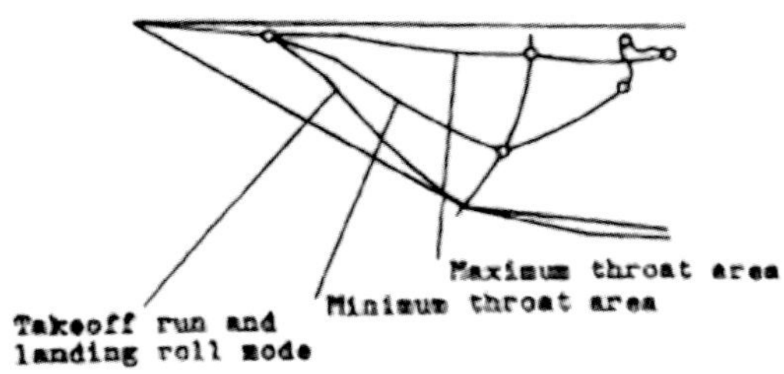

1 - front panel door; 2 - front panel frame; 3 - ramp rear panel; 4 - ramp hydraulic cylinder; 5 - door hydraulic cylinder; 6 - air intake upper doors; 7 - front panel valve; 8 - valve hydraulic cylinder; 9 - pressurization shutter; 10 - ramp rear panel attachment unit; 11 - ramp panel retracted position sequence valve; 12 - door retracted position sequence valve

PECULIARITIES CF POWER PLANT OPERATION

It is al.owed to control the engines by gradually or energetically shifting (for 1 or 2 s) the throttle control lever from the IDLE stop to the FULL REHEAT stop.

At the altitudes above 10,000 m and the airspeed less than 5.50 km/h, the acceleration hangup (the slow acceleration) may occur when shifting the throttle control lever from the idle power setting to the maximum one.

The correct keeping of the control laws by the engine automatic control equipment at the MAXIMUM and REHEAT power settings is checked by the difference in the speed of the left and right engines and by reference to the indication system.

The difference in the temperatures of the left and right engines is not fixed.

When creating the gload, the engine speed may vary up to 5 7% but not more than up to 103%. and the temperature may vary to some extent but not more than up to the assigned maximum value.

The reheat power is reliably selected at the maximum power setting throughout the entire range of the aircraft employment. When the reheat power is selected at the idle power setting, the reheat power may not be engaged at H 14,500 m at the airspeeds close to the minimum ones. If the reheat power rails to be selected, the LH (RH) ENG REHEAT annunciator is not alive, shift the throttle control lever to the MAXIMUM position and when the sustained parameters of the engine are gained, select the reheat power once more by setting the throttle control lever to the reheat position.

Check the reheat power selection by reference to variation in the longitudinal gload and illumination of the LH (RH) ENG REHEAT annunciator.

If the reheat power fails to be switched off when the throttle control lever is shifted to the MAXIMUM stop (the REHEAT annunciator is alive), switch off the reheat power with the aid of the REHEAT EMERG OFF selector switch. In this case, the reheat power of both engines will be switched off.

AIRCRAFT FUEL SYSTEM

The fuel is contained on the aircraft in five integral tanks located in the fuselage and two integral tanks arranged in the wing of the aircraft. Provision is made for suspension of the ventral fuel tank which may be jettisoned in flight.

The hold capacity of the tanks amounts to:

650 ltr in tank No. 1;
870 ltr in tank No. 2;
1810 ltr in tank No. 3;
310 ltr of two tanks No. 3A in total;
660 ltr of two wing tanks in total.

The total hold capacity of the main tanks is 4300 ltr. The amount of unusable fuel in the main tanks is 100 ltr. The ventral drop fuel tank holds 1500 ltr, including 20 ltr of unusable fuel.

The total bold capacity of the aircraft fuel system, with the ventral drop fuel tank attached is 5800 ltr.

When in flight the ventral drop fuel tank may be jettisoned by depressing the TANK JETSN button located on the aircraft control stick.

The tanks may be fuelled by two methods: the singlepoint fuelling (the main method) and fuelling of the tank via the filler neck.

The single point fuelling is carried out with the aid of two fuelling units located in tank No. 3 and one fuelling unit located in the ventral drop fuel tank. The aircraft tank singlepoint fuelling units are coupled with the aircraft connection to which the refueling hose is attached. The aircraft connection is installed in the LH landing gear well.

The fuel is run out of the tanks in the sequence ensuring the aircraft centre of gravity within the assigned limits

Fuel tank No. 2 is a service one. It also serves as a negative gload compartment. The sequence of the fuel transfer is determined by the order of the fuel usage.

The fuel is transferred into the service tank as follows:

- from the ventral drop fuel tank, the fuel is delivered under air pressure into tank No. 1;
- from tank No. 1, the fuel is transferred into tank No. 2 with the aid of the hydraulic turbopump and if the latter fails, the fuel flows into tank No. 2 by gravity due to the difference in the fuel levels in tanks Nos 1 and 2;
- from tank No. 3, the fuel is transferred into the service tank with the aid of the hydraulic turbopump, at the same time the jet pump transfers the fuel from tank No. 3 into tank No. 1 so that, if the hydraulic turbopump fails, the aircraft rear center of gravity does not increase;
- from the wing tanks and tank No. 3A, the fuel is transferred into tank No. 3 with the aid of the jet pumps.

From tank No. 2, the fuel is delivered to the engine pumps with the aid of two hydraulic turbopumps arranged one above the other so that during the negative or zero gloads in two planes, with any gload available in the third plane, as well as during the inverted flight, one of the pumps is immersed in the fuel.

In case of the zero gload in three planes (in case of the weightlessness), the fuel accumulator located in tank No. 3 is Put into operation, which ensures the delivery of fuel to the engine pumps by forcing the fuel out of the fuel cavity of the accumulator by the diaphragm under the air pressure.

From the pressure line, a certain portion of fuel is fed to centrifugal pump PTsN which increases the fuel pressure versus the engine power setting. This fuel is delivered to the jet pumps and hydraulic turbopumps and put them into operation (the active fuel).

Downstream of the centrifugal pump, a certain portion of fuel is delivered to the fueloil cooler for cooling the oil, and then the fuel is returned to the pressure line.

For precluding the intensive separation of the resin from the fuel and contamination of the engine automatic control equipment with this resin, provision is made for the hot fuel bypass system. In response to the fuel temperature pickup installed downstream of the engine pumps, a certain portion of fuel delivered to the engine pumps begins returning into tank No. 2. The fuel pumping rate increases and the fuel temperature decreases.

For cooling the antifreezing compound of the radar liquid cooling system, the fuel is pumped by the lower hydraulic turbopump of tank No. 2 via the fuelliquid cooler. The hot antifreezing compound circulates through the cooler pipes and gives off the beat to the colder fuel flowing through the channels between the pipes. The warmedup fuel 45 fed to the inlet of the upper hydraulic turbopump of tank No. 2.

During the ground functional check of the radar, the fuel is pumped through the cooler with the aid of centrifugal pump =H. In this case, it is necessary to open the shutoff valve installed under the pump. After the functional check of the radar, close the valve.

The fuel system altitude tolerance and venting of the tanks are ensured with the pressurization system.

The fuselage and wing tanks are pressurized with nitrogen or air, whereas the fuel accumulator and ventral drop fuel tank are pressurized only with air.

The sections of the system pressurizing the tanks with nitrogen or air are separated from one another with the aid of the nonreturn valves. If the nitrogen is used up or the electropneumatic valve is shut off, the tanks will be pressurized with air.

For pressurizing the tanks. the air is bled downstream of the fan of each engine and is directly fed to the pressurization unit through the nonreturn valves and to the fuel accumulator and drop fuel tanks through the flow restrictors.

The pressurization unit is used for decreasing the pressure of the air or nitrogen fed to the fuel tanks to a certain value and for maintaining this pressure versus the flight altitude and also for bleeding the air or nitrogen when the pressure in the tank's exceeds the assigned value.

For pressurizing the fuel tanks, four bottles located in the nose landing gear well are charged with nitrogen at a pressure of 320 kgf/cm. though the charging connection.

After cutting in the electropneumatic valve, the nitrogen is fed to the pressurization unit through the reducer decreasing the pressure from 320 to 8 kgf/cm.

The nitrogen begins to be fed to the fuel tanks as soon as the engine highpressure rotor gains a speed of about 5%, with the opening of the electropneumatic valve. For preserving the assigned concentration of the nitrogen in the fuel tanks when the engines are shut down in the air, the electropneumatic valve is interlocked with the lefthand landing gear uplock limit switch.

The pressurization of the fuel accumulator and drop fuel tanks with air begins from the moment the engine is started. The assigned pressurization pressure is maintained by the safety valves. When the pressurization pressure is not available in the drop fuel tank, the pressure switch delivers the DROP TANK no USAGE command to the EKRAN annunciator panel.

The aircraft is provided with the fuel quantity gauge and flowmeter system comprising four sections: fuel flow metering, fuel quantity gauge, fuel automatic control and available flight range computation sections.

The fuel flow metering section of the system measures the total consumption of fuel and instantaneous flow rate, introducing the necessary correction for the temperature and grade of fuel used.

The fuel quantity gauge section measures the amount of fuel available in the fuselage integral tanks, automatically sets the index of the fuel quantity gauge and flowmeter system indicator to the quantity of fuel filled. The fuel quantity gauge section functions with the due allowance for the compensation for the error caused by the change in the temperature and fuel grade.

The fuel automatic control section of the fuel quantity gauge and flowmeter system is used for generating the tank fuel runout signals, controlling the singlepoint fuelling, automatic indication and correction of the readings of the fuel flow metering section, calling the information from the fuel quantity gauge section to the fuel reserve scale on the fuel flowmeter and quantity gauge system indicator, performing the builtin test during the line maintenance.

The available flight range computer is used for computing and indicating the available flight range by reference to the information on the fuel reserve and consumption and true airspeed.

AIRCRAFT FIRE EXTINGUISHING SYSTEM

The aircraft fire extinguishing system is used for extinguishing the fire areas in the engine compartments and in the aircraft accessories gear box compartment and comprises the fire detection system ES and fire extinguishing system.

Tanks used	Amount of fuel used, ltr
Ventral tank	Fully used
Tank No. 1	250
Tank No. 3	200
Wing tanks	Fully used
Tank No. 3A	Fully used
Tank No. 3	600
Tank No. 1	280
Tank No. 3	Remainder
Tank No. 1	Remainder
Tank No. 2	Fully used
Accumulator tank	Fully used

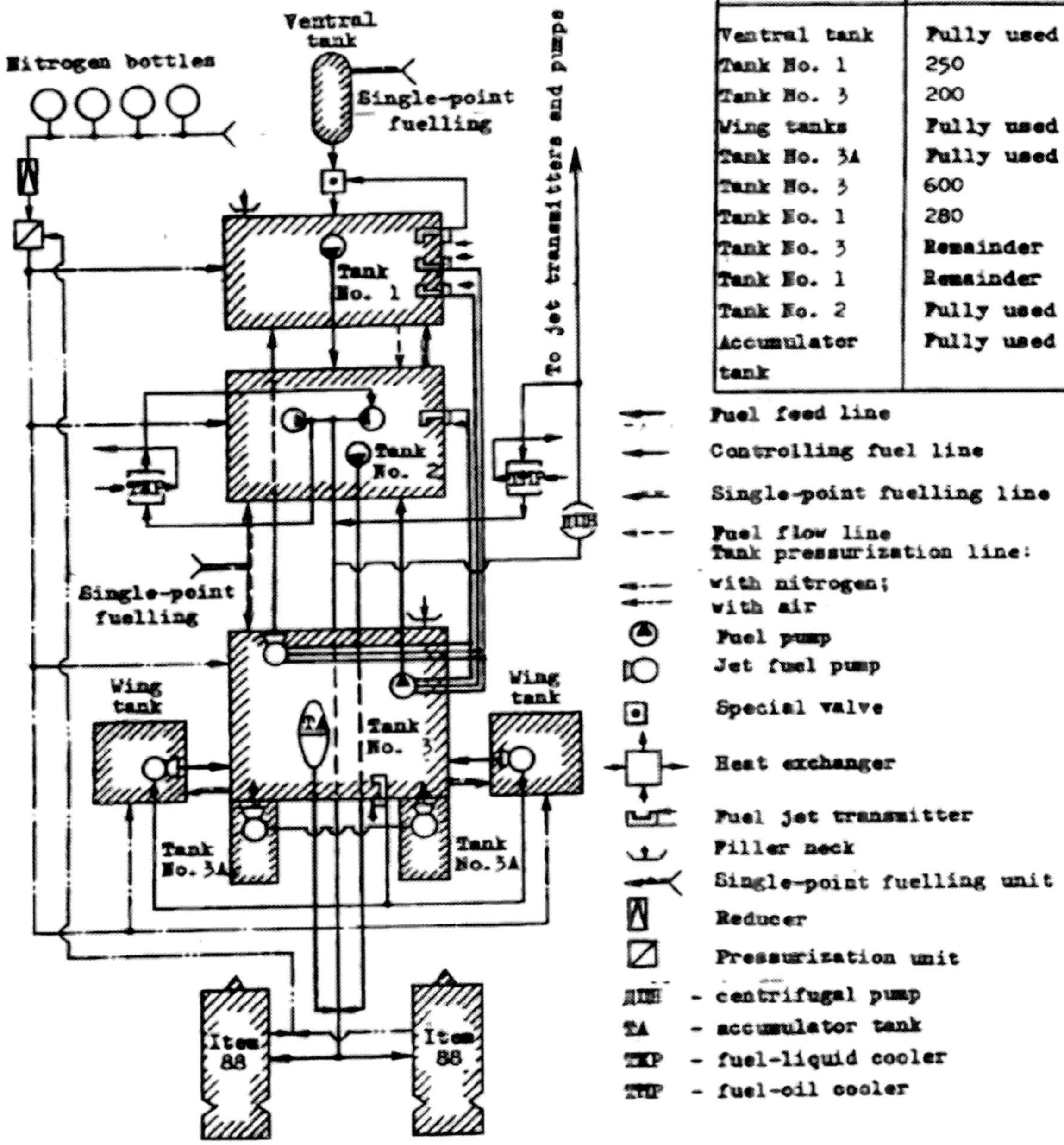

SCHEMATIC DIAGRAM OF FUEL SYSTEM AND FUEL TANK USAGE

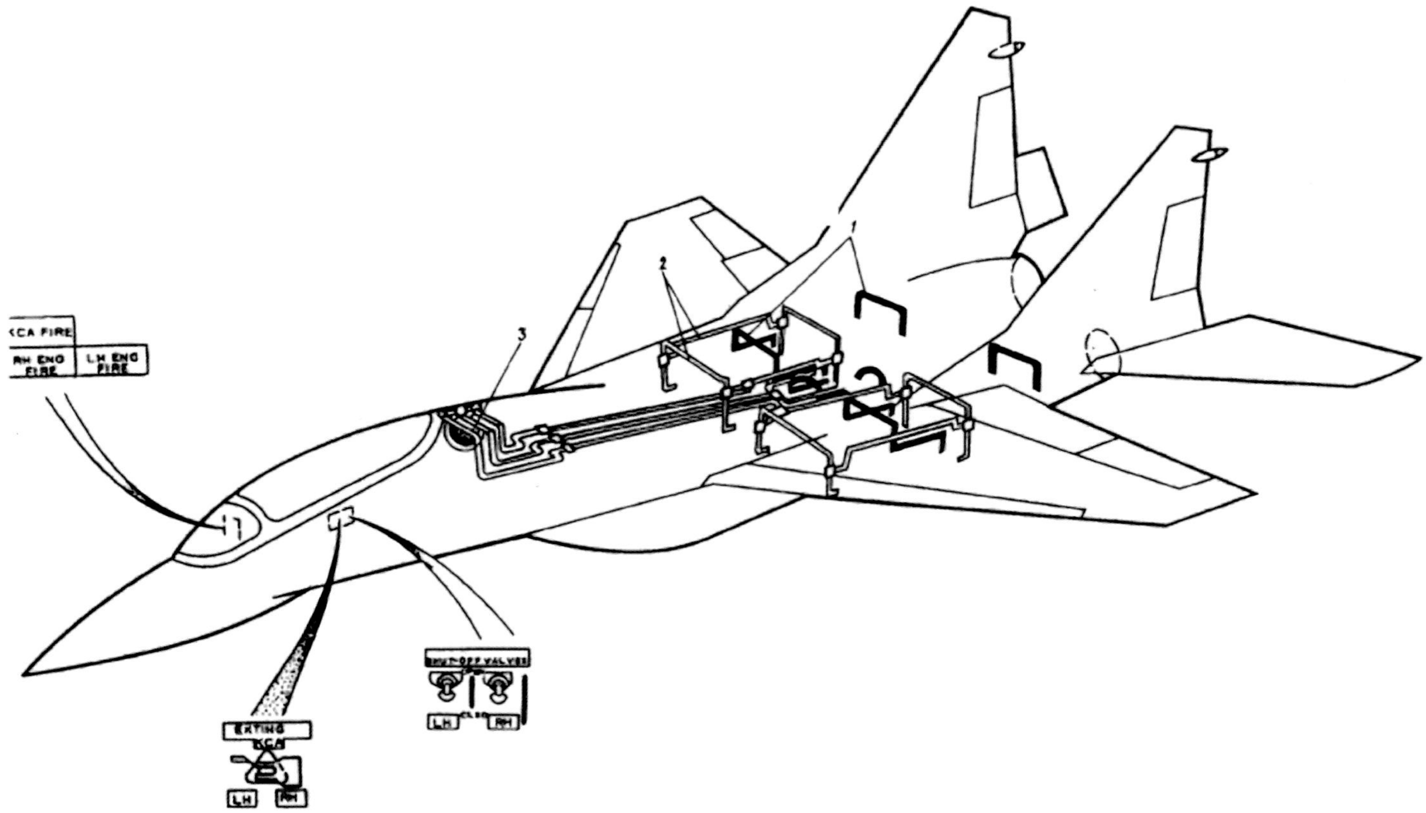

FIRE EXTINGUISHING SYSTEM

1 - fire detector; 2 - fire extinguishing spray rings; 3 - bottle with fire extinguishing mixture

Fire detection system ES functions on the principle of usage of the electric conductivity of the flame when the free electric charges appear therein. The appearance of the flame between the detector (the heat resistant metal tubes installed in the engine compartments and in the aircraft accessories gear box compartment) and the frame makes the space between them electrically conductive.

The fire detector does not respond to the temperature rise. The system precludes the false actuation when the detector is grounded or in case of the decrease in the detector insulation resistance due to the heat or moisture.

Fire detection system ES actuates for not more than 3 s when the flame licks the detector. When the flame disappears, the system returns to the initial position for the time of not more than 3 s.

When the system actuates, lightbutton KSTs begins flickering, the KSA FIRE (LH (RH) ENG FIRE) annunciator illuminates and the aircraft accessories gear box (left (right) engine) fire" voice information is delivered twice to the earphones.

The fire extinguishing system comprises the fire extinguisher and the spray rings located in the engine compartments and in the aircraft accessories gear box compartment. The fire extinguisher is essentially a balltype bottle with three discharge bonnets. Each discharge bonnet mounts two electrically actuated squibs and the operating connection through which the fire extinguishing agent is fed to the spray rings versus the position of the FIRE EXTING: LH (RH, KSA) selector switch located in the cabin.

HYDRAULIC SYSTEM

The aircraft hydraulic system comprises two independent systems the main and booster systems.

The main hydraulic systems ensures:

- the feed of the second chambers of the hydraulic boosters of the stabilizer, ailerons, rudders and the hydraulic cylinder of the control stick pusher of the stall limitation system.
- the retraction and extension of the landing gear, highlift devices, air brakes;
- the control of the air intake ramps, nosewheel steering/damping control valve and the pedal loading.

The booster hydraulic system ensures the feed of the first chambers of the two chamber hydraulic boosters of the stabilizer. ailerons, rudders and the hydraulic cylinder of the control stick pusher of the stall limitation system.

The hydraulic system working fluid is oil AMG-10.

Variable displacement pumps HP, with the operating pressure ranging from 190 to 220 kgf/cm, serve as the sources of the hydraulic energy in both hydraulic systems.

The feed of the working fluid is controlled by the device included into the pump design. Each system employs their own pump installed on the aircraft accessories gear box.

For protecting the pressure lines of the hydraulic systems against the intolerable high pressures, use is made of the safety valves adjusted to an opening pressure of 240 kgf/cm, then this pressure exceeds the rated value, the safety valves bypass a certain portion of fluid into the return line.

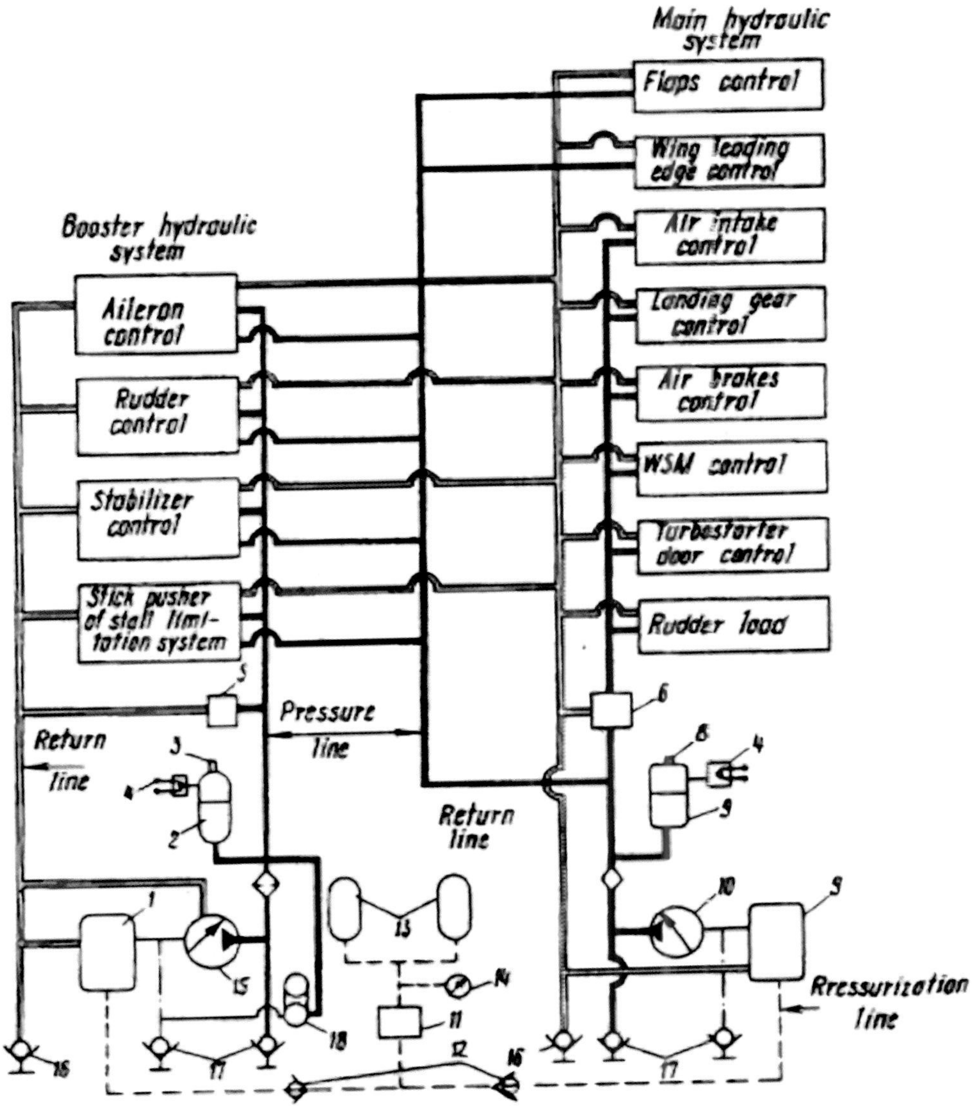

BLOCK DIAGRAM OF AIRCRAFT HYDRAULIC SYSTEM.

1 - booster system hydraulic tank; 2 - booster system hydraulic accumulator; 3, 8 - charging connection; 4 - pressure transmitter; 5 - safety valve; 6 - pressure and flow restrictor; 7 - main hydraulic system hydraulic accumulator; 9 - main hydraulic system hydraulic tank; 10 - pump; 11 - reducer; 12 - non-return valve; 13 - gas bottles; 14 - pressure gauge; 15 - pump; 16 - filling connections; 17 - ground pump coupling connections; 18 - emergency pump unit

For protection against the intolerable high pressure of operating medium, the tanks of the hydraulic systems are provided with the safety valves.

The actuators are controlled by the electrohydraulic distributors which are remotely switched on and switched off from the cabin.

The pressure in the main and booster hydraulic systems is checked against the remote controlled combined hydrogas system indicator installed on the instrument hoard.

The indicator transmitters are connected to the gas chambers of the hydraulic accumulators of both systems, therefore, when there is a zero pressure in the hydraulic systems, the indicator reads the nitrogen charging pressure in the gas chambers of the hydraulic accumulators.

With the engines running, the indexes showing the pressure in the hydraulic systems should be within the green sectors provided with marks.

When starting the engines at the low temperatures and increased flow rates of the .hydraulic fluid, the indicator index may move to the yellow sector (upper or lower one, respectively) for a short time.

When the pressure in one of the hydraulic systems drops below 100 kgf/cm, the BOOST FYD SYS or MAIN HYD SYS signal illuminates on the EKRAN annunciator panel, and when the pressure drops to the same value in both hydraulic Systems, the TWO H/SYS FAIL annunciator illuminates on the right vertical console and the AFCS power supply is automatically cut off. Simultaneously with the illumination of said annunciator, the respective voice information is delivered to the earphones. When the pressure exceeds 100 kgf/cm , the light and voice information is cut off and the AFCS power supply is cut in.

For the emergency supply of the booster hydraulic system, the aircraft mounts the emergency pump unit (HC) fed from the fuel pumps of the automatic control equipment of both engines. The pump unit is automatically cut in when the pressure in both hydraulic systems drops below 100 kgf/cm and ensures the aircraft control with the engines running at the speeds not below 85 %. The pump unit may also be cut in manually with the aid of the EMER PUMP UNIT switch. The pump unit cutin is checked by reference to the EMER PUMP UNIT green annunciator.

AIR SYSTEM

The air system is used for storing and distributing the air under the low and high pressures between the consumer systems. According to its functional purpose the air system is subdivided into three independent systems: the main, emergency systems and the radio unit pressurization system.

The main air system ensures:

- the braking of the landing gear wheels;
- the control of the hinged canopy and cabin pressurization;
- the deployment and release of the brake chute;
- the control of the fuel system shutoff valves.

The emergency air system ensures:

- the emergency extension of the landing gear;
- the emergency braking of the main landing gear wheels.

The radio unit pressurization system maintains the required pressure in the radio unit compartments.

The bottles are charged to a pressure of 150 kgf/cm from the ground compressed air source through the common charging connection located in the LH landing gear well.

Mounted close to the charging connection is the pressure gauge used for checking the charging pressure and air tightness of the main air system on the ground (on aircraft not equipped with hydraulic tank independent pressurization system its charging is effected through the charging connection of the air system).

The air systems are separated by the nonreturn valves preventing the flow of air from one system to the other and to the charging line.

The air is fed to the main air system consumers through the reducer and to the emergency air system consumers and radio unit pressurization system directly from the bottles.

The air reducer decreases the pressure from 150 to 63 kgf/cm.

For increasing the total amount of air and for ensuring the reliable functioning of the brake chute, an additional air bottle is installed thereon. This bottle is separated from the main air system by the nonreturn valve and may be discharged only into the brake chute units.

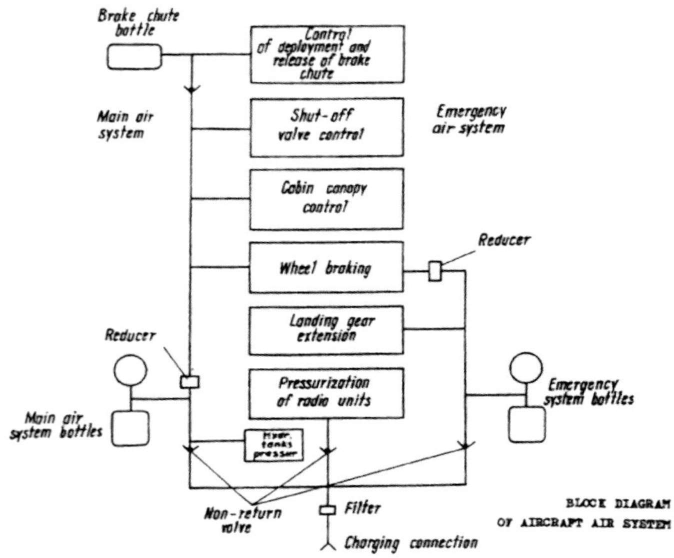

BLOCK DIAGRAM OF AIRCRAFT AIR SYSTEM

The pressure in the air systems is checked against the combined hydrogas system indicator. The indexes showing the pressure in the gas systems should be within the green sector of the scale.

ALIGHTING GEAR

The alighting gear includes the landing gear, brake chute system, flaps and deflectable leading edges of the wing.

The landing gear is retracted and extended with the aid of the main hydraulic system, by means of the actuating struts. The nose landing gear leg is retracted backward, whereas the main landing gear legs are retracted forward in the direction of the flight into the wells provided in the fuselage and are closed with the doors. All the landing gear legs are retracted and extended at a time. When retracted, the landing gear legs are fixed by the mechanical locks located in the landing gear wells. When extended, the landing gear less are fixed by the mechanical locks built into the actuating struts. The landing gear retraction time is 9 to 10 s, and the landing gear extension time is 7 to 8 s.

The turbostarter doors are closed and opened with the aid of the special actuating cylinder simultaneously with the retraction and extension of the landing gear.

The emergency extension of the landing gear is effected from the emergency air system of the aircraft with the aid of the actuating struts and actuating cylinders doors. If necessary, the autonomous emergency extension of only nose landing gear may be performed.

For performing the emergency extension of the landing gear, it is necessaryto pull the EMERG LG handle all the way backward than turn the handle clockwise the main landing gear will be extended. The initial positions for operation of the system are any position of the landing gear and any position of the landing gear valve selector knob. The design of the selector valve provides for the hydromechanical interlock precluding the inadvertent retraction of the landing gear after its emergency extension when the landing gear valve selector knob is placed in the RETRACTED position. When the air pressure is applied for the emergency extension of the landing gear, the switch in looked in the position connecting the leading gear retraction line with the return line. This position precludes the retraction of the landing gear and is retained after bleeding the air from the emergency lines. For restoring the landing gear emergency retraction system, it is necessary to return the landing gear emergency extension handle to the initial position in the reverse order; in this case, the air is discharged from the actuating doors and actuating struts of the landing gear into the atmosphere through the landing gear emergency extension valve.

For retracting the landing gear, after restoring the landing gear emergency extension system, it is necessary first (if the pressure is available in the hydraulic system) to set the landing gear valve knob to the EXTENDED position, if the knob has been placed in the RETRACTED position and, then to the RETRACTED position not earlier than in 10 to 151 s.

The landing gear position is checked by reference to landing gear and flaps position indicator.

Before the takeoff and landing (with the landing gear extended) three green indicator lights of the landing gear are alive. After the retraction of the landing gear, all the indicator lights should extinguish.

The landing gear intermediate position red indicator light burns continuously if even one of the landing gear legs is not downlocked with the valve knob set to the landing gear extended position or uplocked with said valve knob set the landing gear retracted position.

The landing gear wheel braking system is part of the aircraft air system and is subdivided into the main and emergency systems. The compressed air fed from the air system under high pressure is reduced by the reducers of the main and emergency systems.

The main braking system comprises two sections: the control and actuating sections.

In the control section of. the system, the air reduced to a pressure of 63 kgf/cm is fed from the bottles to the reducing valve which in its turn reduces the air versus the braking lever depressing value: to the maximum pressure of 8 kgf/cm 2 if the lineup braking trigger is not depressed or 11 kgf/cm if said trigger is depressed Then, the air is fed to the pneumaticallycontrolled reducing valves through the differential control unit. When moving the pedals, the differential control unit bleeds the air fed to one of the pneumaticallycontrolled reducing valves of the main landing gear wheels thus releasing the brake of the respective wheel of the landing gear. Under the action of the control pressure, the pneumatically controlled reducing valves increase the pressure applied to them from the air reducer three times for the main landing gear wheels and two times for the nose landing gear wheels. Under this pressure the air is fed to the wheel brakes and the latter get braked.

In the nose wheel braking line, the air upstream of the pneumaticallycontrolled reducing valves passes through the shutoff valve serving for abutting off the nose wheel braking line with the aid of the NOSE WHEEL BRAKE knob located in the cabin.

The braking system air pressure is checked against the twopointer pressure gauge connected into the system downstream of the differential control unit and reading the pressure in the control section of the braking system.

For braking the wheels simultaneously with the retraction of the landing gear, the automatic braking cylinder is installed in the hydraulic system. When the landing gear valve is set to the RETRACTED position, the respective command is applied to the automatic braking cylinder rod (in the hydraulic system) which through the linkage of cables and bellcranks opens the reducing valve to a pressure of 2 to 4 kgf/cm . Under the action of this control pressure, the pneumaticallycontrolled reducing valves reduce the pressure applied to them from the reducer and apply the pressure to the wheel brakes. After retraction of the landing gear and actuation of the sequence valves, the cylinder rod retracts, shutting off the reducing valve and, thus releasing the wheel brakes.

The automatic brake release system prevents the skidding in the process of its onset and precludes the complete interlock of the wheel. In case of the skidding origination under the effect of the negative acceleration, the rotating flywheel of the skid detector of the given wheel moves and closes the detector contacts.

The electrical signal generated thereby causes the respective electropneumatic valve .to actuate, bleeding the air from the wheel brake into the atmosphere. The wheel brake is thus released.

If the skidding appears on one of the main landing gear wheels, only this wheel brake is released and if the skidding appears on any of the nose landing gear wheels. the brakes of both wheels of this landing gear are released.

The automatic brake release system provides for braking when the braking lever is fully depressed. which ensures the optimum mode of braking

In this case, the air consumption increases to some extent, but the process of braking is considerably simplified.

The main landing gear wheel braking system is used in case of the failure of the main braking system. The nose landing gear wheels are not provided with the emergency braking system.

The emergency braking is effected by feeding the air from the emergency air system.

The emergency braking system is controlled with the aid of the EMERG BRAKE , knob located on the upper section of the instrument board. When pulling the knob, the air is fed from the reducer at a pressure

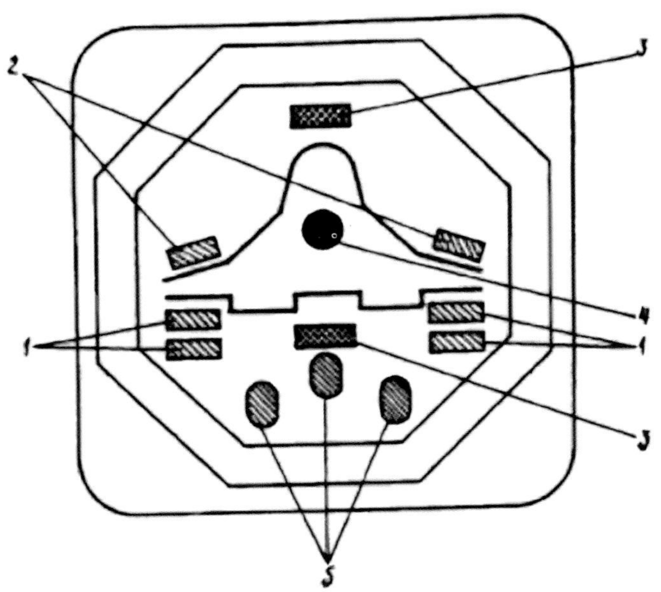

LANDING GEAR AND FLAPS POSITION INDICATOR MП

1 - flaps position annunciators; 2 - leading edges position annunciators; 3 - air brakes position annunciators; 4 - landing gear intermediate position annunciator; 5 - landing gear extended position annunciators

of 63 kgf/cm through the reducing valve to the emergency selectors which actuate and connect the wheel brakes with the emergency air system, shutting off the main air system pipelines. The wheels get braked.

The reducing valve reduces the pressure proportional to the linear travel of the knob to the maximum pressure equal to 18 kgf/cm.

When the EMERG BRAKE knob returns to the initial position, the air is discharged from the brakes into the atmosphere though the reducing valve and the wheel brakes are thus released.

The emergency braking system 'is not provided with the antiskid control unit.

The nose wheel steering system. The aircraft movement on the ground is controlled by the nose wheel steering/damping control valve installed on the nose landing gear. The nose wheel steering/damping control valve is used for counteracting the inadvertent turns of the aircraft during the takeoff run and landing roll and for controlling the aircraft when taxiing. In these cases, the nose wheel steering/damping control valve functions in the control mode.

The damping mode is used as an emergency one. The nose wheel steering/dam ping control valve is changed over to functioning in the damping mode when the WSM EMERG OFF switch located on the left horizontal console in the cabin is turned off or in case of the failure of the main hydraulic system.

The nose wheel steering/damping control valve has the twostage control drive: the first stage the takeoff/landing mode, with the wheel limit steering angle being 8°, the second stage the taxi mode, with the nose wheel limit steering angle being 31°. The nose wheel steering/damping control valve is put into operation when the nose landing gear is downlocked provided the WSM EMERG OFF switch is turned on.

The changeover of the nose wheel steering/damping control valve to the great steering angles is effected by depressing the WSM, LOCKON PF button on the throttle control lever, having first retracted the flaps.

The nose landing gear wheals are interlocked when the nose wheel steering/damping control valve is changed over to the second stage. Simultaneously with the changeover of the drive to the second stage, the electrical signal is applied to the electropneumatic valve of the braking system and the nose landing gear wheel brakes get automatically released.

The wing flaps and leading edges: The aircraft mounts the singlepiece slotted flaps, with the angle of deflection being 25°. Each flap is extended and retracted by one hydraulic cylinder. When the flaps are retracted, they are retained by the working fluid pressure and mechanical locks in the cylinders and when the flaps are extended, they are retained only by the working fluid pressure.

When in flight, the angles of deflection of the extended flaps depend on the dynamic pressure the flaps are "floating". The extension of the flaps should be checked against landing gear and flaps position indicator EP by reference to illumination of four green indicator lights.

Mounted on the wing leading edges are the leading edge flaps comprising two mechanically isolated sections: inboard and outboard. The inboard section is divided into two parts interconnected by the pins. The outboard section is a single piece.

The leading edges are extended at M < 0.75 to 0.9, their extended position is displayed on landing gear and flaps position indicator EP (two green indicator lights come on). The leading edges are retracted with the decrease in the angle of attack up to 7° or at M>0.75 to 0.9.

In the retracted and extended positions, the leading edges are retained by the working fluid pressure and twoway hydraulic locks located in the wing panels one lock per section of the leading edges.

The flaps are extended and retracted from the cabin control panel located on the left horizontal console. in the landing configuration, the leading edges are extended (retracted) automatically when the LANDING (RETRACTED) button is depressed on the flaps control panel.

The depression of the TAKEOFF or LANDING button located on the flaps control panel ensures the extension of the flaps through 25°.

The air brakes are extended and retracted from the main hydraulic system, with the aid of four hydraulic cylinders mounted in pairs on each air brake.

The air brake extension and retraction are controlled by the A/BRAKES button located on the right throttle control lever.

The air brakes should be checked for proper extension by reference to illumination of the yellow indicator lights on landing gear and flaps position indicator EP. The maximum angles of deflection of the air brakes are ensured up to an airspeed of 1000 km/h. The time required for deflection of the air brakes to the maximum angles should not exceed 3 s.

For maintaining the normal functioning of the stabilizer servos in case of the pressure drop in the main or booster hydraulic system, the provision is made for the air brake extension electrical interlock versus the pressure drop in one of the hydraulic systems.

When the pressure drops in the booster system to 100 kgf/cm the air brake control distributor, irrespective of the position of the A/BRAKES selector switch, is changed over to the retraction of the air brakes under the action of the main hydraulic system pressure.

When the pressure drops to the same value in the main hydraulic system, the air brakes are retracted under the action of the dynamic pressure. When the pressure drops in any of the hydraulic systems, with the air brakes retracted, the latter are not extended.

The aircraft provides for the air brake extension interlock versus the landing gear extended position and the suspension of the drop fuel tank.

The brake chute system is used for reducing the landing roll and comprises the container, brake chute and brake chute deployment and release electropneumatic system.

The brake chute container is arranged in the tail section of the aircraft, between the upper and lower air brakes.

The brake chute is connected with the lock hook by means of the shackle ensuring the emergency separation of the brake chute in case of its inadvertent deployment in flight. The brake chute separation speed is 320 km/h.

For the visual check of the brake chute lock position, the mechanical indicator in the form or the red rod is installed thereon. The rod is extended through the upper air brake to a length of 30 mm when the lock is released. When the lock is closed, the upper rod edge is flush with the air brake skin.

For deploying brake chute, it is necessary to depress the BRAKE CHUTE DEPLOYMENT button and for releasing the brake chute, depress the BRAKE CHUTE RELEASE button after landing the aircraft.

For precluding the inadvertent release of the brake chute, the brake chute deployment and release system has the brake chute release interlock circuit which is cut off after depressing the BRAKE CHUTE DEPLOYMENT button.

AIRCRAFT CONTROL SYSTEM

The system controlling the aircraft in flight comprises:

- the longitudinal control of the stabilizer;
- the lateral control of the ailerons and differentiallydeflectable stabilizer;
- the directional control of the rudders;
- the control of the wing leading edges, flaps and air brakes.

The hinge moments due to the aerodynamic forces originated during the deflection of the rudders are taken up by the hydraulic boosters. For simulating the forces applied to the control stick and pedals when they are deflected, the spring feel units are installed in the longitudinal, lateral and directional control channels.

The adjustment of the forces applied and trimming of the aircraft in flight are effected by the trimming mechanisms connected with the spring feel units. Apart from that, the trimming mechanisms serve as the servo units in the automatic flight control system. The neutral positions of the trimming mechanisms are indicated by illumination of the respective lamps.

The load of the control stick in the longitudinal channel and the stabilizer deflection angles versus the amount of control stick deflection with the airspeed and altitude varying, are corrected by the automatic transmission ratio controller system.

The automatic control of the aircraft in all flight regimes is effected by the automatic flight control system.

In all the channels, connected to the control linkage are the rods of the selfcontained servo units which ensure the transmission of the motion to the control surfaces from the control clearances both with the selfcontained servo unit rods being motionless and from the self contained servo units, with the control surfaces being motionless.

The control of the aircraft in the longitudinal, lateral and directional control channels is effected through the central unit combining the control stick and pedals which are connected through the system of the rods and bellcranks with the stabilizer, rudders and ailerons.

The aircraft control stick mounts:

- the gun trigger used for firing the gun and located under the reversible trigger;
- the RKT MSL B trigger used for launching the missiles, rockets and releasing the aerial bombs;
- the fourposition pressure switch for controlling the roll and pitch trimming mechanisms;
- the AFCS MODE OFF red button;
- the LEVELING button;
- the TANK JETS button with protective cap;
- the INTERR button;
- control button KI;
- the trigger for the shorttime disengagement of the automatic modes of the automatic flight control system. (except for the damping mode);
- the braking lever with the lineup braking trigger.

Mounted in the lower section of the control stick is the screw stop for adjusting the control stick to the pilot's palm.

The pedals are adjusted to the pilot's stature by the adjusting mechanism with the aid of the locking ring provided on the control stick, with the free forces applied to the pedals for the smaller stature of the pilot and with the forces applied to both pedals (forward) for the greater stature of the pilot. The adjustment of the angle of deflection of the control stick to the length of the pilot's hand is effected with the aid of the screw installed in the front bottom section of the control stick.

LONGITUDINAL CONTROL

The longitudinal control system includes the control stick; two servos; the spring feel mechanism:

- the trimming mechanism;
- two nonlinear mechanisms;
- two selfcontained servo units;
- the automatic transmission ratio controller system.

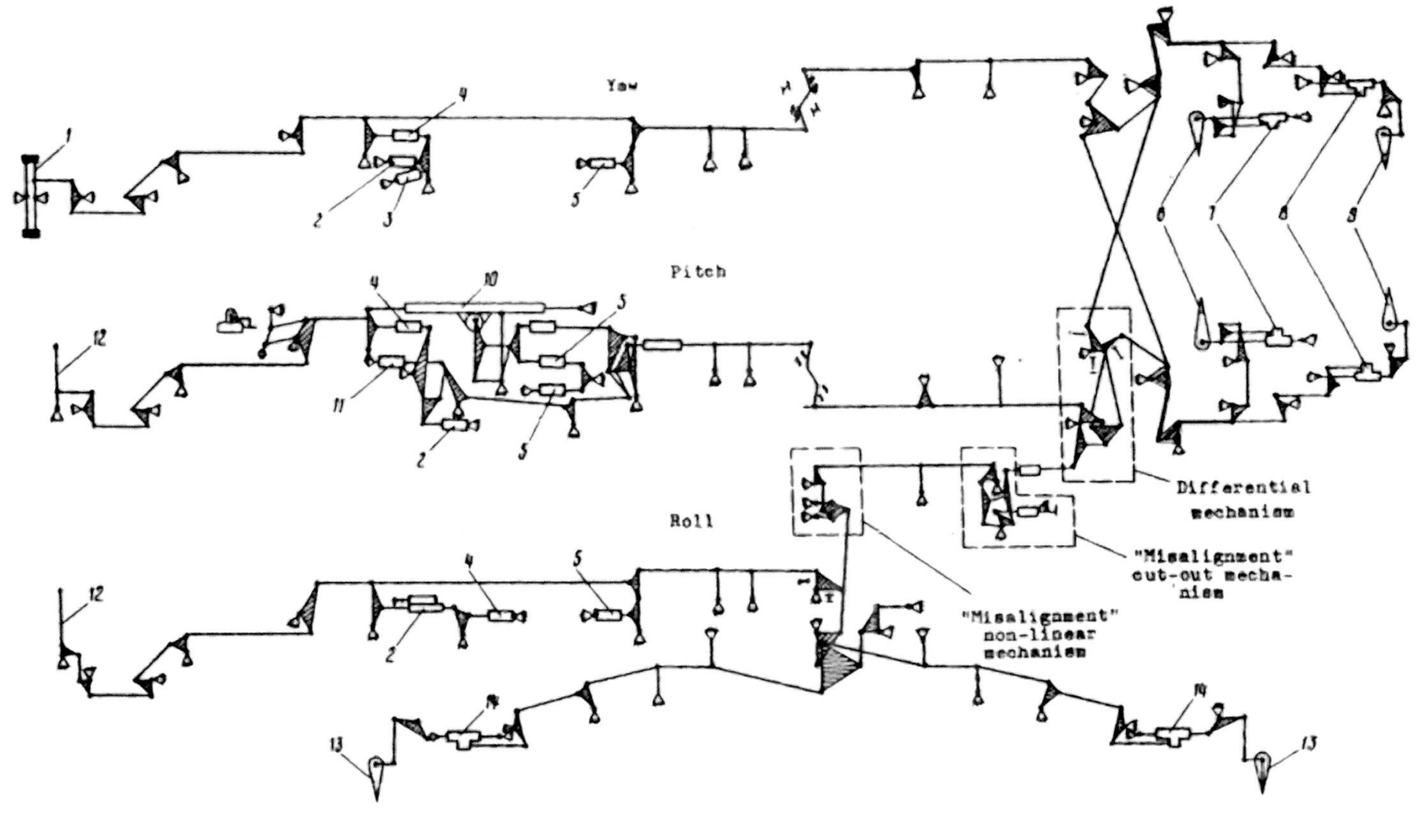

AIRCRAFT CONTROL LINKAGE SCHEMATIC DIAGRAM

1 - pedals; 2 - load feel unit; 3 - rigid spring limiting pedal travel; 4 - trim mechanism; 5 - self-contained servo unit; 6 - stabilizer; 7 - stabilizer boosters; 8 - rudder boosters; 9 - rudders; 10 - non-linear mechanism; 11 - control ratio mechanism of automatic transmission ratio controller; 12 - aircraft control stick; 13 - aileron; 14 - aileron booster

The aircraft control in pitch is effected by the synchronous deflection of the stabilizer panels with the aid of the servos (during the longitudinal travel of the control stick).

The trimming range in the longitudinal control system amounts to 80% of the control stick full travel in all modes. The selfcontained servo units in the pitch channel are used as follows: one servo is used as a damper servo of the automatic flight control system, the other servo is employed by the longitudinal stability augmentor.

The automatic transmission ratio controller (ARI) is used for the automatic correction of the control stick feel and the stabilizer deflection angles versus the airspeed and altitude of the flight.

The automatic transmission ratio controller is put into operation after turning on the BAT GND SUP switch; in so doing, the APY TAKEOFF LAND annunciator illuminates (the automatic transmission ratio controller selector switch is placed in the AUTO position). When the airspeed is less than 400 km/h, the rod of the control ratio mechanism is fully extended, the control stick feel in this case is minimum. whereas the stabilizer deflection angles are maximum; the APY TAKEOFF LAND annunciator illuminates.

At the altitudes of about up to 1000 m, the control ratio mechanism rod moves only versus the dynamic pressure (i.e. , the indicated airspeed): when the airspeed exceeds 400 km/h, the rod starts retracting and when the airspeed reaches 870 km/h, the rod is fully retracted. With the further increase in the airspeed up to 1200 km/h. the rod remains retracted. In this case, the control stick feel is maximum, whereas the stabilizer deflection angles are minimum. With the further increase in the airspeed at the same altitudes the control ratio mechanism rod starts extending, thus reducing the forces applied to the control stick and increasing the stabilizer deflection angles.

The trimming range corresponds to 60% of the full travel of the control stick in roll. The damper selfcontained servo Unit of the singlechannel action deflects the ailerons through maximum angle of 5° in response to the commands delivered from the automatic flight control system.

The aileron setting neutral position is 5° (both ailerons set upward). This is done to improve the efficiency of the lateral control at the great angles of attack and high indicated airspeeds.

For improving the lateral controllability of the aircraft at the high airspeeds. use is also made of the stabilizer differential control system.

The differential control unit installed in the stabilizer control system ensures:

- the stabilizer control in the roll mode when, with the control stick deflected in the lateral direction, in addition to the ailerons, both halves of the stabilizer deflect through the equal angles.
- the stabilizer control in the combined (pitch and roll) mode when moving the control stick over the diagonal. both halves of the stabilizer, deflecting in one direction in pitch, at the same time deflect through one and the same angle in the opposite directions in roll, thus forming the "misalignment".

The stabilizer "misalignment" is eliminated in response to the wing leading edge extension signal. This is done with the purpose of decreasing the yaw moments caused by the stabilizer "misalignment" at the great angles of attack.

DIRECTIONAL CONTROL

The directional control system comprises:

- the control pedals;
- the spring feel unit;
- the trimming mechanism;
- the selfcontained servo unit;
- the power servos;

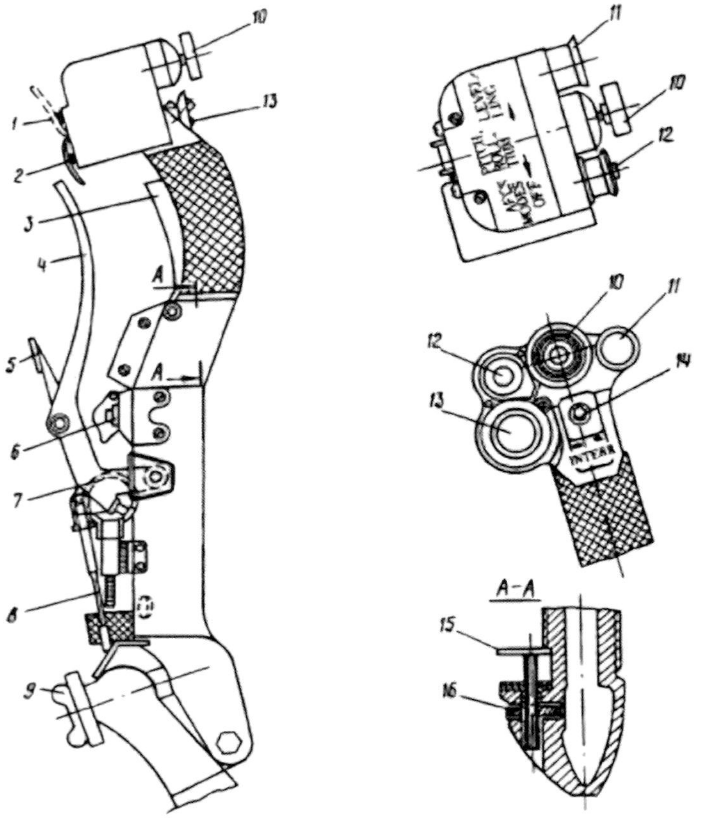

AIRCRAFT CONTROL STICK

1 - GUN trigger; 2 - RKT MSL B trigger; 3 - AFCS modes (pitch and roll) cut-off trigger; 4 - braking lever; 5 - line-up brake trigger; 6 - capped TANK JETSN button; 7 - control ring of pedal adjusting mechanism; 8 - wheel brake control linkage cable; 9 - screw for adjusting and locking control stick lengthwise; 10 - pitch and roll trim mechanism control four-position pressure switch; 11 - LEVELLING button; 12 - AFCS modes cut-off button; 13 - sight and radar control button; 14 - INTERR, RELEASE button; 15 - screw stop adjusting the control stick to palm width; 16 - adjusting wheel

The law of the rudder deflection towards the pedal travel is a linear one.

The trimming range corresponds to 60% of the total travel of the pedals. The single channel damper selfcontained servo unit deflects the rudder through the maximum angles of 5° in damping the shortperiod oscillations.

For simulating the forces applied to the pedals, in addition to the spring feel unit, the twochamber hydraulic cylinder is also installed. With the pedals moving, the forces applied thereto increase Proportionally to the amount of the spring compression in the spring feel unit and in steps when the hydraulic cylinder is connected.

The hydraulic cylinder is an additional feel unit ensuring the increase in the forces required for deflecting the rudders in flight at M >0.8. After retracting the landing gear at M > 0.8, the hydraulic cylinder is connected to the control linkage and due to the pressure of the hydraulic mixture, creates in steps the additional forces (about 30 kgf) when the pedals are deflected through the length exceeding 24 mm which presents a certain difficulty in deflecting the rudders through the angle exceeding 6° with respect to the trimmed position. The limitation of the rudder deflection angle is dictated by the necessity of limitation of the load across the vertical tail.

For precluding the rudder oscillations, a hydraulic damper is installed in the lower section of each rudder.

For checking the pitch, roll and yaw trimming mechanisms for functioning, use the respective pressure switches to deflect the control stick from the neutral position in pitch and roll and the pedals. In this case, the trimming mechanism rods, extending or retracting, move the control linkage through the bellcranks and deflect the respective control elements to depress the switch. In these cases, the spring feel unit functions as a rigid rod. The control stick and pedals are set to the neutral position at the moment the trimming mechanism neutral position annunciator illuminates in the following sequence: the control stick is shifted from the forward to the backward position in pitch; from the left to the right in roll; the right pedal is shifted from the forward to the backward position. This is explained by the peculiarities of arrangement of the areas within which the control stick and pedals neutral position indicator lamps illuminate.

Stall Limitation System

The stall limitation system actuators are the duplicate hydraulic cylinders installed in the control linkage, upstream of the automatic transmission ratio controller. The hydraulic cylinders are fed from the main and booster hydraulic systems. The stall limitation system ensures the increment of the forces applied to the control stick when the aircraft reaches an angle of attack of 26°, with the wing leading edges extended at $M < 0.9$, and the control stick push from the position corresponding to the stabilizer deflection angle of 15°.

If the rod cannot be retracted because of the failure of the stall limitation system, then, after extension of the landing gear, the SOS STOP signal is displayed on the EKRAN annunciator panel. It proves the fact that while landing the aircraft, the forces required for pulling the control stick backward will increase by 17 kgf.

In case of the common failure of the stall limitation system, the SOS FAIL annunciator will illuminate on the right console If one of the system channels fails. the NO RESERVE SOS annunciator illuminates on the right horizontal console.

If the wing leading edges fail to be extended or if the Mach number exceeds 0.9, the stall limitation system ensures the automatic readjustment of the system actuation.

For enhancing the reliability of the stall limitation system operation, the system is duplicated in the hydraulic and electric circuits in the computation and switching unit.

In the hydraulic circuit, the stall limitation system is duplicated by the feed of the working fluid to the control stick pusher cylinder from two hydraulic systems (from the main and booster systems), in the electric circuit, the system is duplicated by the presence of two subchannels in each channel, generating the electrical signals to out in the electrohydraulic distributors. In case of the failure of one of the subchannels or the channel as a whole, the system remains serviceable for the complete or intermediate extension of the control stick pusher cylinder rod.

The system is provided with the builtin test system which makes it possible to check the system for proper functioning when the pressure is available in the main and booster hydraulic systems.

Prior to checking the system for proper functioning with the aid of the builtin test system, it is necessary to set manually the vanes of the angleofattack transmitters to the lower position and turn on the BAT GYD SUP and AC GEN switches.

For checking the stall limitation system for serviceability, it is necessary to pull the control stick backward and depress the SOS CHECK button. The vanes of the angleofattack transmitters will start deflecting from the lower stop to the upper one. When an angle of attack equal to 10° is reached against the indicator the wing leading edges will be automatically extended, and when an angle of attack equal to 16° in reached, the control stick will be pushed forward and the SOS STOP signal will be displayed on the EKRAN annunciator panel. Pull the control stick backward once more and when an angle of attack of 26 or 27° is reached, the control stick will be pushed and the SOS STOP signal will be displayed once more. The control stick will be pushed to the position corresponding to the stabilizer position equal to 15°. Release the SOS CHECK button. In this case, the control stick force is trimmed out, the SOS STOP signal extinguishes and the wing leading edges are retracted.

AUTOMATIC FLIGHT CONTROL SYSTEM

The automatic flight control system (AFCS) is used for controlling the aircraft in the automatic and director nodes. improving the aircraft stability and controllability characteristics during the manual control within the entire operating range of the altitudes, airspeeds and angles of attack.

The system ensures:

- the damping of the shortperiod oscillations of the aircraft during the automatic, director or manual control and improvement of the aircraft stability and controllability characteristics throughout the entire operating range of the altitudes and airspeeds;
- the manual control of the aircraft by one control stick, with the stabilization of the selected roll, pitch and yaw angles;
- the leveling from any attitude with the subsequent stabilization of the altitude and course of the flight;
- the automatic recovery from the selected low altitude in response to the radioaltimeter signal with the automatic modes selected and landing gear retracted, the stabilization of the pressure altitude;
- the automatic and director control in executing the landing approach, during the conjoint operation of the automatic flight control system with the navigation one;
- the trimming during the manual control in roll and pitch;
- the indication of the present roll and pitch angles and deviation of the aircraft from the selected flight path on the combined course and flight director indicators in response to the signals delivered from the navigation system.

The automatic flight control system operating principle is based on measuring the values characterizing the aircraft attitude and converting these values into the signals acting through selfcontained servo unit and trim actuators on the control surfaces of the aircraft.

The DAMPER mode is effected through the selfcontained servo unit and the automatic modes (except the automatic approach), through the trim actuators; the automatic approach mode in carried out according to the combined circuit, i.e. through the selfcontained servo unit and the trim actuators. In the automatic control modes the automatic flight control system shapes the signals determining the amount and direction of the action on the control surfaces required for holding the aircraft on the selected flight path.

In the stabilization, altitude stabilization, recovery and leveling modes; the control the longitudinal channel is effected with due regard for the limitation of the angle of attack of 19+-2°.

When the system functions in the director mode, the aircraft is controlled by the pilot by reference to the command bars on the flight director indicator or to the command circle on the integrated display system screen.

CONTROL AND INDICATION ELEMENTS

The automatic flight control system operation is governed from the control panel located on the cabin port side vertical board and with the aid of the control elements provided on the aircraft control stick.

The automatic flight control system control panel mounts:

- the DAMPER annunciator button with indicator lamp for selecting the DAMPER mode;

• the AP annunciator button with indicator lamp for selecting the attitude (pitch, roll, yaw) stabilization and automatic control modes;
• the RECOVERY annunciator button with indicator lamp for selecting the LIMIT ALTITUDE RECOVERY mode;
• the NIGHT PATH CTL annunciator button with indicator lamp for selecting the director control mode during the landing approach;
• the ALT STAB annunciator button with indicator lamp for selecting the pressure altitude stabilization mode;
• the MISSED APPROACH annunciator button for selecting the missed approach mode (included into the SN29 control element).

The control stick mounts:

• the trigger for switching off the automatic modes of the automatic flight control system (except for the DAMPER mode) for the time the trigger is depressed;
• the LEVELING lightbutton for selecting the automatic recovery to the straight level flight node;
• the AFCS MODE OFF red button for switching off all the modes of the automatic flight control system.

When the AFCS MODE OFF button is depressed, all the modes of the automatic flight control system are switched off, including the DAMPER mode and the longitudinal stability augmentor. If the button has been kept depressed for not more than 3 S, than, after releasing the button, the DAMPER mode and the functioning of the longitudinal stability augmentor are automatically restored. For switching off the DAMPER mode and longitudinal stability augmentor, it is necessary to depress the AFCS MODE OFF button and keep it depressed for more than 3 s.

When the DAMPER mode is switched off, the KSTs lightbutton flickers and the DAMPER OFF signal on the right console is displayed. The LSAR FAILURE red annunciator is installed on the right console of the cabin to indicate the failure of the longitudinal stability augmentor and to set the rod of the selfcontained servo unit of the longitudinal stability augmentor to the neutral position.

CAUTION. PILOT'S INTERVENTION WITH THE AIRCRAFT CONTROL DURING OPERATION OF THE AUTOMATIC FLIGHT CONTROL SYSTEM IN THE AUTOMATIC MODES (EXCEPT FOR THE "DAMPER" MODE) WITHOUT PRELIMINARY COMPRESSION OF THE TRIGGER IS PROHIBITED.

INTEGRATION OF AUTOMATIC FLIGHT CONTROL SYSTEM WITH AIRBORNE EQUIPMENT

The automatic flight control system is integrated with the following airborne: equipment:

1. Information directional/ vertical gyro system MKBK (being part of the navigation system) which generates the present aircraft roll, pitch and yaw angle values to the automatic flight control system, which are used there in shaping the control signals in the stabilization, leveling modes as well as during the director and automatic control. Therefore, when information directional/vertical gyro system EKVK is not ready, said modes cannot be selected.

2. Air data computer system CBC (being part of the navigation system) which delivers the present Mach numbers to the automatic flight control system, which are used therein for correcting the gear ratios of the automatic flight control system.

3. Airborne radio and navigation equipment (being part of the navigation system) which generates the signals of the selected course, azimuth, range, readiness of the localizer and glide slope radio beacons, deviation of the aircraft from the equisignal zones of the landing radio beacons as well as the signals for indication on the instruments and generation of the landing control signals. When the airborne radio and navigation equipment fails, the aircraft cannot be lauded in the automatic or director control mode.

4. Radioaltimeter RV which generates the "Limit altitude" signal to the automatic flight control system for switching on the preselected limit altitude recovery mode.

5. Stall limitation system SOS which generates the true attack angle signal for correcting the gear ratios and also for shaping the longitudinal stability augmentor control law and to limit the angle of attack of 19±2° implemented in the automatic flight control system.

The present angles of roll, pitch, yaw and deviation of the aircraft from the selected flight path are indicated on the flight director and combined course indicators.

BASIC COMPONENTS OF AUTOMATIC FLIGHT CONTROL SYSTEM AND THEIR APPLICATION

1. The selfcontained servo units are used for moving the slides of the boosters of the ailerons, stabilizers, rudders in compliance the control laws. The selfcontained servo units are the independent electrohydraulic servos installed in succession in the aircraft control linkage, between the control stick (pedals) and the booster and have the mechanical device for setting the rod to the neutral position, which actuates when the selfcontained servo unit electrohydraulic valve is de-energized. The electrohydraulic valve is de-energized when depressing the AFCS MODE OFF button located on the control stick, when the servos are filled with working fluid below the standard level, in case of disappearance of even one 200V 400Hz supply phase at the servo electric motor and disappearance of the secondary supply voltage across the automatic flight control system computer.

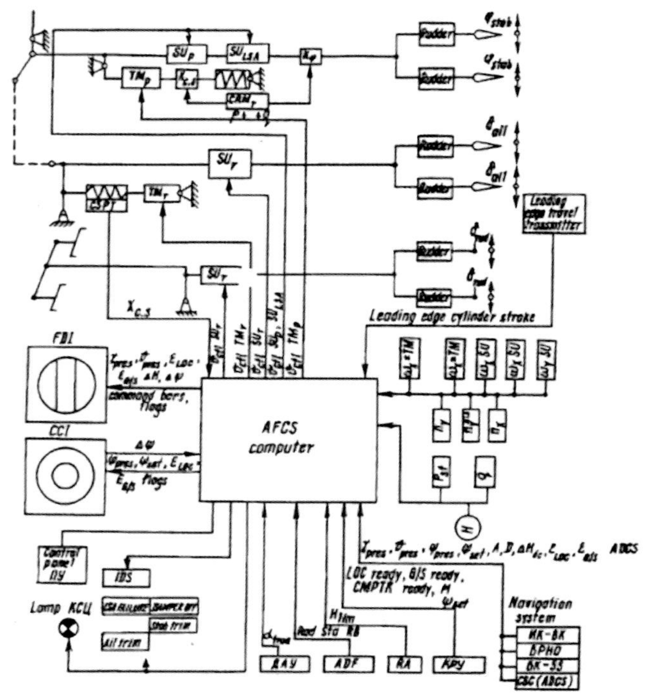

AUTOMATIC FLIGHT CONTROL SYSTEM FUNCTIONAL COUPLING DIAGRAM

2. The trimming mechanisms are used for trimming out the control stick forces during the manual control and also for moving the control surface boost slides in compliance with the control laws during the automatic control. The trimming mechanisms are essentially the electromechanical units with the variable rate of the output rod travel.

3. The computer of the automatic flight control system is used for shaping the control signals and for delivering these signals to the selfcontained servo units and trimming mechanisms and also for generating the director signals to the indicators. In the computer of the automatic flight control system, the control signals are shaped in compliance with the control laws in response to the signals delivered from the rate gyros, gload transmitter, roll control stick transmitter as well as in response to the signals delivered from the inertial directional/vertical gyro system and the altitude corrector. These control signals are corrected with respect to the altitude. dynamic pressure, Mach number and angleofattack.

AUTOMATIC FLIGHT CONTROL SYSTEM OPERATION MODES

When the automatic flight control system switch is turned on, the power is applied to the automatic flight control system and the latter undergoes the testcheck (with the landing gear struts compressed, the inertial directional/vertical gyro system ready, and the pressure available in one of the hydraulic systems). The builtin test is performed for 1.5 minutes in compliance with the cyclogram. The damper section of the automatic flight control system and the other modes are checked within the first 40 s. On a lapse of 40 s after turning on the automatic flight control system switch, the DAMPER mode may be selected manually by depressing the DAMPER annunciator button; in this case, the testcheck is discontinued and the other modes of the automatic flight control system cannot be selected.

For the reliable execution of the testcheck, it is necessary, prior to turning on the automatic flight control system switch, to set the control stick to the neutral position with the aid of the roll and pitch trimming mechanisms till the STAB TRIM) and AIL TRIM annunciators illuminate.

During the testcheck the DAMPER annunciator button flickers at a frequency of 1.5 to 2 Hz on the control panel. When the system is serviceable, on completion of the built-in test, the DAMPER OFF annunciator extinguishes and the DAMPER annunciator button illuminates with continuous light, indicating the serviceability of the automatic flight control system and the DAMPER mode selection.

When the system is under the testclock, the aircraft control surface forces should be trimmed out. In this case, the control surfaces and control stick are actuated from the selfcontained servo units and trimming mechanisms. If, after completing the testcheck, the control stick remains in the released position, it is necessary to set the control stick to the neutral position with the aid of the trimming mechanisms.

While performing the testcheck, all lightbuttons (except for the MISSED APPROACH lightbutton) located on the automatic flight control system control panel may illuminate for a short time; in this case, the testcheck is not discontinued.

If a failure occurs in the automatic flight control system, all the annunciator button lamps on the control panel (except for the MISSED APPROACH lightbutton) start flickering at a frequency of 1.5 to 2 Hz. If the failure is detected at the stage of the DAMPER mode testcheck, all the other modes of the automatic flight control system cannot be selected. If the failure is detected at the stage of check of the other modes of the automatic control, the DAMPER mode in this case may be selected by depressing the DAMPER annunciator button, the DAMPER annunciator button illuminates with continuous light, the other annunciator buttons and the DAMPER OFF annunciator extinguish. The other modes of the automatic flight control system cannot be selected. For repeating the test-check it is necessary to set the control stick to the neutral position. and depress the APCS MODE OFF red button located on the control stick for a short time.

DAMPER Mode

The DAMPER mode is used from the moment of the takeoff to the landing. It is allowed to taxi the aircraft with the DAMPER mode selected.

Flying the aircraft with the DAMPER mode selected substantially improves the stability and controllability characteristics of the aircraft.

The aircraft oscillations are damped by deflecting the control surfaces, irrespective of the pilot's actions for counteracting the angular accelerations of the aircraft around the centre of masses.

When in flight, with the change in the altitude and airspeed, the aircraft oscillation damping changes to some extent as well. For providing the artificial damping (from the automatic flight control system) close to the continuous one with respect to the aircraft oscillation damping, correct the gear ratios in the DAMPER mode relating to the altitude and airspeed of the flight.

For improving the stability and controllability characteristics at the great angles of attack, the cross coupling versus the deflection of the control stick in the lateral direction is introduced into the rudder channel. The amount of the rudder deflection in response to the crosscoupling signals is dependent on the control stick travel and is acted versus the dynamic pressure, flight altitude and the angleof attack.

The DAMPER mode is selected automatically after executing the builtin test, provided the automatic flight control system is serviceable, or by depressing the DAMPER (AP) annunciator button on the control panel with the pressure available in one of the hydraulic systems. The damper mode is automatically selected with the selection of the AP and LEVELING modes.

The DAMPER mode is switched off by depressing the AFCS MODE OFF red button located on the aircraft control stick and keeping this button depressed for more than 3 s; in so doing, the selfcontained servo unit rods are set to the neutral position.

Operation of Longitudinal Stability Augmentor

For improving the aircraft longitudinal stability and controllability characteristics and for maintaining the constant amounts of the control stick displacement per unit of gload in piloting, at the angles of attack of 14 to 25° with the wing leading edges extended, the longitudinal stability augmentor is put into operation when the DAMPER mode of the automatic flight control system is selected, an well as when the landing gear and flaps are retracted.

At the angles of attack equal to 9° when the wing leading edges are extended, the diving moment develops, which is then counteracted by deflecting the stabilizer with the aid of the selfcontained servo unit of the longitudinal stability augmentor.

With the flaps, landing gear retracted and the DAMPER mode selected, in response to the leading edge extension signal the selfcontained servo unit deflects the stabilizer for pitch up through 3.5°, and when the aircraft reaches an angle of attack equal to 14° , said servo unit moves the stabilizer according to the linear law for pitch down versus the angle of attack .

When retracting the wing leading edges or extending the flaps and also when switching off the DAMPER mode, the rod of the longitudinal stability augmentor servo unit moves to the neutral position and the longitudinal stability augmentor gets disengaged.

On the aircraft with improved aerodynamic the longitudinal stability augmentor is additionally engaged at the angles of attack of 2 to +5° with the wing leading edges retracted.

On the aircraft with modified logic of operation of the wing leading edges. the longitudinal stability augmentor operates only when the nose landing gear is retracted irrespective of the flaps position.

In case of failure of the longitudinal stability augmentor. the LSA FAILURE signal illuminates on the right console and the selfcontained servo unit of the longitudinal stability augmentor is set to the neutral position.

Aircraft Attitude Stabilization (STABILIZATION) Mode

The STABILIZATION mode is used for automatically maintaining the aircraft attitude set by the pilot in flight, with the control stick forces trimmed out. The STABILIZATION mode is selected by depressing the AP button located on the control panel of the automatic flight control system and is checked by reference to illumination of the AP button indicator lamp. When this mode is selected, the LEVELING mode is disengaged.

Prior to selecting the mode, it is necessary to trim out the control stick force with the aid of the roll and pitch trimming mechanisms After selecting the mode, release the control stick of the forces and release the trigger.

When the aircraft attains the angle of attack of 19+2°, the changeover from the pitch angle stabilization to this attack angle stabilization is performed in the longitudinal channel.

For changing the roll, yaw or pitch of the aircraft, it is necessary to press the limit switch trigger located on the control stick, set the now roll, yaw or pitch angle values as in case of the normal piloting (in this case, the AP button indicator lamp will flicker), trim the aircraft with the aid of the trimming mechanisms and relieve the trigger of the forces the automatic flight control system will again automatically stabilize the aircraft attitude, the AP button indicator lamp will come on.

In the STABILIZATION mode, only the ailerons and stabilizer are deflected with the aid of the roll and pitch trimming mechanisms. It is allowed to select the STABILIZATION mode directly before the takeoff on the runway or in flight. While taxiing the aircraft, the mode should be switched off. After selecting the STABILIZATION mode during the takeoff run and landing roll, it is necessary to hold the control stick: against its possible drift by the roll and pitch mechanism. In this case, the increment in the forces applied to the control stick to drift the latter is intolerable.

In case of the change in the altitude or airspeed of the flight and when the landing gear, flaps, air brakes are extended or retracted with the control stick forces trimmed out, the roll, yaw and pitch angle values may vary up to 4° from the preset ones, which calls for the pilot a intervention with the aircraft control to restore the previous flight regime.

When in f light with the STABILIZATION modes switched on, at the angles of attack exceeding 17° the aircraft oscillations in roll up to 7° are possible. The initial roll angle may change by 5 to 7° and the aircraft may reach the angle of attack of 19±2° with its subsequent stabilization.

When the STABILIZATION mode is switched off with the aid of the red button located on the control stick, the aircraft may be partially trimmed out.

It is allowed to land the aircraft with the mode selected; in so doing, the automatic flight control system is switched off with the aid of the red button located on the control stick at the end of the landing roll.

The mode is disengaged by depressing the APCS MODE OFF red button located on the control stick for a short time.

The move is automatically disengaged when the LEVELING mode is engaged or when the LIMIT ALTITUDE ZONE RECOVERY mode is selected.

LEVELING Mode

The LEVELING mode is used for the automatic recovery to the level flight in case of the loss of theaircraft attitude by the pilot.

The mode is selected by depressing the LEVELING lightbutton located on the control stick with the limit switch trigger relieved of the forces applied thereto. In this case, all other modes of the automatic flight control system, except for the DAMPER mode, are switched off. The mode switchingon is checked by illumination of the light button.

At the angles of roll exceeding 80°, the leveling is executed only in roll. When the roll is less than 80°, the aircraft is simultaneously leveled in roll and pitch. When the aircraft reaches the angles of roll within +7° and the trajectory angles less than 5°, the altitude corrector is put into operation on a lapse of 3 or 4 s the pressure altitude in stabilized, the ALT STAB annunciator button lights up on the control panel and the flight course is also stabilized. The LEVELING light button remains alive.

If the pilot intervenes with the aircraft control while leveling (compresses the trigger on the control stick or presses on the slide of the trimming mechanism), the LEVELING lightbutton flickers and the automatic leveling is temporarily discontinued. After completing the intervention, the leveling continues and the leveling angular rates, gload, angle of attack as well as the transients may vary to some extent versus the nature and character of the Intervention on the part of the pilot.

The aircraft is leveled at the roll rates within 10 to 45 deg/sec, the pitch leveling is ensured with a gload ranging from 1.0 to 5 g and at the angle of attack exceeding 19+2° depending on the initial angle of pitch, altitude and airspeed of the flight. When leveling the aircraft by means of the automatic flight control system from the positive pitch angles, the longtime (more than 10 s) gload of about 0 or 0 may take place. To stabilize the aircraft attitude, intervene with the aircraft control before completion of the automatic leveling process.

The mode is switched off by depressing the AFCS MODS OFF red button located on the control stick for a short time and by selecting the STABILIZATION mode as well.

When selecting the LEVELING mode, it is necessary to take Into account the following:

- after selecting the mode, it is necessary to set the pedals to the neutral position and keep them in this position during the whole leveling time;
- while leveling, check the airspeed and altitude of the flight and set. the engine power in compliance with their values. To prevent the growth of the negative gload beyond the limits it is not recommended to throttle back the engine;
- when the mode is selected at the descent with the trigger relieved of the forces, the initial period of the leveling may be accompanied first by the negative gload with the subsequent transition to the positive gload if the control stick, prior to selecting the modes, is completely trimmed forward and when the pull forces are applied.

Pressure Altitude Stabilization (ALTITUDE STABILIZATION) Mode

It is allowed to select the ALTITUDE STABILIZATION mode throughout the entire operating range of altitudes and airspeeds of the flight at the vertical speeds close to zero (the aircraft trajectory angle is not more than +5°). For ensuring the stabilization mode, the preselected altitude deviation signal fed from the altitude corrector is introduced into the pitch channel. After conversion, the signal is fed from the altitude corrector to the trimming actuator which, deflecting the stabilizer, ensures the variation of the trajectory angle as a result of which the signal will be decreased to some extent.

The mode is selected by depressing the ALT STAB annunciator button with the AP annunciator button depressed.

The mode selection is checked by reference to illumination of the AP and ALT STAB annunciator buttons. Besides, the mode is selected automatically with the LEVELING mode switched on, 4 or 5 seconds after the aircraft reached the angles of roll less them 7° and the trajectory angles less than 5°.

When the mode is selected, the flight altitude may very with respect to the preset one at the vertical speed of 5 to 8 m/s, after which the altitude stabilization begins. In this case, the roll is stabilized in the lateral channel and when the roll is less than 7°, the course is stabilized therein. It is recommended to pass over the transonic area with the mode switched off (the trigger located on the control stick is compressed) due to the retriming of the aircraft.

CAUTION. PROCEEDING FROM THE FLIGHT PRESSURE ALTITUDE STABILIZATION ACCURACY, IT IS RECOMMENDED TO USE TO MODE AT THE ALTITUDES ABOVE 100 M.

The ALTITUDE STABILIZATION mode is switched off by depressing the AFCS MODE OFF red button located on the control stick for a short time.

Limit Altitude Zone Recovery (RECOVERY) Mode

The RECOVERY mode is used for the automatic recovery from the preselected limit altitude when flying the aircraft with the landing gear retracted at the altitudes below 1500 (1000) m.

The mode is selected by preliminarily depressing the RECOVERY annunciator button at the altitude exceeding that set on the selector of the radio altimeter when any of the modes of the automatic flight control system is selected. The mode readiness is Indicated by the RECOVERY annunciator button located on the control panel.

When descending the aircraft to the altitude less than the selected limit altitude, the radio altimeter warning system operates and the "limit altitude" voice information is delivered to the pilot's earphones. The aircraft is automatically brought Into a climb at the trajectory angle equal to 8°, bringing the aircraft at the same time to the zero angle of roll. In this case, the LEVELING lightbutton located an the control stick will illuminate.

The limit altitude zone recovery" is executed with a vertical gload of 1.5 to 5.0 g.

After climbing to the altitude exceeding the limit one (the "limit altitude" command is cancelled), the LEVELING mode is automatically selected and after , the aircraft gains the level flight at the angle of roll less than 7° and the trajectory angle less than 5°, on a lapse of 4 or 5 s, the pressure altitude and flight course stabilization mode is automatically switched on.

After the automatic selection of the RECOVERY mode, the repeated selection of this mode may occur at the small angles of descent and for precluding this action, the pilot should intervene with the aircraft control and climb to the altitude exceeding the limit one by 50 m.

When the pilot intervenes with the aircraft control during the recovery from the limit altitude. the trimming mechanism automatic control is switched off, the LEVELING lightbutton located on the control stick flickers and the DAMPER annunciator button remains alive. On completion of the pilot's intervention with the aircraft control, the LEVELING mode is switched on again and the LEVELING lightbutton again burns continuously.

If the aircraft has entered the limit altitude zone with the trigger compressed or the trimming mechanism slide pressed, then, in response to the radio altimeter signal the stabilizer deflects only the servo of the pitch channel of the selfcontained servo unit for the pitchup; the pilot perceives the increase in the positive gload for 3 or 4 s with the control stick motionless and the forces applied to it invariable. The automatic entry of the aircraft into a climb is not executed in this case.

The mode is switched off during the repeated depression of the RECOVERY annunciator button located on the control panel of the automatic flight control system, the RECOVERY annunciator button extinguishes. The DAMPER mode remains selected or when the APCS HOME OFF red button located on the control stick is depressed and kept in this position for more than 3 s (until the DAMPER annunciator button extinguishes).

AUTOMATIC AND DIRECTOR CONTROL OF AIRCRAFT DURING LANDING APPROACH

When the automatic flight control system functions together with the navigation system, it is allowed to perform the landing approach in the automatic or director control mode.

The difference between the automatic and director control modes consists in the following:

- during the automatic control, the control signals generated by the automatic flight control system are delivered to the trim actuators and selfcontained servo units which, deflecting the aircraft control surfaces (without intervention on the part of the control the aircraft flight along the descent path on the runway landing heading. Besides, the control signals deflect the command bars on the flight director indicator and in the command circle on the screen on the integrated display system, whereas the aircraft deviation from the glide slope is determined by the pilot by reference to deflection of the deviation bars on the flight director and combined course indicators;
- during the director control, the control signals shaped in the automatic flight control system are delivered to the command bars on the flight director indicator and to the command circle on the screen of the integrated display system. For maintaining the assigned flight regime, it is necessary for the pilot, deflecting the control surfaces of the aircraft, to hold the command bars within the circles, thus maintaining the assigned glide slope and check the correctness of the glide slope execution as compared with the assigned, one by reference to the deviation bars. In aligning with the assigned glide slope, with the deviation bars reading zero, the command bar of the respective channel may be deflected to some extent. The deflected command bar shows the direction to which it is necessary to deflect the control stick whereas the deflected deviation bar shows the direction to the equisignal zone of the lending radio beacon system with respect to the aircraft attitude, i.e., the assigned glide slope.

The director control mode is selected by depressing the FLIGHT PATH CTL annunciator button located on the control panel (after receiving the "LOC beacon ready command, disappearance of the localizer flag on the combined course indicator, with information directional/vertical gyro system EKVK being serviceable); in so doing, the LOC (K) and pitch (T) flags disappear on the flight director indicator.

The landing approach automatic control mode is selected by depressing the AP annunciator button after switching on the director control mode. In so doing. the entry of the aircraft in the equisignal zone of the localizer is accomplished in the lateral channel till the interception of the equisignal zone of the glide slope radio beacon, the altitude stabilization is effected in the zone of trajectory angles of less than 5° (the ALT STAB annunciator button lights up).

When the automatic landing approach mode is selected in case of taking the runway heading and on the glide slope the aircraft control stick oscillations in roll and in pitch with the amplitude of up to 5 mm and the frequency of 1.5 to 2.0 Hz are possible. In this case, the aircraft oscillations in roll amount to 1° maximum.

The automatic and director control modes are selected by depressing the APCS MODE OFF red button located on the control stick for a short time.

FAILURES LIKELY TO OCCUR IN AUTOMATIC FLIGHT CONTROL SYSTEM

The failures likely to occur in the automatic flight control system may be divided into the failures of the servos and failures of the control computer.

1. The failures of the servos of selfcontained servo units and trimming mechanisms:

- the break of the feedback coupling of the selfcontained servo units in the pitch, roll, yaw channels or the longitudinal stability augmentor;
- the fast travel of the selfcontained servo unit rod to the stop in pitch, roll and yaw or the longitudinal stability augmentor;
- the fast travel of the rod of the pitch and roll trimming mechanisms.

The extension of the self-contained servo; unit rod to the stop in the straight level flight causes:

- the aircraft pitchup or pitchdown at the g load differing from the initial one by not more than +2.5 g (in case of the failure in the pitch channel);
- the rotation of the aircraft around the longitudinal axis (in case of the failure in the roll channel) at the maximum angular rates of up to 60 deg/s;
- the aircraft slip (in case of the failure in the yaw channel) up to +3° with the simultaneous rolling at the rate of up to 10 deg/s.

In case of failure of the selfcontained servo unit of the automatic flight control system, the overrun or spontaneous increase of the initial. angle of attack up to 2° and the gload of up to 1.5 g in the horizontal flight takes place. In case of maneuvering at the angles of attack exceeding 13°, the angle of attack spontaneously increases till the operation of the stall limitation system while the gload increases by 2g as compared with the initial value.

The above failures hinders the aircraft piloting and the failure of the selfcontained servo unit of the automatic flight control system in the modes close to the limiting gload modes may result in the excess of its maximum permissible value.

When the selfcontained servo unit rod extends to the stop in the pitch or roll channel in the automatic modes of operation of the automatic flight control system, the respective trimming mechanism will deflect the control linkage. In this case, the failure will lead to the delayed changes of the aircraft attitude.

The break of the feedback coupling of the selfcontained servo unit in any of the channels (the pitch, roll or jaw channel) results in the longitudinal, lateral or directional steadystate oscillations of the aircraft, with a frequency of 1.2 to 1.6 Hz and amplitude of +1 .5 g, a roll of 3° and a rate of rotation around the longitudinal axis up to 7°/s.

When the oscillations are detected, do not counteract them, but switch off the automatic flight control system with the aid of the APCS MODE OFF red button located on the control stick, keeping the button depressed for more than 3 s.

The fast travel of the trimming mechanism rod in roll and pitch in the automatic control mode causes the deflection of the respective control surfaces with the drift of the aircraft control stick.

This failure does not result in the considerable change of the flight parameters.

The forces applied to the aircraft control stick are tolerable and enable the pilot to safety complete the flight.

2. The failures of the automatic flight control system control computer may result in the fadeout of the signals included into the control laws (the present and selected signals). These failures with the automatic modes selected lead:

(a) in the longitudinal channel:

- to the maintaining of a 1g load and gradual drift from the flight path in the vertical plane;
- to the longitudinal maneuver at the gload of up to 5 g;

(b) in the lateral channel:

- to the maintaining of the zero angle of roll and, consequently, to the drift from the selected flight path in the horizontal plane;
- to the aircraft turn at the constant roll;
- to the roll reversal;
- to the deflection of the ailerons with the aid of the selfcontained servo units and trimming mechanism and to the deflection of the rudder with the aid of the selfcontained servo unit and, consequently, to the rotation of the aircraft around the longitudinal and lateral axes.

CABIN AIR CONDITIONING SYSTEM

The air conditioning system is used for maintaining the preset air temperature and pressure in the cabin and for cooling the electronic equipment.

The air fed to the air conditioning system is bled from the compressor of each engine and delivered through the common line to two successively installed pressure differential regulators which limit the pressure of the air fed to the systems. The second regulator duplicates the operation of the first regulator.

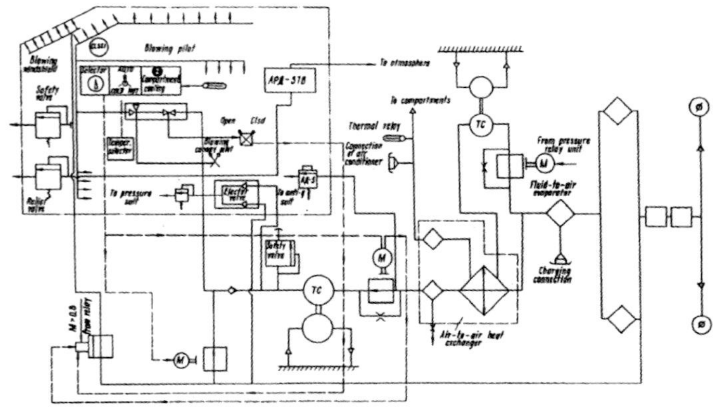

AIR CONDITIONING SYSTEM SCHEMATIC DIAGRAM

First, the air is fed to two (according to the number of the engines) airtoair heat exchangers located in the wall of the air intake ducts.

From the air-to-air heat exchangers, the air is delivered to the fluidtoair evaporator whose cooling agent is a 50 % alcohol and water mixture. Then, the airflow is separated into two flows: one flow is used for cooling the cabin and the other flow is used)r cooling the instrument compartments.

The air used for cooling the instrument compartments is cooled off in the turbocooler and, then passes on its way through the airtoair heat exchanger provided with the airblown moisture separators where it cools off the air coming to the cabin.

Installed in the pipeline feeding the air to the compartments is the thermal relay for checking the air temperature at the compartment inlet. When the temperature reaches +80°C, the NO COOL signal illuminates on the EKRAN annunciator panel.

The air fed to the cabin is additionally cooled in the airtoair heat exchanger and in the separate turbocooler.

The air cooled off in the turbocooler is divided into two flows: one flow is used for blowing the canopy glazing and pilot and the other flow is used for ventilating the suit. The air used for blowing the canopy glazing and pilot is mixed with the hot air fed from the electricallyoperated shutter controlling the air supply in response to the commands delivered from the automatic air temperature control system and is fed to the CABIN AIR SUPPLY valve and, then to the BLOWING valve.

The selected temperature of the air in the cabin is maintained by the automatic air temperature control system within 15 to 25° C in all flight regimes in response to the signals delivered from the temperature selector (installed in the cabin at the bottom, on the instrument board) and the temperature sensor with the aid of the electrically-operated shutters varying the flow rate of the cold and hot air.

The cabin air temperature may also be changed manually with the aid of the CABIN HEAT selector switch having four fixed positions: AUTO HOT COLD OFF When the selector switch is set to the AUTO position, the air temperature set by the temperature selector is maintained automatically; when the selector switch in set to the HOT position, only the hot air shutter is open; when the selector Witch is set to the COLD position, only the cold air shutter is open; when the selector switch is set to the OFF position, the system and shutters are deenergized; in so doing, the shutters will remain in the positions in which the system is deenergized.

CABIN CANOPY

The cabin canopy comprises the front fixed section the windshield rigidly secured on the fuselage and the hinged canopy hinged up and aft (as looking aft) in service

The canopy is provided with the hinged canopy control service system and the emergency hinged canopy jettison system independent on the service one.

The hinged. canopy control service system is actuated by the compressed air fed from the main air system and ensures:

- the lifting and lowering of the hinged canopy depending on the position of the canopy control valve handle located in the under canopy panel, on the port side of the cabin;
- the fixing of the hinged canopy in the closed position by the locks;
- the hinged canopy sealing;
- the canopy inadequate looking indication by illumination of the LOCK CANOPY red annunciator and after starting the engines, by the display of the LOCK CANOPY signal in the priority on the EKRAN annunciator panel and delivery of the "Look canopy" voice information to the pilot's earphones.

The canopy control handle has three positions: CANOPY OPEN CANOPY PARTIAL OPEN, CANOPY CLOSED.

The canopy is opened by shifting the canopy control handle to the CANOPY PARTIAL OPEN position and, then to the CANOPY OPEN position.

In the closed position, the hinged canopy is secured on the fuselage by means of four pins which enter the hinges of four mating emergency looks of the hinged canopy. The retaining of the locks are interconnected through the bellcranks with the aid of the rods of the service system actuated by the rod of the power cylinder driven by the compressed air fed through the canopy control valve.

When open, the hinged canopy is retained by the lifting cylinder (if the pressure in available in the air system).

If the smoke appears or the canopy glazing becomes fogged, the cabin is depressurized with the aid of the CAB EMER DEPRESSOR lever located on the cabin port side. The valve lever may assume two positions: the working position PRESSUR (the lever is locked) and the emergency position CAB EMER DEPRESSOR. After airing the cabin, the latter should be pressurized. For this purpose. the valve lever should be set to the PRESSURE position.

The hinged canopy may be automatically jettisoned with use of emergency jettison system during the ejection from the bifurcated jettison handle or autonomously, without ejection, from the CANOPY EMER JETSN individual handle located on the cabin starboard.

During the emergency jettison of the canopy, the canopytofuselage emergency locks are released (in this case, the hinges with the service lock pins inserted through them remain in the undercanopy panel) and the forced removal of the hinged canopy is effected by the powder gas energy of the pyromechanical system which actuates both as in case of ejection and in case of the autonomous emergency jettison of the canopy.

For opening the canopy if air is not available in the main air system, it is necessary to shift the canopy control handle to the extreme backward position and, after opening the looks, lift the hinged canopy with hands and fix the hinged canopy with the canopy open position rod.

The canopy is furnished with the electrical system used for heating the front section of the windshield glass to prevent the latter from icing on the outer side. The electric heater effectively functions on a lapse of 5 to 7 minutes after turning an the PITOT GLASS HEAT switch.

For precluding the glass panel fogging inside the cabin, the canopy glazing is blown with hot air bled from the engine compressor when the BLOWING CANOPY PILOT air flow shutter control front lever is set to the BLOWING CANOPY position.

For ventilating the cabin while taxiing, it is allowed to partially open the hinged canopy by shifting the canopy control handle to the middle position.

For precluding the cabin windshield fogging in flight at the low airspeed (M< 0.8), provision is made for blowing the windshield with hot air which in bled down stream of the pressure differential regulator and is fed to the separate manifold through the electromagnetic valve.

The feed of the hot air risen the common air temperature in the cabin, therefore, for maintaining the preset air temperature in the cabin, provision is made for the automatic cutout of the windshield blowing line: with the increase in the cabin air temperature, the shutter installed in the cold air line becomes open progressively. When the shutter gets completely open, the limit switch cuts off the electromagnetic valve and the windshield blowing line is shut off.

The windshield blowing in also discontinued when the Mach number exceeds 0.8.

The windshield blowing way also be discontinued manually by closing the cabin air supply valve. In so doing, the microswitch mounted on the supply valve lever will de-energize the electromagnetic valve.

AIRCRAFT EMERGENCY ESCAPE SYSTEM

The emergency escape system of the aircraft comprises the ejection seat, hinged canopy and pyromechanical system controlling the hinged canopy jettison and pilot's ejection. The aircraft emergency escape system ensures the safe ejection of the pilot practically at all the altitudes at the airspeed of up to 1300 km/h as well as during the takeoff run and landing roll at the airspeed not less than 75 km/h.

The protection of the pilot against the gloads onset during the ejection and against the effect of the dynamic pressure is ensured by the highaltitude outfit, the pilotto-seat restraint with arm guard and leg soft straps, steady stabilization of the seat while ejecting, and when the ejection occurs at the airspeed more than 900 km/h, the pilot is additionally protected by the wind deflector against the air flow.

The ejection is effected by the pilot by pulling up the bifurcated ejection system control handle located on the front wall of the seat pan, after which the hinged canopy emergency jettison and seat ejection systems actuate in the required sequence until the lifesaving parachute is put into operation.

The hinged canopy jettison control pyromechanical system ensures the release of the emergency looks of the hinged canopy, its throwing up and eliminates the interlock between the hinged canopy and ejection seat both when pulling the ejection handles and the hinged canopy autonomous jettison handle.

For executing the autonomous jettison of the hinged canopy, it is necessary to deflect the handle in the horizontal plane from the side towards the axis of the aircraft symmetry and, then energetically turn it fully down backward. During the first motion the handle gets unlocked while during the second motion the pin is extracted from the pyromechanical system gas generator.

The main gas generator of the pyromechanical system is actuated just after operation of the pilottoseat restraint when using the jettison handles or directly the hinged canopy emergency jettison handle. The gases are directed through the pipelines to the right and left canopy remover cylinders, emergency lock actuating cylinder (cabin part side) and twocavity emergency look actuating cylinder (cabin starboard) ensuring their release even in case of the mechanical damage to the coupling between the left and right locks. The locks are released and the hinged canopy is thrown up. The travel of the hinged canopy through the static line secured thereon will unlock the firing mechanism of the seat.

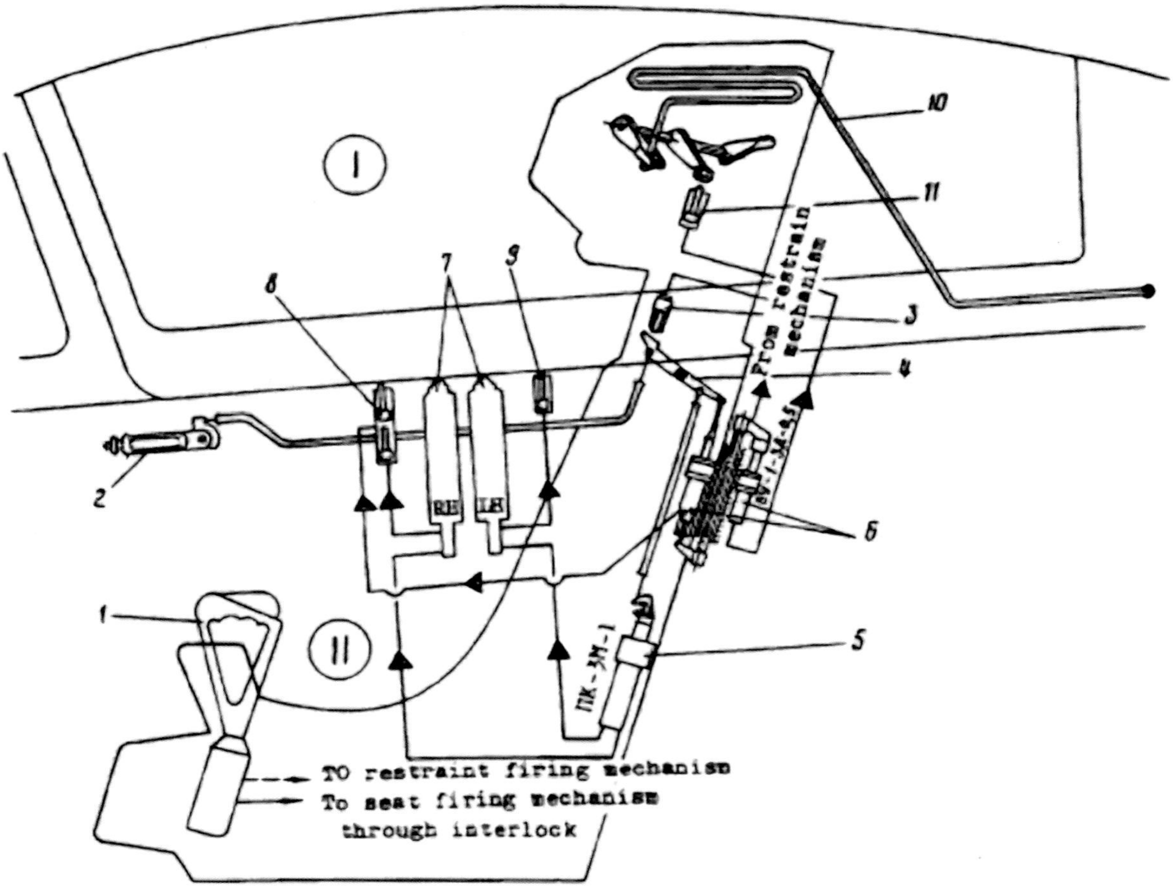

EJECTION SYSTEM SCHEMATIC DIAGRAM

1 - ejection handle; 2 - canopy autonomous jettison handle; 3 - seat restraint firing mechanism actuating cylinder; 4 - canopy jettison gas generator and time delay mechanism actuating bellcrank; 5 - canopy jettison gas generator; 6 - time delay mechanism; 7 - canopy remover cylinder; 8 - canopy lock release two-chamber cylinder; 9 - canopy actuating cylinder; 10 - canopy separation-to-seat unlocking line; 11 - seat unlocking duplicating power cylinder

The standby canopy jettison system is put into operation simultaneously with the actuation of the main canopy jettison system. In case of the failure of the main gas generator, the standby system, on a lapse of 0.5 s, opens the canopy locks and on a lapse of another 0.5 s unlocks the firing mechanism of the catapult.

The ejection seat consists of the seat proper, combined firing mechanism, mechanism box, parachute harness system with the canopy stowed in the headrest, survival kit NAZ7M, emergency (seat) oxygen system, service systems ensuring convenience for the crew member's work in the seat and the emergency systems ensuring the safe ejection.

For precluding the inadvertent actuation of the catapult, the ground safety pins are inserted into the catapult members.

The seat is adjusted in height with the aid of the stepless electromechanical system adjusting the seat according to the pilot's stature with use of the SEAT Up DOWN selector switch. The maximum adjustment value (from the lowermost to the uppermost position of the seat pan) amounts to 135 mm.

When in flight, the pilot is held in the seat by means of the normal operation restraining system.

The normal restraining is effected by the shoulder and waist restraint harness coupled with the parachute harness system and controlled with the aid of the SHOULDER LOCK and WAIST LOCK handles. The shoulder restraint limiting the harness motion, prevents the free displacement of the pilot in the direction of the flight and holds the pilot from the impact against the instrument board. If the backbreast gload exceeds 2 g, the shoulder harness restraint automatically locks the harness. The waist restraint, reeling in the straps, presses the pilot to the seat.

While ejecting under action of gases of the ejection gun cartridge, the log uplifting mechanism, shoulder restraint, arm guards, waist restraint actuate. After actuation, all said mechanisms get locked.

The first stage ejection gun of the combined firing mechanism is also put into operation in response to the ejection control system command.

The combined firing mechanism is essentially a power generator which ensures the acceleration and lift of the seat to the height required for the seat to pass over the fin, actuation of the wind deflector of the system ensuring the additional protection of the pilot against the air flow, deployment of the lifesaving parachute and separation of the seat from the pilot. The power generator comprises the telescopic ejection gun (the first stage), powder booster (the second stage), and parachute actuator. The first stage (the ejection gun) of the combined firing mechanism accelerates the seat in the guide rails to the speed of at least 13.6 m/s. The second stage of the combined firing mechanism consists of the powder booster which ensures the safe ejection trajectory of the seat. The parachute actuator ensures the firing of the headrest for initiating the lifesaving parachute.

The ejection is controlled by pulling the bifurcated ejection control handle. In this case, the ejection control mechanism supplies the commands to actuate the restraining system firing mechanism as well as the canopy emergency jettison system firing mechanism, electricallyignited cartridge of the crash helmet light filter, the first stage ejection gun of the combined firing mechanism, closes the circuit delivering the signals to the airborne emergency regimes and flight data recorder and a certain portion of the circuits. of the electricallyignited cartridge of the pyrotechnic valve used in the system ensuring the additional protection of the pilot against the air flow. The actuation of the first stage of the combined firing mechanism (KSM) is interlocked by the interlocking mechanism till the canopy is jettisoned for the time not exceeding one second. The interlocking mechanism is out off by the static line while jettisoning the canopy. One end of the static line is coupled with the canopy, whereas the other and of this line is coupled with the interlocking mechanism through the cable running via the guide tube of the seat headrest.

If the canopy fails to be jettisoned or if the static line is not coupled, the seat ejection gun cutin interlock is automatically released by the standby gas generator on a lapse of 1 s after actuation of the restraint. When the canopy cannot be jettisoned, the ejection may occur through the canopy glazing.

When the seat moves in the guide rails of the cabin at the various rates of the seat travel, the airborne equipment, including the oxygen equipment set, is actuated, the wind deflector is initiated under the action of gases of the first stage of the combined firing mechanism (at the airspeed of 900 km/h and more). The additional protection system is coupled with the first stage of the combined firing mechanism with the aid of the pyrotechnic valve disabling the system in response to the command delivered from aircraft pressure relay unit EKDR operating versus the dynamic pressure value. When the seat is ejected at the airspeeds of 900 km/h and more, the pressure relay unit does not deliver the command to close the pyrotechnic valve and the latter remains open and passes the gases of the cartridge of the combined firing mechanism first stage to the deflector hoses pushing the deflector proper and its apron into the air flow.

The seat ejection trajectory is dependent on the altitude, attitude and airspeed of the aircraft at the moment of ejection. When the seat is ejected at the low airspeed during the landing roll or takeoff run, the position of the seat along the trajectory is not stabilized: the parachute harness system is actuated when the seat approaches the apex of the trajectory active section. If the ejection occurs at the high airspeeds at the altitude of up to 5000 m, the ejection seat motion is stabilized till the airspeed tolerable for actuating the lifesaving parachute is gained.

When the seat is ejected at the altitude above 5000 m, the stabilized motion of the ejection seat and its descent to an altitude of 5000 m are ensured. The descent to an altitude of 5000 m occurs with inconsiderable rotation of the pilot with seat, with the pilot facing down.

The stabilization of the seat after it clears the cabin till the lifesaving parachute actuates and the seat separates from the pilot, is ensured by two telescopic rods with stabilizing parachutes rotating on the swivels by the air flow, thus ensuring the seat stabilization.

The lifesaving parachute is actuated in response to the combined parachute release control units, type PPK and catapultparachute control unit KPA. Often the seat gains the speed and altitude tolerable for actuation of the lifesaving parachute, the parachute actuator operates in response to the automatic control equipment command. The headrest fired from the seat ensures the unlocking of the chamber and pulls the lifesaving parachute canopy out of the cover. The recoil force originated in firing the headrest is imparted to the seat, the pilot gets rid of the seattopilot harness, i.e., the pilot separates from the seat; in this case, the ejection control handles remain in the pilot's hands, the survival kit clears the seat pan and the pilots hangs on the lifesaving parachute.

After separation of the pilot from the seat, the descent and touchdown (alighting on water) of the pilot are ensured by the parachute recovery system.

The survival kit is used for supporting the pilot's life after abandoning the aircraft or forced landing. The seat is provided with survival kit NAZ7M, automatic radio beacon KOMAR and life raft PSN-1. The life raft, automatic radio beacon and the survival kit pack are interconnected with a 13 m static line stowed in the shaped cover of the seat pan and are automatically released during the lifesaving parachute deployment.

The survival kit pack accommodates the food ration with water, the camping outfit, medicines, signaling means and instructions on use of the survival kit. The food ration is sealed in the PVC bags and water (1.5 litres) is poured in the tank located in the pack pocket. The camping outfit comprises dry combustible and windresistant matches, machete, box with cartridges for the gun, light filter spectacles. The signaling means comprise 10 red signal flares and flare pistol, light signal mirror and orange stain for painting the water or snow to mark the location of the crew member.

OXYGEN EQUIPMENT AND HIGHALTITUDE OUTFIT

The aircraft oxygen equipment and the pilot's highaltitude outfit are used for ensuring the activity of the pilot in executing the highaltitude flights and ejection.

The oxygen equipment comprises the main (airborne) and emergency (seat) oxygen systems.

The main oxygen system includes:

- the gaseous oxygen reserve contained in three bottles having the holding capacity of 4 ltr each and operating pressure of 150 kgf/cm;
- oxygen breathing equipment set KKO (the oxygen volume and reducer, oxygen demand regulator RPK, helmet ventilation unit VISh, remote control panel, oxygen set KP);
- combined lifesupport system indicator EKSh.

The emergency oxygen system is located on the ejection seat and comprises:

- the survival oxygen reserve in the 0.7ltr bottle;
- oxygen equipment unit BKO with the pressure gauge;
- common connector ORK.

The oxygen for the main oxygen system and the engine feed oxygen delivered into the respective systems through the common charging connection up to a pressure of 150 kgf/cm. The connection is installed on the left side, in the nose fuselage, in the access hatch bearing the OXYGEN CHARGE inscription.

The quantity of oxygen in oxygenbreathing equipment set KKO is checked against combined lifesupport system indicator KP with the ground DC and AC power sources connected,

the STBY circuit breakers and BAT GND SUP AC GEN. TOW switches turned on.

The emergency oxygen system is charged with oxygen through the charging connection provided in the seat up to a pressure of 180 kgf/cm . The quantity of oxygen in oxygen equipment unit BKO is checked against the pressure gauge installed in the seat.

The set used aboard the aircraft as part of the oxygen equipment and the highaltitude outfit of the pilot ensures the normal activity of the pilot and supply of the pilot with oxygen in the following conditions:

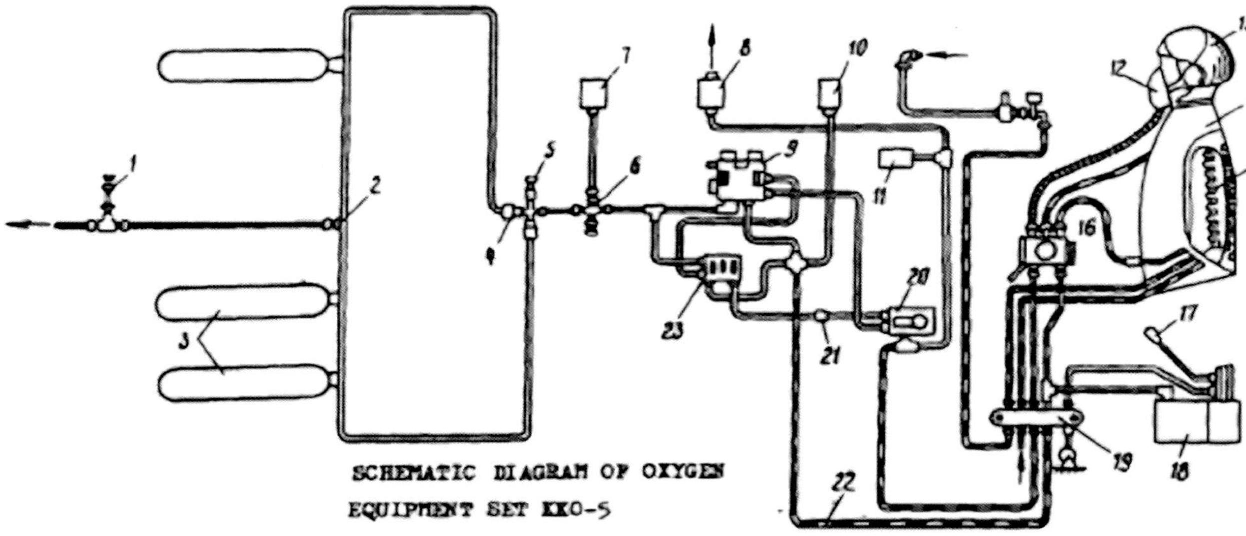

SCHEMATIC DIAGRAM OF OXYGEN EQUIPMENT SET KKO-5

• for a long time, when flying in the pressurized cabin at the altitude not exceeding 20,000 m and when flying in the depressurized cabin at the altitude not exceeding 12,000 m;
• for a short time, when the cabin is depressurized with the subsequent emergency descent from the maximum possible altitude of the flight to the altitude less than 12,000 m;
• for a short time, in case of the ejection at the altitudes not exceeding 20,000 m with the simultaneous automatic changeover to the oxygen supply from the emergency oxygen system.

When flying in the pressurized cabin at all the altitudes as well as at the altitudes up to 12,000 m in the depressurized cabin, the breathing oxygen is fed by the oxygendemand regulator; in so doing, a mixture or oxygen and air is fed to the cabin up to an altitude of 8000 m, and when the altitude exceeds 8000 m, a pure oxygen is fed thereto.

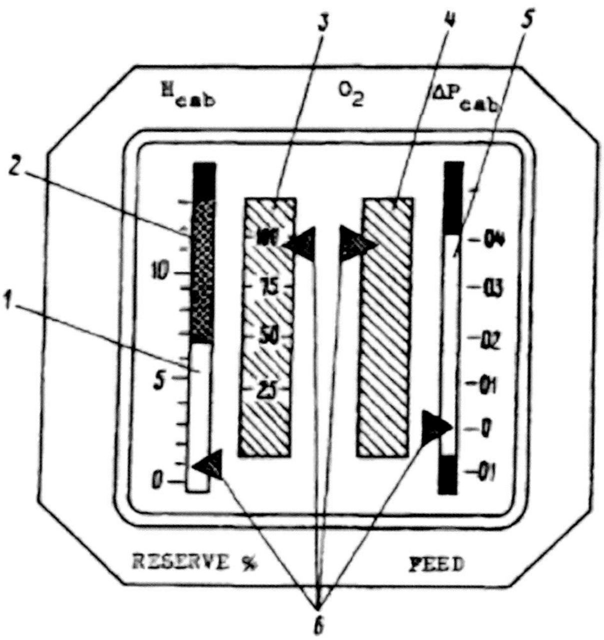

COMBINED LIFE-SUPPORT SYSTEM INDICATOR ИКШ

1, 2 - "cabin (oxygen mask) altitude" scale; 3 - oxygen reserve scale; 4 - scale of oxygen equipment set functioning in "inhalation - exhalation" phases; 5 - cabin pressure differential scale; 6 - scale indexes (indicators)

The percentage of oxygen in the inhaled mixture is regulated automatically versus the flight altitude.

When the cabins are depressurized at the altitudes exceeding 12,000 m, the oxygen is fed to the oxygen mask and highaltitude partial pressure suit under the excessive pressure whose value is controlled automatically versus the flight altitudes.

For more effective removal of the carbon dioxide from the oxygen mask and decreasing the resistance to the breathing, provision is made for the ventilation of the undermost space with the aid of helmet ventilation unit VISh.

The modes of operation of the oxygen equipment set are changed over with the aid of the knobs located on remote control panel Mr. The control panel mounts three knobs: MIXTURE 100 % 0 2 (CMECL 100 % 02) (blue knob), EMERG ON OFF (red knob), BYM ON OFF(black knob).

It is necessary to change over to the pure oxygen supply (100 % 02) in performing the combat flights irrespective of the flight altitude and in case of the smoke in the cabin.

It is necessary to cut in the emergency oxygen feed system in came of malfunction of the oxygen equipment (the shortage of oxygen, the great resistance during inhalation) or the worsening of the state of health. In this case, it is necessary to intensify the check of the oxygen reserve since the oxygen consumption in increased two or three times.

It is recommended to cut in helmet ventilation unit VISh in executing the flights at the low altitudes.

The oxygen equipment serviceability is checked against combined lifesupport system indicator IKSh installed on the instrument board in the cabin. The indicator has four scales with movable indexes (indicators) reading:

- the cabin altitude;
- the oxygen reserve (0 2 RESERVE %)
- the operation of the oxygen equipment in the inhalation exhalation phases (FEED);
- the pressure differential between the pressure cabin (the oxygen mask) and the atmosphere

The "cabin altitude" is checked against the scale in executing the flight in the pressurized cabin at all altitudes and in the depressurized cabin at the altitudes below 12,000 m. When flying the aircraft at the altitudes above 12,000 m in the depressurized cabin, this scale is used for checking the "oxygen mask altitude" which should not exceed 12 km (irrespective of the flight altitude) if the oxygen equipment is serviceable. If the index is within the yellow sector of the scale (from 7 to 12 km), the cabin is depressurized. The location of the index within the red sector of the scale (above 12 km) is indicative of the oxygen equipment failure (the PH ALT ALERT signal is delivered to the EKRAN system; in this case, it is necessary to urgently descend to the altitude below 12,000 m.

The quantity of oxygen (in per cant) available in the main oxygen system is checked against the 02 RESERVE % scale. When the oxygen reserve decreases to 15% (the index is within the red sector of the scale), the respective signal is applied to the TESTER system and the EKRAN system delivers the EMER OXYGEN RESERVE signal, the information reporting system generates the "Switch on oxygen supply" command. In this case, it is necessary to descend to an altitude below 4000 m and do not exceed it until the flight is accomplished.

The FEED scale is used for checking the serviceability of the oxygen equipment set (oxygen demand regulator) in the breathing phases during the inhalation the index moves down and during the exhalation the index moves up. In case of the oxygendemand regulator failure, it is necessary to actuate the emergency oxygen feed system, for which purpose, set the red knob located on the control panel to the ON position.

Note. When the cabin altitude is below 2000 m and the 100% 0 2 MIXTURE knob is set to the MIXTURE position, the breathing oxygen is not fed and the index does not respond to the inlet and outlet.

The pressure differential between the pressure cabin and the atmosphere is checked against the scale. The location of the index within the red sectors of the scale (above the 0.4 kgf/cm division or in the lower section of the scale) is indicative of the failures in the pressure cabin pressure control system.

When the seat is ejected, common connector ORK is automatically released, thus disconnecting the service lines coupling the pilot's highaltitude outfit with the airborne equipment and the oxygen is automatically fed from the emergency oxygen system provided in the seat.

The oxygen reserve in the emergency oxygen system is quite sufficient for the pilot to breath:

- when the ejection is executed from the maximum altitude of the flight, for three or four minutes;
- when the ejection is executed at the low altitudes above the water surface, when the pilot comes to the surface and stays afloat thereon, for three minutes from the moment the emergency oxygen system is switched on;
- in case of the failure of the main oxygen system, for the urgent descent to an altitude of 4000 m, for six minutes.

When in flight, it is necessary to change over to the oxygen supply from the emergency oxygen supply system provided in the seat in case of even one of the following symptoms of main oxygen system failure:

- the sharp decrease in the oxygen reserve(less than 15%) the discontinuation of the oxygen feed when the cabin altitude exceeds 2000 m and the oxygen equipment set functions in the MIXTURE mode and at any altitudes in the 100% mode;
- the absence of the excessive oxygen pressure in the oxygen mask and in the high altitude partial pressure suit during the cabin depressurization at the altitudes exceeding 12,000 m.

For cutting in the emergency oxygen system manually, it is necessary to Pull UP sharply the red knob on the right side of the catapult shaped seat.

Before the flight, check common connector ORK for the proper locking against the common connector locked position indicator (the bright orange pin extending through a length of 3.0 to 3.5 mm over the upper block plane with the common connector lock closed) and check setting of the safety device on the lower block of the common connector in position, "1".

It is allowed to use the following pilot's high altitude outfit:

- crash helmet ZSh 5A or ZSh7A
- oxygen mask KM-34D or KM34 (in oxygen set KKO15 complete with the KP-52M oxygen set).
- high altitude partialpressure suit VKK-6M (in oxygen set KKO15);
- ventilated suit VK3M (in oxygen set KKO15);
- anti g suit PPK3;
- high altitude marine survival suit VMSK4 (in oxygen set KKO15);
- airborne life belt ASP-74;

The pilot should use only the individual highaltitude outfit which is adjusted to the pilot's size and stature with the help of the doctor of the flying unit and the high altitude outfit technician (mechanic).

Before the flight, make certain that the highaltitude outfit is serviceable, check for the radio communication and the oxygen mask tightness.

In executing the flights when the mass destruction weapons are employed use the pilot's mask which forms part of oxygen set KKO15 and comprises:

- oxygen mask KM35P;
- oxygen set KP112;
- gas filtering box KPL2

The pilot's gas mask ensures the protection of the respiratory system and eyes against toxic and radioactive agents, and bacteriological aerosols.

The provision is also made aboard the aircraft for ventilation of the pressure suit underclothes space. The ventilation system is switched on and controlled with the aid of the ejector valve installed in the cabin on the left horizontal console. For switching on the ventilation system, set the SUIT VENT knob to the OPEN position, whereas the temperature is controlled with the aid of the HOT COLD knob.

ELECTRICAL EQUIPMENT OF AIRCRAFT

The aircraft electrical equipment comprises the power supply and electrical automatic control systems.

The power supply system comprises the DC and AC generators, storage batteries, inverters and the power distribution line (mains). The electrical automatic control system ensures the control of the electricallydriven systems of the aircraft (the engines, fuel system, fire extinguishing system, etc.).

The electrical equipment is cut in through the cabin power system.

In case of the failure (cutout) of the main power sources, the respective placard is displayed on the annunciator panel of the EKRAN system, lamp KSTs starts flickering and in case of the failure of two generators, the respective voice information is delivered to the earphones.

DC POWER SUPPLY

The main aircraft DC Power supply source is the DC starter generator operating together with the startandspeed control and protection equipment ensuring a voltage of 28.5 V in the aircraft electrical system. The emergency DC power supply source is presented by two storage batteries.

The DC generator is mounted on the aircraft accessories gear box and generates a voltage of 28.5 V maintained within the specified limits by protection and regulation unit BRZ. Besides, protection and regulation unit BRZ protects the generator and aircraft electrical system from the over voltage as well as from the short circuit in the generator proper. In case of the over voltage or short circuit, the switch operates in protection and regulation unit BRZ, which discontinues the generator excitation winding circuit. After de-energizing the windings, the switch contacts remain open. The switch may be returned to the initial position only on the ground by depressing the BRZ COCKING button. The generator cutout is accompanied by the DC GEN placard displayed on the EKRAN annunciator panel.

If the DC generator fails, the aircraft electrical system is changed over to the emergency supply from the storage batteries. In this case, the powerful consumers such as the 3502 command radio link, integrated radar/sighting system, IFF interrogatorresponder, optoelectronic navigation/sighting system (except for the navigation system) are automatically cut out. The common capacity of two storage batteries amounts to 80 110 Ah, the rated voltage is 22 V.

The storage batteries are installed in the fiber glass laminate container provided with a DC heater.

For cutting in the generator into the aircraft DC electrical system, use is made of the circuit breaker which ensures:

- the connection of the generator to the aircraft electrical system if the generator voltage exceeds that of the system;
- the automatic cutout of the generator if the reverse current flows from the aircraft electrical system through the generator winding.

AC POWER SUPPLY

The main aircraft AC Power supply source is the generator with the startandspeed control and protection equipment feeding the aircraft electrical system with a threephase 115/200 V 400 Hz power. The inverter is an emergency AC power source.

The generator drive is mounted on the aircraft accessories gear box and is used for feeding the threephase stabilized AC power to all airborne AC consumers and comprises the hydraulic drive and AC generator. The hydraulic drive output shaft imparts the rotary motion to the generator shaft. The generator drive output shaft speed and, consequently, the speed of the generator are maintained to be constant with the aid of the builtin control system, irrespective of the drive input shaft speed or the load across the generator. The control system consists of the centrifugal regulator and servomechanism. The rated power of the generator drive is 30 kVA.

The generator is automatically out into a load on gaining the assigned rotary speed and AC voltage as well as 55 % of speed of one of the engines. In this case, the AC GEN signal is cancelled in the EKRAN system priority annunciation circuit.

For maintaining the voltage across the generator within the preset limits, use is made of the voltage regulation unit.

If the generator fails or is automatically cut off by the protection and control unit, the essential consumers are fed from the inverter. In this case, the AC GEN signal is delivered to the EKRAN system and the following equipment is deenergized:

- the heaters of the main and standby direction/vertical gyros;
- the integrated radar/sighting system, optoelectronic navigation/sighting system (except for • the navigation system); the heater of the LH angleofattack transmitter and slip angle transmitter;
- the KRY command red o link;
- the heater of the glass panels (the upper and middle sections);
- the IFF interrogatorresponder, automatic flight control system;
- the external store

To prevent the generator drive breakdown if the failures onset therein, provision in made for uncoupling the generator drive input shaft from the aircraft accessories gear box shaft both automatically and manually.

When the oil pressure in the generator drive drops below the rated value or when the AC frequency exceeds the tolerable value, the voltage is applied to the electric actuator which cuts out the generator drive and the DISCON GEN DRIVE signal is applied to the EKRAN system. After operation of the electric actuator (which cuts off the drive when the AC frequency and voltage drop below the rated values), the generator in disconnected from the aircraft electrical system and the AC GEN signal in applied to the EKRAN system.

In the case the generator drive automatic cutout circuit fails to operate (the DISCON GEN DRIVE signal remains on the EKRAN annunciator panel) or the generator fails (the AC GEN

signal is applied to the MAN system), the drive is cut out manually by depressing the GEN DRIVE EMERG DISCON switch located in the cabin for a short time.

The generator drive protection against the cutting of the ground power supply source with wrong phase sequence is effected by the phase sequence unit.

The consumers are fed with a 36V AC power through the stepdown transformer, whereas the emergency changeover of these consumers from the transfer to the inverter in effected through the transformer protection unit.

The inverter is essentially a threephase, singlephase device with the 1000VA three phase and 1500VA singlephase generator powers is used for converting the 27V W into the threephase 36V AC and singlephase 115 V 400 Hz.

The ground AC power connector in located in the LH landing gear well.

CENTRALIZED BUILTIN TEST AND WARNING SYSTEM "EKRAN"

The EKRAN centralized builtin test and warning system is part of the aircraft warning and caution system. It combines the builtin test circuits into one system and ensures the automated test of the serviceability of the units, accessories and systems of the aircraft during its preparation for the flight and accomplishment of the flight as well as the record of the information on the failures occurred.

After landing the aircraft, the EKRAN system records the failures occurred in the flight. The information on the condition of the systems, accessories and units, whose failures and excessive parameters do not endanger directly the flight safety, without delivering the information to the EKRAN annunciator panel as well as the information on the systems, units and accessories, which is delivered to the red annunciators, are recorded.

The system comprises the logic and control unit and the universal annunciator panel.

For controlling and checking the EKRAN system, the latter is provided with the EKRAN CALL and EKRAN CHECK buttons.

The logic and control unit successively "interrogates" the builtin test circuits and separate transmitters checking the parameters of the operation of the aircraft accessories, units and systems. If any parameter of the aircraft unit (accessory) or system exceeds the rated value or one or several objects under check fail, the failure signal is generated by the respective builtin test circuits or by the respective transmitters. The logic and control unit receives this signal and generates the information on the failure.

When the EKRAN CALL button is depressed and the priority signals are not applied, the frame disappears on the EKRAN annunciator panel and the MEMORY annunciator illuminates and will be alive until the failure signal disappears. The signal is called from the memory by depressing the EKRAN CALL button if the priority signals are not applied.

In this case, lamp KSTs is not turned on. In case of application of the higher priority signal, the lower priority signal will be put on the queue the

PRIORITY annunciator will illuminate whereas the higher priority signal will be applied to the annunciator panel. When several failure signals are applied, the higher priority signal is displayed on the EKRAN annunciator panel, whereas the other signals are put on the queue the PRIORITY annunciator illuminates. The information is called from the queue by depressing the EKRAN CALL button. In this case, lightbutton KSTs is turned off and the signal applied to the annunciator panel till it is called from the queue goes into the memory which is called by depressing the EKRAN CALL button only after scanning the whole queue and illumination of the MEMORY annunciator.

If the information obtained by the logic and control unit from the builtin test system and the transmitters of the items under test is not related to that delivered to the EKRAN annunciator panel, then said information is stored only in the logic and control unit for the subsequent record of the information after landing the aircraft.

The builtin test system and transmitters checking the operation of the aircraft accessories units and systems are "interrogated" by the logic and control unit according to the assigned test algorithms.

INFORMATION REPORTING SYSTEM

The information reporting system is used for delivering the voice information to the pilot and to the ground control posts on the emergency situation in flight.

The system executes the multichannel magnetic record on the ferromagnetic tape. When a signal is delivered from the transmitter, the head corresponding to the given channel is connected and the voice information signals picked off the reproducing head are applied to the reproducing unit output and, then to the pilot's earphones through the aircraft interphone system. Voice messages 1, 2. 3, 5 are transmitted on the air through the radio set.

In case of the simultaneous application of the signals from the several transmitters, the information in processed successively in the channel number increasing order.

The information reporting system is fed with a voltage of 27 V from the DC generator and if the latter fails, the system in fed from the storage battery.

The information reporting system is controlled with the aid of the IRS REPEAT and IRS CHECK buttons with the BAT GND SUP and ASFT SYS switches turned on.

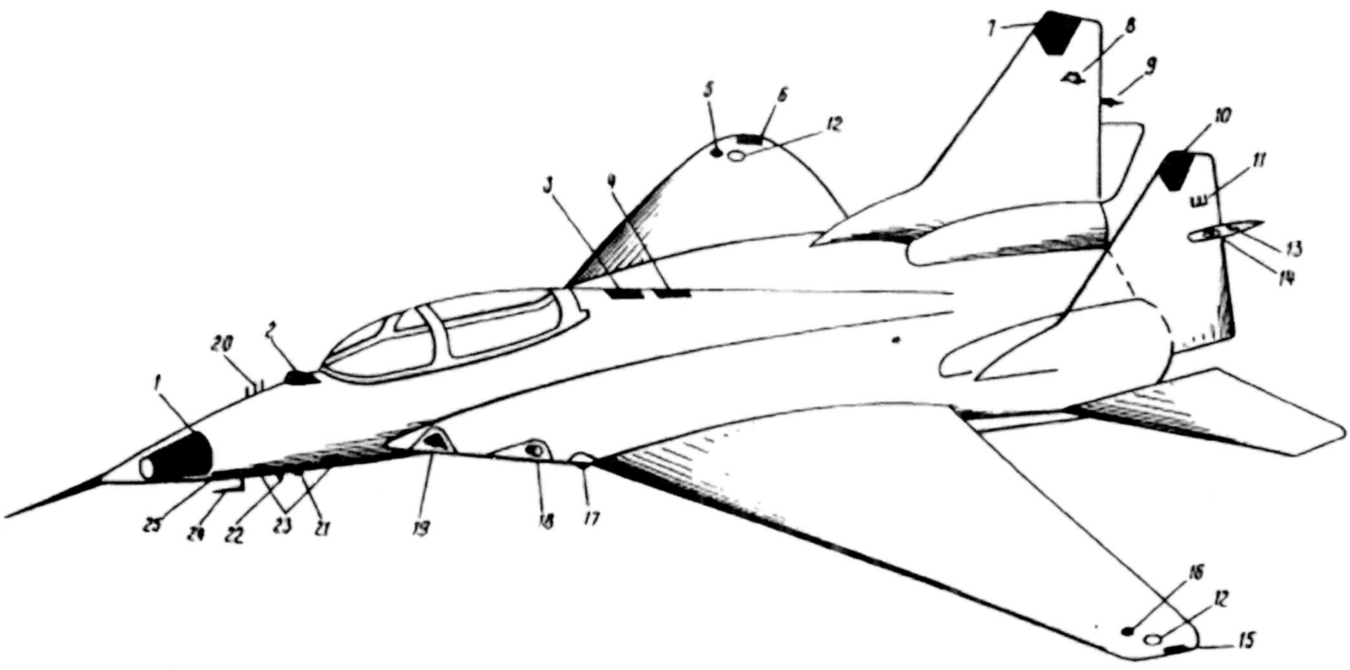

ARRANGEMENT OF ANTENNAS ON AIRCRAFT

1 - radar antenna; 2 - optical locator laser station antenna; 3 - automatic direction finder omnidirectional antenna; 4 - automatic direction finder loop antenna; 5 - СПО-15ЛМ station elevation antenna ("top"); 6, 15 - СПО-15ЛМ station azimuth (rear) antennas; 7 - Р-862 radio set antenna; 8, 21 - СО-69 ATC transponder band III (international communication band) antennas; 9 - "Пион-НМ-02" antenna feeder system rear antenna and СО-69 ATC transponder horizontally polarized band III antenna; 10 - airborne radio link antenna; 11 - IFF transponder band III antenna; 12 - IFF transponder band II antenna; 13 - СО-69 ATC transponder bands I and II antenna; 14 - IFF transponder band I antennas (one antenna, arranged on left side and one antenna, arranged on right side in fin); 16 - СПО-15ЛМ station elevation antenna ("bottom"); 17 - one IFF transponder band I antenna in left wing extension and one IFF transponder band I antenna in right wing extension; 18 - two СПО-15ЛМ station azimuth front antennas (one antenna, in left wing extension and one antenna, in right wing extension); 19 - two СО-69 ATC transponder bands I, II antennas (one antenna, in left wing extension and one antenna in right wing extension); 20 - IFF interrogator-responder antenna; 21 - СО-69 ATC transponder band III antenna; 22 - IFF transponder band III antenna; 23 - radio altimeter transceiving antenna; 24 - "Пион-НМ-02" antenna feeder system and СО-69 ATC transponder band III front antenna; 25 - marker radio receiver antenna

When the IRS CHECK button is depressed, the voice information is delivered from the test channel, and the supply circuit of all relays of the information reporting unit gets open, after the button has been released, the circuit gets closed and if the signals are applied from the transmitters, the repeated reproduction of the commands in the priority order occurs.

When the GND START button is depressed, in response to the automatic control unit signal the information reporting system is set to the initial position.

When the IRS CHECK button is depressed, the information reporting system is being checked and if said system is found serviceable, "This is aircraft No . .... Information reporting system is serviceable" voice information is delivered to the earphones.

The list of voice information and instructions delivered by the information reporting system is given in Table 12A.

RADIO SET R862

Change over to fuel quantity gauge
Decrease temperature
Set to EASY before landing

The radio set ensures the fixed tuned communication within the lineofsight distance between the aircraft, the aircraft and the ground radio stations and the constant reception of the signals delivered by the rescue service on an emergency frequency of 121.5 mHz.

The radio set may function on any of twenty channels tuned to the fixed frequencies preselected in the memory.

The radio set operates on two wave bands:

- on the metric wave band from 100 to 149.375 mHz;
- on the decimetric wave band from.220 to 399.975 mHz.

The communication frequencies are uniformly allocated over the bands every 25 kHz (2000 frequencies on the metric wave band and 7200 frequencies on the decimetric wave band).

The radio set is fed from the DC generator and if the latter fails, the radio set is fed from the airborne storage batteries. The radio set serviceability is ensured even if the voltage in the aircraft electrical system drops to 18 V. The consumption of the power by the radio set in the TRANSMIT mode is not more than 550 W and in the RECEIVE mode, not more than 50 W.

The range of communication between the aircraft and the ground control centres established by the radio set, amounts to at least 120 km when flying at an altitude of I000 m, at least 250 km at an altitude of 5000 m and at least 350 km at an altitude of 10,000 m.

The time required for changing over from the transmission to the reception and vice versa is not more than 0.5 s. The time required for changing over to another channel is not more than 1.5 s.

The radio set is ready for operation just after the power is applied thereto.

The radio set antenna is installed in the right fin tip.

When executing the 360° turns at a bank exceeding 45°, the antenna may be shaded and the radio communication may be lost at some relative bearings.

The radio set may operate in the frequency modulation and amplitude modulation modes. The modulation mode is selected in compliance with the assigned flight mission with the aid of the 01 FM -AM selector switch. During the training flights, the selector switch is placed in the AM position and looked.

The radio set is turned on with the aid of the RADIO switch.

The following operations can be performed from the radio set control panel:

Table 12

No.	Information	Instruction
1	Left engine on fire (to be repeated twice)	
2	Right engine on fire (to be repeated twice)	
3	Aircraft accessories gear box on fire (to be repeated twice)	
4	550 kg remain	
5	No fuel boost	Cut out afterburner
6	Left engine overheat	Decelerate left engine
7	Right engine overheat	Decelerate right engine
8	Left engine RPM exceed permissible ones	Decelerate left engine
9	Right engine RPM exceed permissible ones	Decelerate right engine
10	Vibration of aircraft accessories gear box above permissible	Abort flight mission
11	Limit altitude (to be repeated two times)	
12	Left engine oil pressure low	Decelerate left engine
13	Right engine oil pressure low	Decelerate right engine
14	Low oil pressure in aircraft accessories gear box	Abort flight mission
15	Left engine oil temperature above permissible	Decelerate left engine
16	Right engine oil temperature above permissible	Decelerate right engine
17	Booster hydraulic system failure	
18	Main hydraulic system failure	Extend landing gear in emergency way
19	Vibration of left engine	Decelerate left engine
20	Vibration of right engine	Decelerate right engine
21	Failure of DC generator	Flight time - 15 minutes
22		
23		
24	Launch permitted	
25	Air intake upper inlet	Do not exceed Mach number 0.8
26	Failure of left air intake automatic control system	
27	Failure of right air intake automatic control system	
28	Left engine supplied from stand-by system	Watch temperature and RPM
29	Right engine supplied from stand-by system	Watch temperature and RPM
30	Air data computer system failure	
31	Chips in left engine oil	Decelerate left engine
32	Chips in right engine oil	Decelerate right engine
33	Left engine fuel pressure above permissible	Decelerate left engine
34	Right engine fuel pressure above permissible	Decelerate right engine

Table 18, continued

No.	Information	Instruction
35	Check fuel load	Change over to fuel quantity gauge
36	No compartment cooling	Decrease temperature
37	Extend landing gear (to be repeated twice)	
38		
39	Limit stagnation air temperature	
40		
41	Lock canopy (to be repeated twice)	
42	Drop tank fuel consumption failure	
43	Switch on oxygen supply	
44	See EKRAN	
45	EKRAN failure	
46	ARI controller failure	Set to EASY (LEGKO) before landing
47	Emergency oxygen remainder	
48	Information reporting system serviceable	

- the selection of any of twenty communication channels by setting the channel selector switch to the required position;
- the volume control by turning the volume control knob;
- the cut in and cut out of the noise suppressor with the use of the NS switch;
- the cut in and cut out of the emergency channel monitoring circuit with the aid of the AJ (active jamming) switch.
- when the AJ lamp flickers, turn on the AJ switch and listen to the information delivered by the crew in distress;
- the application of the call signals generated by the automatic direction finder or short range radio navigation system simultaneously with the radio set signals with the aid of the DF switch.

The NARROW WIDE band pass selector switch is installed in the aftofcabin compartment, an the receiving driver front panel. Setting the selector switch to the NARROW position makes it possible to exclude the noise during the radio communication with the similar radio set.

RADIO SET R855YM

The small size transceiving radio set is part of the KOMAR radio beacon and is used for the two way radio communication between the pilot of abandoned distressed aircraft and the aircraft of the search group. This radio set serves also as a radio beacon operating in the TONE mode.

The KOMAR radio beacon comprises:

- the radio set (the transceiver);
- the inflatable envelope with antenna filled with carbon dioxide from the bottle during the emergency escape;
- the PRIBOY power supply battery (the storage term is not more than 12 months);
- the telescopic rod antenna.

The power supply battery ensures the continuous operation of the radio set in the following mode: one minute transmission and three minutes reception for 60 hours and in the radio beacon TONE mode for at least 70 hours.

The range of two way radiotelephone communication with the similar radio set, with the antenna being 1.5 m high over the ground , amounts to at least 800 m.

The range of radio communication with the aircraft and the hosing range are dependent on the flight altitude and the sensitivity of the airborne receiver and at an altitude of 1000 to 3000 m it amounts to:

(a) 16 to 60 km for the two way radio communication;
(b) 30 to 70 km for the homing radio communication in the TONE mode.

The radio set remains serviceable after being immersed into the water to a depth of up to 1 m for one hour.

The reception and transmission are exercised through one dynamic microphone mounted into the transceiver casing (or through the headset connected via the patch cable available in the survival kit).

The radio set is controlled with the aid of the RECEIVE and TRANSMIT buttons located on the side wall of the transceiver casing. The buttons are locked and unlocked when moving the retainer towards the antenna and in the reverse direction, respectively.

The casing upper section is provided with the receptacle for connecting the antenna, whereas the lower section of this casing is provided with the screw used for depressurizing the radio beacon when the atmospheric pressure or ambient temperature varies considerably.

The radio beacon TONE mode is automatically selected during the ejection, drop and inflation of the antenna and is switched off by depressing the RECEIVE or TRANSMIT button.

AIRCRAFT interphone SYSTEM

The aircraft interphone system is used for establishing the twoway telephone communication between the pilot and the aircraft technician, for listening to the radio communication signals through the radio set receiver, the homing radio station call signs through the direction finder, the signals delivered from the airborne information reporting system and marker receiver.

The aircraft interphone system is fed with a voltage of 27 V DC from the DC generator andif the latter fails, the system is fed from the aircraft storage batteries. The maximum power consumption through the supply circuit is 13.5 W.

The aircraft interphone system is put into operation with the aid of the BAT GND SUP switch.

AUTOMATIC DIRECTION FINDER

The automatic direction finder is used for the air navigation with the aid of the homing and broadcasting radio stations and ensures:

- the flight towards the radio station and from the latter, with the indication the relative bearing;
- the determination of the radio and aircraft bearings against the course scale on the combined course indicator;
- the execution of the landing approach with the aid of the homing radio station;
- the reception of and listening to the mediumwave radio stations within the direction finder frequency range.

The automatic direction finder operates in two modes:

- the COMPASS automatic direction finding mode. In this mode the automatic direction finder automatically reads the relative bearing of the radio station the direction to which is being taken;
- the ANTENNA mode. This mode is used for listening to the call signs of the homing radio stations or the monitoring of operation of the broadcasting radio stations. 'When this mode is selected, the automatic direction finder operates as a standard medium wave receiver having the rather high sensitivity and noise immunity.

The automatic direction finder is cut in with the aid of the ASFT SYS switch.

The automatic direction finder is fed with a voltage of 27 V DC from the DC generator and with a voltage of 36 V 400 Hz from the AC generator through the stepdown transformer.

When the AC generator fails, the automatic direction finder is fed with an AC power from inverter PTO. 'When the DC generator fails, the automatic direction finder is automatically changed over to the power supply from the storage batteries.

The power consumed over the AC circuit is not more than 10 VA and over the DC circuit, not more than 50 W.

During the landing approach, the changeover from the outer homing radio station to the inner one is effected automatically in response to the marker receiver signal (with the landing gear extended) and when the radio station relative bearing pointer deflects through 90-50° from the zero position. When the change over occurs, the ADF-IHRS lamp illuminates. The changeover from the inner homing radio station to the outer one is effected automatically when the radio station relative bearing pointer deflects through 90-45° from the radio station relative bearing equal to 180°. The INNER-OUTER selector switch is used for the manual changeover to the inner homing radio station.

When the inner homing radio station number selector switch is set to the RB position and the ADF-IHRS selector switch is set to the IHRS position, the automatic direction finder is checked with the aid of the builtin test circuit. In this case, the radio station relative bearing pointer on the combined course indicator settles within 16°, and when the VOICE selector switch is set to the CW position, the 800Hz current will be applied to the earphones.

The switchingon of the manual rotation of the loop antenna (and deflection of the radio station relative bearing pointer, respectively) is effected by depressing the LOOP button located on the automatic direction finder control panel. When the button is released (in the COMPASS position), the radio station relative bearing pointer will again read the direction to the radio station and when the automatic direction finder is set to the ANTENNA, the radio station relative bearing pointer will remain motionless.

When the radio set (receiver) fails, the commands delivered by the flight control officer may be listened through the automatic direction finder. For this purpose, it is necessary to turn on the DF switch located on the radio set control panel, and set the COMPASS ANTENNA selector switch located on the automatic direction finder control panel to the ANTENNA position. In this case, the automatic direction finder functions as a highsensitive mediumwave receiver.

RADIO ALTIMETER

The radio altimeter is used for continuously measuring the absolute altitude of the aircraft flight and automatically delivering the altitude value to the indicator. The range of the altitudes measured is 0 to 1000 m with an accuracy of +1 m at the altitudes of 0 to 10 m and 10 % H at the altitudes exceeding 10 m (against the indicator). The limit altitude of 0 to 20 m is warned with an accuracy of _+O,5 m, and higher, with an accuracy of +3 % H.

The radio altimeter is essentially a radar continuously irradiating the centimetric wave band radio waves with frequency modulation.

The radio altimeter is fed with a voltage of 27 V BC from the DC generator and with a voltage of 115 V 400 Hz from the AC generator. If the DC generator fails, the radio altimeter is fed from the storage batteries, and if the AC generator fails, the radio altimeter is fed from inverter PTO with the INV START switch turned on.

The radio altimeter is cut in with the aid of the ASFT SYS switch; in this case, the unreliability flag appears on the altitude indicator. On expiration of the time required for the warmup when the radio altimeter is serviceable, the unreliability flag disappears whereas the altitude indicator pointer reads altitude of 0+1 m.

The proper functioning of the radio altimeter in flight is continuously checked in the automatic mode by the builtin test system.

When flying the aircraft at the altitudes below 1000 a, with the radio altimeter failed, the unreliability flag appears on the altitude indicator, and the pointer goes to the dark sector, just the same occurs at the altitude exceeding 1000 m when the radio altimeter is serviceable.

The radio altimeter generates the limit altitude signals to the information reporting system, EKRAN system, automatic flight control system and the radio altimeter serviceable and present altitude signals to the integrated radar/sighting system. The limit altitude is set with the aid of the ALT SET. CHECK knobbutton by aligning the limit altitude index with the preselected altitude value on the radio altimeter indicator scale. When descending the aircraft below the preselected limit altitude, the LIMIT ALTITUDE yellow lamp illuminates on the indicator, the LIMIT ALTITUDE signal is displayed on the EKRAN annunciator panel, and the "Limit altitude" voice information is delivered to the earphones.

The radio altimeter reads the altitude values with the errors not exceeding those mentioned above at the angles of roll of up to 15° ; with the increase in the angles of roll, the errors grow respectively, and when the angle of roll reaches 30° and more, the readings of the radio altimeter should not be used.

It is not recommended to use the radio altimeter when flying the aircraft above the mountainous terrain when the sharp changes in the altitude may exceed the range of altitudes being measured.

IFF EQUIPMENT

The IFF equipment is used for identifying the nationality of the aerial targets (the friend or foe identification). The IFF equipment comprises the SRO15 aircraft radar IFF interrogator and the SRO2 aircraft radar IFF transponder.

The IFF interrogator is used for interrogating the aerial target in identifying its nationality. The interrogator irradiates the highquality coded signals, receives the coded reply signals, decodes them and, if the reply code corresponds to the acting one, delivers the identification mark to the integrated display system screen.

The IFF radar transponder receives and decodes the interrogation signals and, then codes and irradiates the reply signals corresponding to the received interrogation and effective table (program) of the codes.

The IFF transponder ensures:

- the "friendorfoe" identification of the aerial targets;
- the generation of the "Distress" emergency signal.

The presence of the identification mark alongside with the target blip of the detected object on the radar indicator indicates the friendly object.

The IFF interrogator and IFF transponder are fed with a voltage of 27 V DC from the DC generator and with a voltage of 115 V 400 Hz from the AC generator. If the DC or AC generator fails, the IFF interrogator supply is automatically cut off and the IFF transponder is connected to the storage batteries and to inverter PTO respectively.

The controls and serviceability test elements are arranged on the horizontal section of the right console in the cabin:

- the code selector switch for selecting the code according to the table of codes;
- the TRANSM TRIG CODE annunciator for indicating the code triggering;
- the RESPONSE lamp for checking the transponder serviceability when interrogating the target by the aircraft, ground or ship IFF interrogator;
- the DISTRESS switch for sending the information on the emergency situation.

The code is set with the aid of the code selector knob. The code is checked by reference to the number displayed in the port and by reference to illumination of the TRANSM TRIG CODE annunciator which will remain dead if the code has not been triggered.

If the emergency situation occurs aboard the aircraft, it is necessary to turn on the DISTRESS switch.

The transponder and interrogator supply is out in with the aid of the ASPT SYS and ACS selector switches.

The interrogation mode of the IFF interrogator is selected with the aid of the button located on the aircraft control stick.

In the SCAN mode, the identification result is delivered in the form of a mark over the target blip, spurious identification marks are likely to appear over the target and noise blips.

In the LOCKON mode, the identification result is displayed in the form of the "C" symbol.

ATC TRANSPONDER

The ATC transponder is used for operation together with the searching, air traffic control and groundcontrolled approach radars, types P35, PSP in the active mode.

The transponder ensures:

- the increase in the operating range of the ground radars in determining the aircraft coordinates;
- the automatic transmission of the additional information on the aircraft number (the pilot's call sign), flight altitude and fuel remainder in per cent to the ground radars;
- the discrimination of the target blips on the screens of the ground radars in the presence of the noise caused by the weather factors and clutters.

The operating principle of the ATC transponder consists in reception of the interrogation signals delivered from the ground radars and automatic delivery of the reply coded signals.

The LTC transponder has three modes of operation: the groundcontrolled approach (RSP) air traffic control (IVP) and P35 modes selected with the aid of the function selector switch.

In the groundcontrolled mode, the ATC transponder ensures the delivery of the coordinate mark of the aircraft to the screens of the air traffic control and groundcontrolled approach radars operating in the active radio location modes.

In the air traffic control mode, the ATC transponder ensures the delivery of the coordinate mark of the aircraft to the screens of the air traffic control and groundcontrolled approach radars and the automatic delivery of the additional information on the aircraft number (the pilot's index), flight altitude and fuel remainder in per cent in response to the interrogation of the air traffic control radars provided with the information display equipment, type HOMER-TM.

In the P35 mode, the ATC transponder ensures the delivery of the coordinate mark of the aircraft to the screens of the scanning radars, type P35, operating in the active mode.

For coding the aircraft number, use is made of the selection head in which one fivedigit code have been preselected, the change of the number transmitted by the transponder is effected by changing the head before the flight.

For transmitting the data on the flight altitude, the ATC transponder is integrated with the air data computer system which delivers the signals proportional to the pressure altitude to the ATC transponder.

For transmitting the data on the fuel remainder in per cent, the transponder is conjugated with the fuel quantity gauge and flowmeter system.

The ATC transponder is fed with a voltage of 27 V DC from the DC generator and with voltages of 115 V 400 Hz and 36 V 400 Hz from the AC generator.

If the DC generator fails, the ATC transponder is changed over to the supply from the storage batteries and in case of the failure of the AC generator drive, the ATC transponder is automatically cut off.

For controlling the ATC transponder, provision is made for the following control elements:

- the GCA, ATC, P35 function selector switch;
- the IDENT button (for delivering the individual identification signals to the ground radar in response to the command sent from the ground control post);
- the CO CHECK button and indicator lamp;
- the DISTRESS switch (common for the ATC transponder and IFF transponder).

The ATC transponder is cut in with the aid of the ASFT SYS switch. The ATC transponder is checked for serviceability on a lapse of two or three minutes after depressing the CO CHECK button; in so doing, the CO CHECK lamp burns continuously if said units are serviceable. When the aircraft is within the radar coverage zone, the lamp flashes and when the aircraft leaves the radar coverage zone, the lamp extinguishes.

COMMAND RADIO LINK EQUIPMENT

The equipment is used for receiving control and cooperation commands transmitted from the guidance post.

The command radio link equipment is cut in with the aid of the ASFT SYS switch.

The equipment is coupled with the integrated radar/sighting and integrated display systems. The target range, selected true airspeed, target altitude, selected course deviation commands as well as the discrete commands: reheat , rear hemisphere, front hemisphere, re-aiming, homing, vertical are displayed on the integrated display system annunciator panel. The time required for storing the previous commands, when the new commands are not applied, amounts to 30 s. The selected course commands are indicated on the combined course indicator.

The following commands are applied to the integrated radar/sighting system:

- the azimuth and elevation commands (for controlling the antenna position);
- the target range and closing speed (in case of the noise set up for the range measuring channel of the integrated radar/sighting system);
- the discrete commands: the target ranges front (rear) hemisphere attack, irradiation cutin commands (the commands are applied automatically in response to the commands delivered from the guidance post).

The following may be set manually or in the semiautomatic mode from the control panel: 20 numbers of the waves in mode IV; 3 numbers of the code; 8 numbers of the wave separation.

Notes: 1. The modes are automatically switched over depending on the selected number of the code (mode IV with the code numbers being 1 to 3).

2. The numbers of the waves are set versus the numbers of the wave separation (with the wave separation numbers being 1 to 4, the wave number corresponds to the preset one, with the wave separation numbers being 5 to 8. the wave number is increased by 20).

- the check for the proper setting of the radio data by reference to the display of the respective digits on the VOLKA, SHIFR, RAZNOS indicators after depressing the MANUAL button;
- the indication of serviceability of the command radio link equipment when the latter is checked by means of the builtin test system;
- the check of serviceability of the VOLNA, SHIFR and RAZNOS indicators.

When the MANUAL and AUTO buttons are depressed at a time, digits 88, 08. and 8 should be displayed on the VOLNA, SHIFR and RAZNOS indicators, respectively.

The equipment provides for the semiautomatic changeover to the now radio data transmitted from the guidance post. Said changeover is effected when the cooperation commands are applied (a continuous 1000Hz audio signal is applied to the earphones, the VOLNA, SHIFR, and RAZNOS indicators start flickering).

For changing over to the new radio data, it is necessary to depress the AUTO button; in so doing, the new values of wave, code and separation will be transmitted from the guidance post. When the commands are delivered from the new guidance post, the audio signal is not applied and the VOLNA, SHIFR and RAZNOS indicators flicker.

For changing over to the old radio data, the pilot should depress the MANUAL button.

The equipment is checked for serviceability with the aid of the builtin test system.

The built in test mode may be switched on both from the aircraft builtin test control system (by the EKRAN system) and from the equipment control panel. For cutting in the built in test system, it is necessary to depress the BIT button on the control panel. When the equipment functions normally, the VOLKA, SHIFR and RAZNOS indicators will start flickering on the control panel, a 1000Hz continuous signal is applied to the pilot's earphones.

When the BIT button is released, the VOLKA, SHIFR and RAZNOS indicators stop flickering and the audio signal is not applied to the pilot's earphones:

- the "R (reheat)" command to switch on the full reheat without delay. The disappearance of the R command on the integrated display system screen informs the pilot on cessation of the command transmission from the guidance post;

- the "!" command to inform the pilot on the forthcoming guidance of his aircraft at another target or the repeated guidance at the same target. The fadeout of the signal is indicative of the execution of the new guidance commands;
- the "Homing" command to inform the pilot on accomplishment of the indicated guidance at the target. When the "Homing" command is applied, the information on the target range is removed from the integrated display system screen (the range scale in not illuminated). When the "Homing" command is delivered from the guidance post, the landing airfield approach may be performed in response to the course, altitude and airspeed commands delivered by the equipment;
- the "Altitude" command to inform the pilot on the transition to the altitude required for attacking the target.

The equipment is fed with a voltage of 27 V DC from DC generator GSR and with a voltage of 115 V 400 Hz from generator SGK. If one of said generators fails, the equipment is automatically cut out.

RADAR ILLUMINATION WARNING STATION

The radar illumination warning station (SPO) is used for warning the pilot of the aircraft illumination by the radars of the antiaircraft missile or antiaircraft artillery system or by the airborne radars of the enemy fighters.

The radar illumination warning station ensures:

- the detection and direction finding of the ground and airborne radars illuminating the aircraft;
- the approximate determination of the range and speed of the closure with the illuminating radar, the moment of the aircraft approach to the effective zones of the antiaircraft missile system or the enemy fighter missiles for the most dangerous attacking means;
- the determination of the most dangerous attacking means;
- the light and audio warning.

The detection range of the operating radars (airborne radars) is at least 120 % of the range within which the weapon can be employed. The radar illumination warning station scanned area amounts to 360° in azimuth and +30° in elevation.

According to the operating principle, the radar illumination warning station is a passive fixedtuned amplitude direction finder with the developed processing of the received signals, which makes it possible, in compliance with the set program, to determine the type and degree of the threat presented by the illuminating means.

The antenna system forms the lobe fanshaped (in the azimuth plane) and funnelshaped (in the vertical plane) directional patterns.

The direction to the illuminating radar is determined by reference to the angular position of the sector in which the radar signals have the maximum power. The type of the radar is identified by comparing the parameters of the signal received with the parameters set in the programming device.

When the aircraft is illuminated by several radars, provision is made for the automatic selection and display of radar presenting the maximum threat to the aircraft, taking into account the radar operating mode. The "Nike Hercules" antiaircraft missile system radar is considered to be the most threatening one and is selected as the main one in all the cases since the illumination of the aircraft by the signals of this radar mainly occurs when the attacking missile approaches the aircraft, when the missile and aircraft prove to be within the main beam of this radar at a time. In absence of the illumination by the "Nike Hercules" antiaircraft missile system radar, the most threatening radar is considered that operating in the tracking mode. If the radar operating modes are similar, the selection is effected depending on the priority of the antiaircraft missile systems employing the illuminating radar. The data on the priority of the radar types are put into

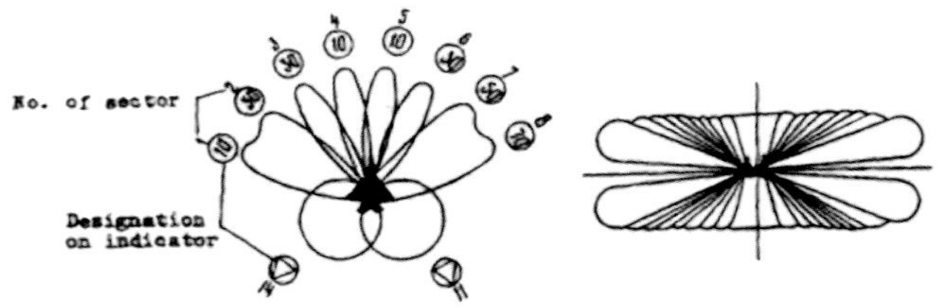

Directional patterns of antenna system of radar illumination warning station in horizontal and vertical planes

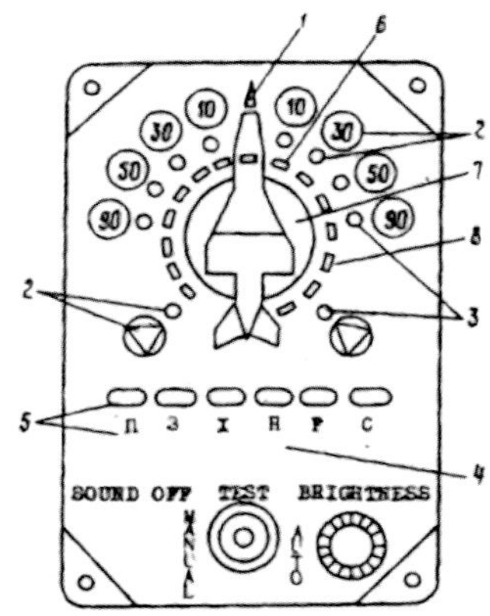

RADAR ILLUMINATION WARNING STATION INDICATOR
1 - serviceability signal; 2 - main target bearing; 3 - target bearing; 4 - type of target; 5 - type of main target; 6 - power graduation marks; 7 - lock-on mark; 8 - effective zone boundary

the program of the radar illumination warning station. The priority of the radar types decreases clockwise: P, Z, X, N, G, S.

All the information on the illuminating radars is presented by illumination of the various marks. The respective audio signal is applied to the pilot's earphones simultaneously with the light indication. The application of the marks is described below.

The "Serviceability signal" mark is illuminated when the radar illumination warning station is serviceable, just after switching on the station.

The azimuth position (bearing) of the illuminating radars is periodically presented by illumination of the bearing green marks at the scanning rate of this radar. If the aircraft is illuminated in the direction between the adjacent sectors, two adjacent bearing marks will be simultaneously illuminated for one radar.

The radar types are indicated by illumination of the radar type green marks designating:

- P (aircraft F4, F104 ship antiaircraft missile "Terrier");
- Z (sixbarrel selfpropelled antiaircraft gun mount "Vulcan" antiaircraft missile "Chaparral", ship antiaircraft missile "Seawolf");
- X (antiaircraft missile "Hawk", modified antiaircraft missile "Hawk");
- N (antiaircraft missiles "Nike Hercules", antiaircraft "Talos", "Patriot");
- F (aircraft F14, F15, F16, F18);
- S (aircraft F4, F104, F5, F111, "Mirage", "Lighting", "Jaguar", "Super Fledermaus", ship anti-aircraft missiles "Terrier" and "Talos").

When the aircraft is illuminated only by one radar or if this radar presents the maximum threat to the aircraft, the marks of the radar main type and main bearing appear above the radar type and bearing marks, with the angular values of this radar in degrees still displayed for 8 to 12 s after illumination, and, if the scanning rate of the most threatening illuminating radar antenna is less than 8 s, the information on this radar will be continuously displayed m this indicator.

When the aircraft is illuminated in the front hemisphere within the sector from 50 to 80°, index 50° appears from the left and right sides, indexes 50° and 90° appear at a time within the sector from 80 to 90°, and the rear sector index appears in the sector above

90° (up to 180°). The dynamics of closure with the radar, the radar operating mode (the number of the power gradations) and the effective zone relate only to the radar presenting the maximum threat.

The radar signal power is presented by alternate illumination of the power gradation yellow marks counterclockwise.

When the illuminating radar is changed over to the tracking mode, the LOCKON red mark will be displayed. When the "Nike Hercules" missile is launched at the aircraft, the LOCKON mark starts flickering at a frequency of 2 Hz/s.

CAUTION. 1. THE MODE OF OPERATION OF THE AIRBORNE QUAZICONTINUOUS ILLUMINATION RADARS OPERATING IN THE TRACKWHILEFLYBY MODE IS NOT DETERMINED.

WHEN THE AIRCRAFT IS ILLUMINATED BY THE CONTINUOUS AND QUAZICONTINUOUS ILLUMINATION RADARS (WEEK UP TO 5 GRADATIONS OF THE POWER ARE DISPLAYED), SECONDARY TYPE MARK 'X" FLICKERS. AT THE SMALLER RANGES (WHEN THE POWER EXCEEDS 5 GRADATIONS) WHEN THE AIRCRAFT IS ILLUMINATED BY THE HAWK ANTIAIRCRAFT MISSILE RADAR, MARK "X" IS CONTINUOUSLY DISPLAYED, AND WHEN THE AIRCRAFT IS ILLUMINATED BY THE AIRBORNE RADARS OF AIRCRAFT F14, F15, F16, AND F18, THIS MARK DISAPPEARS AND MARK TYPE "F" IS DISPLAYED THEREON.

WHEN THE AIRCRAFT IS ILLUMINATED BY THE AIRBORNE QUAZICONTINUOUS ILLUMINATION RADAR, THE NUMBER OF THE POWER GRADATIONS VARIES FROM THE FIFTH EIGHTH TO THE FIRST GRADATION FOR A SHORT TIME AT THE MOMENT MART TYPE "F" APPEARS, WITH THE SUBSEQUENT INCREASE IN THE POWER GRADATIONS.

The following audio signals are applied simultaneously with the application of the light indication:

- when the illuminating radar operates in the scanning mode, the intermittent low pitch signal (having the frequency similar to the radar antenna scanning rate);
- when the radar operates in the tracking mode, the continuous highpitch signal;
- when the "Nike Hercules" missile is launched, the audio signal variable in pitch and volume.

The radar illumination warning station is controlled from the control panel:

- the cutin of the radar illumination warning station is effected with the aid of the RIWS switch;
- the application and control of the audio warning are effected with the aid of the VOLUME knob. When the audio warning is switched off, the SOUND OFF placard is displayed on the indicator;
- the display of the information on the indicator in the full volume is effected by setting the SCAN OFF selector switch to the SCAN position. When the selector switch is placed in the OFF position, the indication on the aircraft illumination by the radars operating in the scanning and track while fly by modes is cancelled on the indicator.

The radar illumination warning station is checked for serviceability with the aid of the builtin test system, by setting the TEST springreturn switch to the AUTO position. At the moment the switch is depressed, the serviceability mark disappears, whereas the other marks are displayed. On a lapse of 5 to 7 s, the serviceability mark should again appear thereon; if said mark does not appear, it means that the radar illumination warning station is unserviceable.

After checking the station and placing the TEST switch to the neutral position, all the marks, except for the serviceability mark, should disappear.

Placing the TEST switch to the MANUAL position and then alternately depressing it, it is possible to check and evaluate the sensitivity of each sector separately by comparing the number of the power gradations displayed on the indicator with the number of gradations entered in the special table. The switch is met to the MANUAL position when the ground checks are performed.

AIRCRAFT ARMAMENT SYSTEM

The aircraft armament system ensures:

- the interception and destruction of the aerial targets with the employment of the missiles and guns;
- the destruction of the aerial targets in the close maneuvering combat with employment of the missiles and guns;
- the destruction of the ground (surface) targets when they are visible, with employment of the rockets, bombs and gun (the "airtoair" missiles provided with infrared homing head may be launched against the heatcontrast targets).

The armament system comprises:

- armament control system SIV29;
- the means for destroying the aerial and ground targets and their suspensions.

ARMAMENT CONTROL SYSTEM

The armament control system is used for selecting the sighting equipment and weapons the most efficient for the given situation. for controlling them when executing the combat employment and navigation missions.

The operating principle of the armament control system consists in determining the coordinates of the target with respect to the aircraft with the aid of the airborne sighting equipment included into the armament control system, and delivering the respective target designation to the missile homing head in accordance with these coordinates or the computing the lead angles for the unguided weapons and computing the most effective zones of the weapon employment, ensuring the maximum probability of the target hit.

The basic components of the armament control system are:

- the integrated radar/sighting system (RLPK);
- the integrated optoelectronic navigation/sighting system (OZPRNK).

Apart from the sighting systems, the armament control system also comprises the systems and devices ensuring the proper functioning of said systems:

- the pylon antennafeeder assembly ensuring the tuning of the radar homing head of missile R27R to the frequency employed by the integrated radar/sighting system;

• the data conversion units (DPE) converting the digital data into the analog ones in the lines coupling the armament control system with the missiles, type R60 ;
• the input/output device (IVV) ensuring the coupling of the airborne computer, integrated radar/sighting system and integrated optoelectronic navigation/sighting system with the analog transmitters and the data receivers, the multiplex and digital coding of the commands from the control panel;
• the coupling and distribution unit (BSR) ensuring the switching and conversion of the commands and signals in the lines coupling the integrated optoelectronic navigation/ sighting system with the other systems;
• the linear acceleration transmitters (BDLI), and rate gyros (BDIS) delivering the data to the computers.

Integrated Radar/Sighting System

The integrated radar/sighting system ensures the interception of the aerial targets in the day tine and at night, in the VFR and IFR conditions and when the jazzing is applied as well as the conduct of the close maneuvering combat with employment of the missiles and the aimed gun firing, and executes the following operations:

• the search, detection and identification (together with the IFF identification equipment) of the aerial targets;
• the tracking while flyby of up to ten targets and automatic lockon of the target presenting the maximum threat to the aircraft;
• the manual selection and lockon of one target;
• the tracking of one target with the accurate determination of the coordinates of this target;
• the delivery of the radio correction commands to missile R27R until the target is locked on by the radar homing head of the missile;
• the illumination of the target being tracked for operation of the radar homing head of missile R27R;
• the delivery of the target designation to the missiles and computation of the missile launch permissible zones.

The operating principle of the integrated radar/sighting system is based on the shaping of the powerful SHF pulses and irradiation of these pulses into the space, reception of the signals echoed from the targets, their processing, determination of the coordinates of the aerial targets and delivery of the measured coordinates to the airborne systems.

Basic Specifications of Integrated Radar/Sighting System

1. The altitude and airspeed of the aerial targets being attacked. The integrated radar/sighting system ensures the detection and autotracking of the targets flying at the altitudes from 30 50 to 23,000 m with a stepup vertical separation of the aircraft being up to 10,000 m and stepdown vertical separation being up to 6000 r on the headon and followon courses.

The detection of the target and the changeover to the autotracking depending on the attack hemisphere and the mode of operation of the integrated radar/sighting system are limited by the minimum component of the target airspeed to the fightertotarget line of sight, which amounts to:

• on the headon courses, in the ENCOUNTER mode 230 km/h, at the range of detection not exceeding 20 km (250 km/h at the edge of the scanned area) and 325 km/h when the target is detected at a range of 60 km;
• on the headon courses, in the PURSUIT mode 270 km/h;
• on the follow on courses in the PURSUIT mode 210 km/h;
• on the headon and followon courses, in the CLOSE COMBAT mode, at the ranges of at least 10 km, when attacking the target from below (the angle of the antenna tilt in elevation is more than 50) 60 km/h and when attacking the target from above 200 km/h. At the ranges exceeding 10 km, it is possible to lock on and track the targets only when the closure (lag) speed is at least 30 km/h.

The maximum airspeed of the targets lockedon for tracking amounts to:

• on the headon courses in the ENCOUNTER mode up to 2500 km/h;
• on the headon courses in the PURSUIT mode up to 1110 km/h;
• on the followon courses in the PURSUIT mode up to 2200 km/h;
• on the headon and follow on courses in the CLOSE COMBAT code up to 1200 km/h.

For the AUTO and TRACKWHILEFLYBY modes, the conditions for detecting the aerial targets are similar to those in the ENCOUNTER and PURSUIT modes.

2. The coverages of the integrated radar/sighting system, when attacking the aerial target, type MIG21 in the VFR conditions at the various altitudes of the fighter flight, with the equipment operating in the respective modes.

The ranges of detection and lockon of the target are mainly dependent on the fighter flight altitude and type of the underlying surface. The substantial reduction of these ranges is observed when decreasing the interceptor flight altitude to less than 3 km. For the PURSUIT mode, the coverage, when flying the aircraft above the floodplain, waterlogged forest or in the IFR conditions, is reduced by 10 to 15 %. When attacking the target against the ground, the detection and lookon ranges practically do not decrease with respect to the free space.

3. The angular characteristics of the searched areas versus the mode of operation of the integrated radar/sighting system.

The scanning cycle of the integrated radar/sighting system depends on the mode of operation of the system and amounts to 3 to 5 s.

The limit tracking angles are:

• +56 to 36° in vertical;
• +65° in horizontal.

4. The maximum aspect angles of the aerial target attack. The integrated radar/sighting system ensures the detection and lockon of the targets up to the aspect angles defined by the value of the target airspeed radial component whose values for the various modes are given above.

When the attack is launched in the rear or front hemisphere in the free space (3 km), the integrated radar/sighting system ensures the super aspect tracking of the targets maneuvering in the horizontal plane with the interception of the reciprocal course at the closing speed of at least 50 km/h.

When the target is attacked against the ground, the targets are tracked when they execute the maneuvers with the interception or the reciprocal course up to the course angles defined by the ratios:

• for the ENCOUNTER mode;
• for the PURSUIT and CLOSE COMBAT modes.

At the great aspect angles and at the fighter altitudes below 3 km, the system is changed over to the tracking of the signals echoed from the ground.

5. The resolving power in range and angular coordinates. The resolution of two targets in range and speed is not ensured if the targets are not resolved in angular coordinates. The integrated display system screen displays one blip or its division into 2 or 3 blips versus the target range.

The separate display of two target blips on the indicator is ensured when they are spaced through 5 to 6 km in range when the attack is launched in the front hemisphere and through 3 to 4 km when the attack is launched in the rear hemisphere, if said targets are resolved in angular coordinates.

The integrated radar/sighting system ensures the lockon and trackingof two targets (without their resolution on the indicator) flying at an interval of 500 m at the same range. From the range less than 8 km, the steady transition to the tracking of one target is ensured; in so doing, the angular resolution occurs at the angles of 3.5°.

6. Target autotracking transition time. The time from the moment of depression of the WSM LOCKON. PF button, provided the target blip is located within the gate, till the range computation and replacement of the scanning information by the aiming one, amounts to:

- 2 to 7 s for the ENCOUNTER mode;
- 1 to 4 s for the PURSUIT mode;
- 1.5 to 2 s for the CLOSE COMBAT mode.

However, the target lockon time is increased up to 710 s when gate moves in jumps or the aircraft executes the changes in roll and pitch while gating the blip and depressing the WSM, LOCKON PF button due to the wrong setting of the centre of the antenna additional search zone with respect to the target,

7. During the operation under the interaction interference conditions to ensure the simultaneous attack by a pair of aircraft it is recommended to set the maximum frequency spacing.

In the dual situation of the headon missile combat conducted between fighter MiG-29 and the fighter provided with the radar operating in the same frequency range (3 cc) in the scanning mode the interference is not observed. During the attack the fighter, whose radar has a greater operating range, possesses an advantage and this advantage allows this aircraft to be first in detecting and locking on the target. Owing to the precise direction of the illumination, the aircraft being tracked has the false range readings, resulting in the false lockson (the quick displacement of the present range mark towards the greater or smaller ranges).

However, when attempting to lock on by the escorting aircraft, the lockon by the attacking aircraft may abort.

The integrated radar/sighting system comprises the airborne radar (BRLS) and airborne digital computer.

The airborne radar is essentially a pulseDoppler system operating in the QUAZICONTINUOUS mode with employment of the channel time gating and frequency selection on the base of a large number of the Doppler filters .

The receiving device filter system discriminates the signals echoed from the target.

The reference frequency signal generated by the master oscillator, reinforced by the powerful output amplifiers operating in the pulse mode, is shaped into a main pulse and is applied to the antenna system presenting a twomirror antenna.

The antenna system shapes the narrow directional pattern 3.5° wide, by means of which the transmitter energy is irradiated into the space and moves therein according to the certain laws. The transmitter has the main and illumination channels

The transmitter main and illumination channels are combined in one band, but are spaced in the illumination frequency and time of operation. For ensuring the operation Of the equipment without the interaction interference when flying in the group formation, provision is made for changing over the indexed frequencies of the main and illumination channels on the ground.

The beam is shifted line by line with the aid of the antenna control channel according to the preset program, with the beam stabilized up to 120° in roll and up to +40 to 30° in pitch. There are two antenna control modes (scanned area control modes): manual and automatic.

In case of the antenna manual control in elevation the assigned stepup (stepdown) vertical separation of the target with respect to the fighter is set with the aid of the selector switch and in range, bythe gate from the control knob. The discrete correction turn of the antenna in azimuth through an angle of 120° from the aircraft axis is executed by depressing the selector switch located on special mode panel PSR. The scanned area attitude is checked by reference to the' displacement of the marks of the scanned area attitude in azimuth and elevation on the integrated display system screen.

The antenna automatic control (the automatic control of the scanned area) in azimuth and elevation is ensured with the aid of the ground automatic control system through the Command radio link when the GUID switch is turned on. In searching the target, the antenna scans the space by four lines (by six lines during the instrument guidance).

When in the SCANWHILEFLYBY mode, the radar executes the tracking of up to 10 targets, with determination of their angular coordinates, range and closing speed. During such a tracking, the target presenting the maximum threat to the aircraft is selected and automatically gated and locked before reaching the maximum launch range. In this case, the airborne digital computer generates the "Lockon of target being attacked" command in response to which the beam stops moving in the line, the beam is directed to the target area, beginning the additional search of the target within the "box pattern" sizing 8 x 40. When the target is detected, the ' Target signature" signal appears. In response to this signal the beam halts in this angular position and the radar begins operating in the continuous direction finding mode. When the "Target signature" signal is not applied, the beam continues its moving within the 'box pattern" for three cycles and if the target is not detected, the radar is changed over to the normal scanning for this time.

Apart from the automatic lookon of the target, the target may be locked on manually with the aid of control button KI located an the aircraft control stick. In this case, the centre of the additional search within the "box pattern" will be the centre of the gate superimposed on the target.

Depending on the mode of operation of the integrated radar/sighting system, the main pulses way have the high repetition frequency (HRF) for detecting the targets in the front hemisphere and the medium repetition frequency (HRF) for detecting the targets in the rear hemisphere.

Besides, in absence of the hemisphere signature, the radar employs the combined HRF-MRF mode.

The ERF mode makes it possible to obtain the reception zone on the headon courses in the frequency range free of the signals echoed from the ground. The use of the airborne radar with quazicontinuous illumination limits the possibility of launching the attacks at the aspect angles close to 4/4. The tracking of the target against the ground in the front and rear hemispheres is ensured up to the aspect angles dictated by the radial speed of the target. In separate cases, the tracking may be executed at an aspect angle of 4/4. The radar is controlled with the aid of the control elements of the .armament control system, located on special mode panel PSR, missile control panel PIR, control panel PI, on the throttle control lever and aircraft control stick.

The airborne digital computer executes the secondary processing of the radar signals, solves the airborne radar control problems, executes through the airborne digital computer of the integrated optoelectronic navigation sighting system the interaction of the sighting systems and coupling of the airborne radar with the aircraft systems, computes the armament control and employment problems. The sighting information generated by the airborne digital computer is delivered to the integrated display system.

Structurally, the integrated radar/sighting system is made in the form of a monoblock and separate units. The monoblock is arranged in the nose section of the aircraft, whereas the other units are located in the aftofcabin compartment and wing extensions.

The units are cooled off with air used for air conditioning of the pressurized cabin with the subsequent discharge of the air into the integrated radar/sighting system monoblock and for blowing off the radio equipment.

The pressurization system ensures the feed of the air to the antennawaveguide channel and to the pressure seal of the transmitter under the excessive pressure for precluding the breakdown of the RF energy. The airborne air bottle filled with the air compressed to 120 to 150 kgf/cm is the air source, the test pressure gauge is located from. below in the starboard wing extension.

The integrated radar/sighting system may be cut in on the ground without cooling the equipment for up to 10 minutes at the air temperature not exceeding +40° C.

The equipment is automatically cut off if the pressure drops, the cooling fluid gets overheated, and the pressurization is not available.

Modes of Operation of Integrated Radar/Sighting System

Depending on the position of the RDR MODES selector switch, the equipment has its own peculiarities in operation, which should be taken into consideration while intercepting the aerial target, proceeding from the tactical situation and the ground support of the fighter.

When scanning the space and searching the targets, the radar operates in the HRF or MRF.

The illumination modes are selected from the missile control panel with the aid of the RDR MODES selector switch versus the commands delivered from the ground. When launching the attack in the front hemisphere, the selector switch is set to the ENCOUNTER position and when launching the attack in the rear hemisphere, the selector switch is set to the PURSUIT position.

The ENCOUNTER mode ensures the detection, lockon and autotracking of the targets on the headon courses both in the free space and against the ground. The equipment functions in the 1W mode which ensures the adequate filtration of the noise signals from the lowmoving ground targets and radio contrast cloud cover. The range measuring method in the HRF mode does not ensure the detection of the target at the range less than 5 km. Apart from that, the range measuring error may reach 8 km, which will be displayed on the screen as the difference between the target range mark and the guidance navigator's information or in the form of the separate range mark jumps. The indicator screen is practically clear, the detection and gating of targets do not offer any difficulty, however, at the high closing airspeeds, it is necessary to ensure the displacement of the gate centre with the lead in range with respect to the target blip. In thin case, it is necessary to perform, the lockon when the target blip gets into the gate during the subsequent scanning cycle, which reduces the time to some extent for transition to the autotracking.

The PURSUIT code ensures the detection, lockon and autotracking of the targets on the pursuit courses, as well as against the ground with the target flight airspeed and closing speed limitations indicated below. The equipment operates in the MRF mode. When the equipment functions in this mode, the targets are detected even at the short ranges by changing the repetition frequencies, however, the target detection is hampered because of the presence of the false clutters at the ranges. equal to the fighter flight altitude.

When this mode is selected, the integrated display system screen displays the separate false marks at the various ranges and azimuths hampering the detection of the target blip without the ground target designation. The number of the false marks grows when flying the aircraft above the marshland, marshy forest or floodplain. In separate cases, the error of up to 8 km in measuring the target range is observed. When changing over to the autotracking mode with such an error, the target lockon abortion may occur at the small ranges. In this case, it is necessary to compare the computed range with that transmitted by the navigator, reset the abortive lockon and lock on the target once more.

When the equipment functions in the PURSUIT mode, the targets are detected and attacked on the headon courses when they are flying at the airspeeds from 270 to 1100 km/h.

The AUTO mode is employed when the target flight course is unknown. This mode ensures the simultaneous detection of the targets flying both on the headon and followon courses and the transition to the autotracking of the gated target. The equipment functions during the automatic changeover of the antenna high and medium repetition frequencies line by line.

If, during the given scanning cycle, the illumination mode linebyline sequence will occur in the MRF HRF MRF HRF succession, then, during the next scanning cycle, the illumination nodes will be presented in the MRF HRF MRF HRF succession. During the antenna additional search, the signals are followed up in one and half cycles of the box pattern at the high repetition frequency and in one cycle, at the medium repetition frequency. In this mode, the coverages are similar to those in the ENCOUNTER and PURSUIT modes, i.e., when the target flies on the headon courses, the target detection occurs at the ranges similar to those in the ENCOUNTER mode. From the range loss than 30 km, the target blip in observed in each scanning cycle due to the measurements and at the medium repetition frequency (at the range less than 10 km, only at the medium repetition frequency). After locking on the target, the target steady tracking in ensured on the headon and followon courses.

In the AUTO mode, when the target is within the scanning zone, it in possible to detect the target even after the target has reversed its flight direction, excepting only for the came when the target blip disappears on reaching the blind aspect angles (from 3/4 to 4/4, depending on the target airspeed).

The TRACK-WHILE-FLYBY mode in employed for the concealed accomplishment of the attack, when the fightertotarget approach in executed, with the equipment operating in the scanning node, and the most dangerous target in automatically looked an at the ranges close to those permissible for launching the missiles in attacking the target from the front and rear hemispheres. The TRACK-WHILE-FLYBY mode is selected only when the integrated radar/sighting system is used in the ENCOUNTER and PURSUIT modes (in the AUTO mode, the equipment is inoperative). The mode ensures:

- the tracking up to 10 targets detected in the scanning mode and selection of the most dangerous target (a group of targets). When the attack in launched in the front hemisphere, the dangerous target is determined by the maximum ratio of the closing speed to the target range. When the target is attacked in the rear hemisphere, the criterion is unsteady. In the most cases, the closest target is gated;
- the automatic control of the centre of the zone scanned in elevation and of the target displayed on the screen of the integrated/sighting system with respect to the coordinates of the dangerous target or a group of targets in compliance with the resolving power of the integrated radar/sighting system in range and angular coordinates. During the consecutive detection of the targets, the first target blip is taken for the automatic gating after displaying the blip in two or three scanning cycles, without the subsequent transition to another target being more dangerous, with the air situation varying;
- the automatic transition to the autotracking of the gated target at the ranges exceeding the maximum missile launching range by 05 to 5 km versus the closing speed.

The TRACK-WHILE-FLY-BY mode assists the pilot in executing the tracking operations with the aid of the target range gate in response to the navigator's command and in looking on the target especially at the high closing speeds an the attack headon courses. Besides, after setting the equipment to the automatic gating, it is possible to attack the target performing altitude maneuver at the vertical speed of up to 50 m/s. However, in case of appearance of the false marks and disappearance of the target blip, observed in more than three scanning cycles are the gate abortions calling for execution of the subsequent manual gating and lockon of the target.

In all cases, for getting the target into the scanned zone, the respective stepup (stepdown) vertical separation should be set with the aid of the selector switch and the target range with the use of the control button according to the information delivered by the navigator.

The CLOSE COMBAT mode is used for attacking the visually visible targets in the front and rear hemispheres, including the maneuvering one, with the employment of the close combat missile and gun. In said mode, the target is locked on when the vertical search zone is superimposed on the closest visible target and the WSM,LOCKON PF button is depressed. While launching the attack, the target is steadily tracked at the equalized airspeeds as well as in case of the delay in delivering the target designation to the homing head of missiles R27R and R60.

The CLOSE COMBAT mode has the following characteristics:

• the twoline scan in case of the vertical beam scanning;
• the azimuth scanning 6°;
• the elevation scanning from 13° to 4.5°;
• the scanning cycle time 2.5 s;
• the maximum lockon range 10 km;
• the minimum lockon range 450 m;
• the minimum tracking range 250 m.

The COOPERATION mode is selected with the aid of the COOP RETARD NO RETARD selector switch and is employed at the medium repetition frequencies irrespective of the RDR MODES selector switch position.

Integrated Optoelectronic Navigation/Sighting System

The integrated optoelectronic navigation/sighting system is used for solving the navigational problems, controlling the armament in the maneuvering aerial combat autonomously, or together with the integrated radar/sighting system, for conducting the aimed fire and bombing an the ground targets.

The integrated optoelectronic navigation/sighting system ensures the following operations at all combat altitudes, including the operation carried out against the ground, in the day time and at night, in the optical visibility conditions, in the case the jamming are set up:

• the search, detection and lockon of the aerial targets in the rear hemisphere autonomously or in response to the target designation signals delivered from the integrated radar/sighting system as well as the target range measurement with the aid of the laser range finder;
• the computation of the possible launching zones, the autonomous delivery of the target designation and other command required for aiming and launching missiles R60 and R27R;
• the aimed gun fire in the rear hemisphere at the nonmaneuvering and actively maneuvering aerial targets with the use of the information delivered from the optical locator laser station as well as the fire against the visually visible targets when no information is delivered from the optical locator laser station;
• the measurement of the distance to the ground surface in the direction defined by the aiming angle;
• the delivery of the sighting angles, range value and the "Target lockon" command with respect to the angles to the integrated radar/sighting system;
• the manual delivery of the smooth target designation to the missile homing head and optical locator laser station within the field of view of the windshield indicator;
• the aimed gun and rocket firing against the ground targets in the visually visible conditions;
• the precision level, dive, divetoss bombing and toss bombing against the ground targets in the conditions of the target visual visibility;
• the determination of the navigation and flight parameters required for controlling the aircraft;
• the shaping of the control signals for the automatic flight control system at the stage of the navigation, intermediate approach and landing approach;
• the shaping and display. of the radar and infrared radar situations, signals and commands required for controlling the aircraft and armament on the integrated display system screen;
• the photo recording of the visually visible targets and information displayed on the windshield indicator in all nodes of operation of the armament control system.

The range of the aerial target lockon by the optical locator laser station (0 3 m with respect to the angular coordinates in the rear hemisphere at an aspect angle of 3/4, with the engine running at the maximum power, the fighter and target flying at the altitude above 5 km, in the VFR conditions and in the free space, with the probability being 0.9, is at least 12 km.

The minimum sighting angle of the optical locator laser station towards the sun, when tracking the aerial target, is 3 to 15°.

The aerial target range measured by the optical locator laser station is 200 m to 6.5 km.

The maximum range depends on the target aspect angle, meteorological visibility, the characteristics of the transceiving channel of the laser range finder of the optical locator laser station.

The maximum range to the ground target, measured by the laser range finder of the optical locator laser station at the dive angles of 15 to 25° is 5 to 8 km.

The maximum error in determining the target range by the optical locator laser station is not more than 10 m. The zone for delivering target designation to missiles from the infrared homing head, optical locator laser station and integrated radar/sighting system with the aid of helmet-mounted target designation system:

• +600
• +60 to 140

The zone for delivering target designation to missiles from the infrared homing head, optical locator laser station and integrated radar/sighting system by use of optical method, with the aid of control button 17 (manual target designation) in azimuth and elevation in +12°.

Opto-electronic Sighting System

The opto-electronic sighting system comprises:

• the optical locator laser station (KOLS);
• the helmetmounted target designation system (NSTs).

The optoelectronic sighting system functions in the following modes:

• the scanning;
• the guidance and lockon;
• the target designation of the missile homing head and integrated radar/sighting system with the aid of the optical locator laser station or helmetmounted target designation system;
• the asynchronous aiming against the aerial targets during the operation of the optoelectronic sighting system together with the optical locator laser station;
• the measurement of the range to the aerial and ground targets with the aid of the laser range finder;
• the aimed firing and bombing against the ground targets.

The information on aerial targets is delivered to the optical locator laser station.

The optical locator laser station measures the angular position of the sighting line at the target with respect to the aircraft datum line, the absolute angular rates of the sighting line and the aerial and ground target ranges at the moment of sounding. The other parameters required for conducting the fire (the linear accelerations of the sighting line, the present target range and its derivatives) are not measured by the optical locator laser station, but computed by the aircraft digital computer of the integrated optoelectronic navigation/sighting system.

The optical locator laser station consists of the scanning and tracking infrared radar and laser range finder. The infrared radar is used for searching, detecting and locking on the nonmaneuvering and maneuvering aerial targets with the subsequent tracking of said targets in response to their heat irradiation. The laser range finder measures the aerial target range when the target is automatically tracked by the infrared radar or the range to the ground surface when attacking the ground target in the direction preset by the aiming angle.

All the components of the infrared radar and laser range finder are installed rigidly in the station casing, and the direction of the optical axis is changed with the aid of the mirror mounted in the gimbal which is controlled by means of two independent drives.

The optical locator laser station may execute only the successive operation in each of the scanning and lockon modes and, then may accomplish the tracking of the infrared target simultaneously with the range measurement.

In the scanning mode, the optical locator laser ensures the scanning of the zone which is not stabilized in the space and amounts to +15° in elevation and to 30° in azimuth the great zone scanned by the optical locator laser station; apart from that there are two small scan zones.

When the optoelectronic sighting system functions in the scanning mode, the initial position is the great field scan zone.

The manual selection of the search zones is effected with the ZONE selector switch with indexes on special modes panel PSR, which correspond to the shift of the small zone centre in azimuth to positions: 15°, 0°, +15°. When the selector switch is set to the neutral position the great zone is selected.

For operation of the station against the complicated background, provision is made for adjusting the sensitivity of the photo receiver. When turning the IRR AMPL knob counterclockwise on the special modes panel, flickering symbol PP appears on the windshield indicator field and on a lapse of 5 s, this symbol will illuminate continuously, whereas the number of the targets and amount of the noise will reduce on the screen of the windshield indicator versus their irradiation power.

The guidance and lockon mode is a transient one employed by the optical locator laser station for ensuring the main (tracking) mode provided for solving the aiming problems.

The guidance to the target and its lockon may be accomplished by:

- the manual target designation when superimposing the gate on the target blip with the aid of the control button the IRR (GATE) mode;
- flying the aircraft with the visually visible target kept within the lockon zone. the CLOSE COMBAT (IRR) mode;
- the target designation from the integrated radar/sighting system and helmetmounted target designation system;
- the manual target designation in aligning the aiming mark with the visually visible target with the aid of the control button the OPTIC mode.

When the IRR mode is selected, the transition to the lockon begins from the moment the WSM, LOCKON PF button located on the throttle control lever is depressed; in this case, the scan ceases and the antenna mirror is set in the shortest way to the position corresponding to the angular coordinates of the gate centre where the search is executed within the gate (4 x 6°) with the subsequent lockon of the target for being automatically tracked with respect to the angular coordinates. During the target lockon time (from the moment the WSM,LOCKON PF button is depressed to the moment the scanning image is changed by the lockon one on the integrated display system screen), the information on the target attitude is not updated on the integrated display system screen and corresponds to the moment of the lockon command delivery. The target is not looked on, if the target moves respectively in azimuth and elevation, which leads to the fact that the target leaves the additional IRR zone during the target lockon.

When the target designation is delivered to the optical locator laser station from the integrated radar/sighting system or from the helmetmounted target designation system, the target is locked on similarly to the target lockon in the IRR mode; in this case, the centre of the additional search zone during this time corresponds to the present target designation angle.

When the CLOSE COMBAT (IRR) mode is selected, the optical locator laser station looks on the target (when the WSM LOCKON PF button is depressed) for automatic tracking with respect to the angles of one of the targets available in the zone of +2° in azimuth and +16 to 14° in elevation.

After locking on the target and its tracking, the integrated optoelectronic navigation/sighting system generates the target designation to the missile infrared homing head. The purpose of the target designation is to align the axes of the missile infrared homing heads parallel to the sighting line of the aerial target for ensuring the target lockon by the infrared homing head. The target designation signals should also ensure the transmission of the target sighting angles from the optical locator laser station (or from the helmetmounted target designation system) to the radar and vice versa.

After locking on the target for its autotracking with respect to the angular coordinates, the laser range finder irradiates the pulses at a repetition frequency of 0.25 Hz for measuring the target range (the standby mode). The target range is measured (which is indicated by appearance of the return pulses) from 6.5 to 3 km. The maximum coverage of the laser range finder independent an the target aspect angle, meteorological visibility and transceiving channel characteristics.

From the range of 1500 m, the laser range finder is automatically changed over to the main mode of operation (the pulse repetition frequency is 2 Hz). The changeover to the main mode of operation of the laser range finder is dictated by the necessity to increase the accuracy of the present target range measurement during the aimed employment of the gun and missiles in the close maneuvering aerial combat.

When attacking the ground targets, the laser range finder irradiates in the main mode. The direction of the laser range finder irradiation is set by the aiming angles computed by the aircraft digital computer of the integrated optoelectronic navigation/sighting system.

The laser range finder irradiation is applied automatically in attacking the ground targets from the dive (in response to the command generated by the aircraft digital computer) when reaching the dive angle exceeding 10° and the target range less than 3500 m (for bombing) and maximum permissible range +500 m (for firing the gun and rockets). When the ground targets are attacked from the level flight, the laser range finder irradiation is applied by depressing control button KI. The command to apply the irradiation in attacking the ground target is accompanied by illumination of symbol A (attack) on the integrated display system screen. The illumination of this symbol is not the reliable symptom of the laser range finder triggering. The laser range finder irradiation is switched off automatically on a lapse of 30 s or by depressing the INTERR button located on the aircraft control stick.

The time of the laser range finder operation in the main mode is not more than 3 minutes per flight, and in the standby mode 12 minutes. On a lapse of 3 minutes of the laser range finder operation in the main mode, the laser range finder is automatically changed over to the standby mode of irradiation irrespective of the aiming conditions and mode of operation of the armament control system.

The helmetmounted target designation system is used for determining the sighting line coordinates of the visually visible target tracked by turning the pilot's head in the conditions of the close maneuvering combat. These coordinates are delivered by the helmet mounted target designation system to the aircraft digital computer for computing the target designation signals and delivering the latter to the missile infrared homing head and optical locator laser station, and when the COOPERATION mode is selected (the COOP RETARD NO RETARD selector switch is placed in the RETARD position), said coordinates are also delivered to the integrated radar/sighting system.

The helmet mounted target designation system comprises the helmetmounted sighting device, scanning units A and B and electronic unit.

The helmetmounted sighting device mounts three reference points (the lightemitting diodes of infrared spectrum), collimating device and reflector on which the aiming and signal marks are displayed. The brightness of the signal and aiming marks is controlled with the aid of the IRR AMPL HEL BRIGHT knob and with the use of the brightness automatic control system.

The irradiation of the reference points (the lightemitting diodes) is taken up by the scanning units (located on the sighting head of the windshield indicator) scanning the cabin space. The aircraft digital computer determines the angular position of the target sighting line by reference to the known distances between the scanning units and reference points on the crash helmet and the measured bearings of the reference points.

The helmet-mounted target designation system serviceability is ensured when the helmetmounted sighting device is located within the space the sections in the horizontal and vertical planes.

If one of the light-emitting diodes leaves the given zone, this results in generation of the signal switching off the information delivered from the helmetmounted sighting device.

The helmet-mounted sighting device is secured on the crash helmet with the aid of the latches. The construction of the helmetmounted sighting device ensures the easy removal from and installation of the device on the crash helmets of both sizes as well as the employment of said device with any set of the helmetmounted target designation system.

However, the employment of the helmetmounted sighting device with any other set of the helmetmounted target designation system is not recommended since this may lead to the increased errors in measuring the target designation angles. The helmetmounted sighting device is kept in the special container. The aircraft technician is responsible for storage of this device.

The reflector is adjusted in the horizontal and vertical planes at the level of the pilot's eyes with the aid of the respective knobs (handwheels) located on the helmet-mounted sighting device, the reflector is also adjusted in the vertical plane with the aid of the gaskets available in the helmet set. For ensuring the reliable observation of the aiming and signal marks in the gload conditions, the crash helmet should be thoroughly fit to the pilot.

Weapon Control System

The weapon control system is used for controlling the preparatory operations and employment of all types of the weapons carried by the aircraft, in all the variants of the weapon combat employment.

The weapon control system ensures the following modes of operation of the integrated opto-electronic navigation/sighting system when the aircraft weapons are employed:

- the semiautomatic mode of operation, when the weapons and the weapon employment variant are selected by the pilot, whereas the preparation of the weapons for employment and the sequence of its employment are ensured by the weapon control system in response to the commands generated by the aircraft digital computer;
- the manual mode of operation with simplified preparation and employment of the weapons without use of the information generated by the aircraft digital computer, accomplished by the pilot manually with the aid of the controls located in the cabin;
- the emergency mode of operation realized through the emergency launch and release independent circuits, ensuring the emergency launch of the missiles and emergency release of the aerial bombs and rocket pods.

The commands ensuring the assigned mode of employment of the weapons are delivered from the controls to the weapon control system. The aircraft digital computer delivers the "Preparation", "Inrange", "Inadmissible aiming error" commands to the weapon control system and in the automatic bombing mode, the "Automatic release" command.

The weapon control system delivers the signals on the weapons selected for employment, missile infrared homing head readiness, present consumption of the gun ammunition load, the LP signal, as well as the signals on serviceability of the units of the weapon control system and the system as a whole to the aircraft digital computer and integrated display system. In the combat modes, the integrated display system ensures the display of the presence and type of the selected weapons, readiness of the missile infrared homing head, LP signal, the availability of the builtin gun ammunition load on the windshield and direct vision indicators located in the cabin.

When the power is applied to the aircraft electrical system and when the weapon control system is energized, the latter functions in the builtin test mode for 30 s from the moment it is cut in, during which the system is checked for serviceability, the system serviceability signals are shaped and delivered thereto. On completion of the testcheck, the weapon control system is charged over to the operating mode.

The safety of the ground work with the weapons in ensured by interlocking the respective launch, release and firing circuits, by turning off the MASTER ON.

Camera Gun FKP

The camera gun is located in the cabin. The camera gun ensures the simultaneous and combined photorecording of the external space (the target) and parameters displayed on the windshield indicator.

The camera gun comprises the camera with feed magazine and electronic unit. The camera has two optical channels:

- the external channel for photographing the target and the outside world;
- the internal channel for photographing the parameters displayed on the windshield indicator.

The camera gun functions in the automatic mode. The camera gun is controlled in response to the signals delivered from the aircraft digital computer of the integrated optoelectronic navigation/sighting system in compliance with the logic ensuring the record of the information for each mode of operation of the weapon control system. When one of the firing buttons is depressed, the pictures are taken on the internal and external channels at a frequency of 9 frames/second.

The depression of one of the firing triggers or WSM LOCKON PF buttons is recorded on the photographic film in the form of the spoiled strip on the right side in the perforated area. The length of film is 20 m (780 frames).

The builtin test system of the integrated optoelectronic navigation/sighting system is used for evaluating the technical condition of the integrated optoelectronic navigation/sighting system during the ground preparation and automatic continuous monitoring of the subsystems and transmitters in flight with the delivery of the respective information on the failures to the EKRAN system.

AIRCRAFT ARMAMENT SYSTEM CONTROL AND TEST ELEMENTS

The weapon control system equipment control, the test and indication of the modes of operation of the weapon control system are effected with the aid of the test panels and controls located in the cabin.

The special modes panel mounts:

- the WCS MODES wafer selector switch having the following positions:

• the PITCHUP position, for selecting the toss bombing mode;
• the NAVIG position, for switching off the combat modes of operation of the weapon control system and switching on the navigation information circuit of the integrated display system;
• the RDR position, for selecting the mode of operation of the integrated radar/sighting system;
• the IRR position, for selecting the mode of operation of the optical locator laser station, with the manual gating of the target blip during the target lockon;
• the CLOSE COMBAT position, for selecting the mode of operation of the optical locator laser station in the close combat (with the accomplishment of the lockon of the visually visible target without preliminary gating);
• the HELMET position, for selecting the mode of operation ensuring the target lockon by the radar, optical locator laser station or by the missile infrared homing head when the target designation is delivered from the helmetmounted target designation system;
• the OPTIC position, for selecting the mode of operation of the integrated optoelectronic navigation/sighting system f or the visually visible aerial and ground targets with the employment of the missiles (the manual target designation with the aid of control button KI), rockets, aerial bombs and gun;
• the F, position, for selecting the missile launch mode, with the infrared homing head set along the weapon axis, and for ensuring the target lookon by the missile infrared homing head by executing the respective maneuvers of the aircraft;
• the SPAN potentiometer, with the numbering in meters and indexes M, C, B (small, medium, large targets), f or setting the target size in measuring the range with use of the external base method and for introducing the target size character into the weapon control system and missiles R27R for computing the maximum permissible launch range and adjusting the fuzes of these missiles. When, the potentiometer is not to position "N", the computation in made for the zone of permissible launch of the missiles against the cruise missiles, types ILA17, KPM1; in position "C", for the missiles launched against the aircraft, types F15, F16; and in position "B", for the missiles launched against the aircraft, types F111, SR71, B1;
• the IRR AMPL HEL BRIGHT potentiometer, for controlling the sensitivity of the receiver of the optical locator laser station in the scanning mode and f or controlling the brightness of the aiming and signal marks of the helmetmounted sighting device in the HELMET mode;
• the ZONE selector switch, for controlling the position of the zone seamed by the optical locator laser station and integrated radar/sighting system in azimuth;
• the PREP: MANUAL AUTO selector switch for the manual delivery Of the commands for preparation of the missiles for launching and for removing the permissible launch ran e interlock in the modes of operation in which this range is not computed in the aircraft digital computer of the weapon control system, and for ensuring the release of the aerial bombs when the firing button is depressed;
• the MASTER OFF switch for connecting the RKT MSL B and GUN firing buttons to the fire control circuits.

The missile control panel is used for selecting the modes of operation of the integrated radar/sighting system. The panel mounts:

• the RDR MODES selector switch for selecting the modes of operation of the radar;
• the ILLUM DUMMY OFF selector switch f or cutting in the radar in the DUMMY or ILLUMINATION mode of operation;
• the COMP selector switch for switching on the compensation channel of the radar manually;
• the TWF FHS RHS selector switch for selecting the TRACKWHILE-FLYBY mode in the FHS position. When the present range signal is not applied in the HELMET, OPTIC and FLIR modes, leave the FHS RHS selector switch in the position corresponding to the attack hemisphere;
• the AJ OFF CAJ selector switch for changing over the modes of operation of the integrated radar/sighting system in the conditions of the active jammings. When the selector switch is set to the AJ position, the active jamming. is displayed on the screen of the integrated display system. The CAJ position is not used.
• control panel PI is used for selecting the modes of the gun fire and bombing. The control panel mounts: the coop: RETARD No RETARD selector switch f or delivering the character of employment of the aerial bombs with retarding devices to the aircraft digital computer of the integrated optoelectronic navigation/sighting system and for selecting the main information channels of the integrated radar/sighting system and integrated optoelectronic navigation/sighting system in the RETARD position. When the selector switch in set to the NO RETARD position, the character of employment of the aerial bombs without retarding devices is delivered to the aircraft digital computer.
• the AIR GROUND selector switch for selecting the modes of employment of the gun and missiles furnished with infrared homing head against the aerial and ground targets. In this case, when the selector switch is placed in the GROUND position, the proximity fuzes of the missiles are not disconnected.

Note. In case of the employmentof the rockets and aerial bombs, the selection of the mode of operation of the weapon control system against the ground targets in ensured by the presence of the respective weapon character at the external store stations irrespective of the position of the AIR GROUND selector switch.

• GUID switch for switching on the mode of the instrument guidance from the ground automatic control system (the automatic application of the irradiation and control of the zone seamed by the integrated radar/sighting system);
• the RESET button for delivering the commands to reset the target lockon by the radar or optical locator laser station,
• the ARMED SAFE selector switch for selecting the emergency mode of bomb release;
• the EMERG RELEASE capped button for the emergency release of the aerial bombs, rocket pods and rockets S24B.

The windshield Indicator mounts the following controls:

• the RETICLE DAY NIGHT wafer selector switch for displaying the collimating reticle on the windshield indicator in the RETICLE position; when placed in the DAY or NIGHT position, the selector switch enables the display of the sighting and navigation information of the respective colour (through the red filter at night). If the windshield indicator is supplied with the puton light filter, to ensure the camera equipment operation in the night conditions, place the RETICLE DAY-NIGHT wafer selector switch in the DAY position;
• the BRIGHT knob for controlling the display brightness on the windshield indicator and the collimating reticle brightness;
• the TEST button for cutting in the testcheck mode of the integrated display system;
• the light shield actuating lever.

Mounted on the top of the instrument board to the left side is the STBY LAUNCH button for executing the standby (unguided) launch of the missiles. Located at the same place, but at the bottom, is the VOLUME LOCKON potentiometer for controlling the volume of the armament audio warning system.

The right throttle control lever mounts:

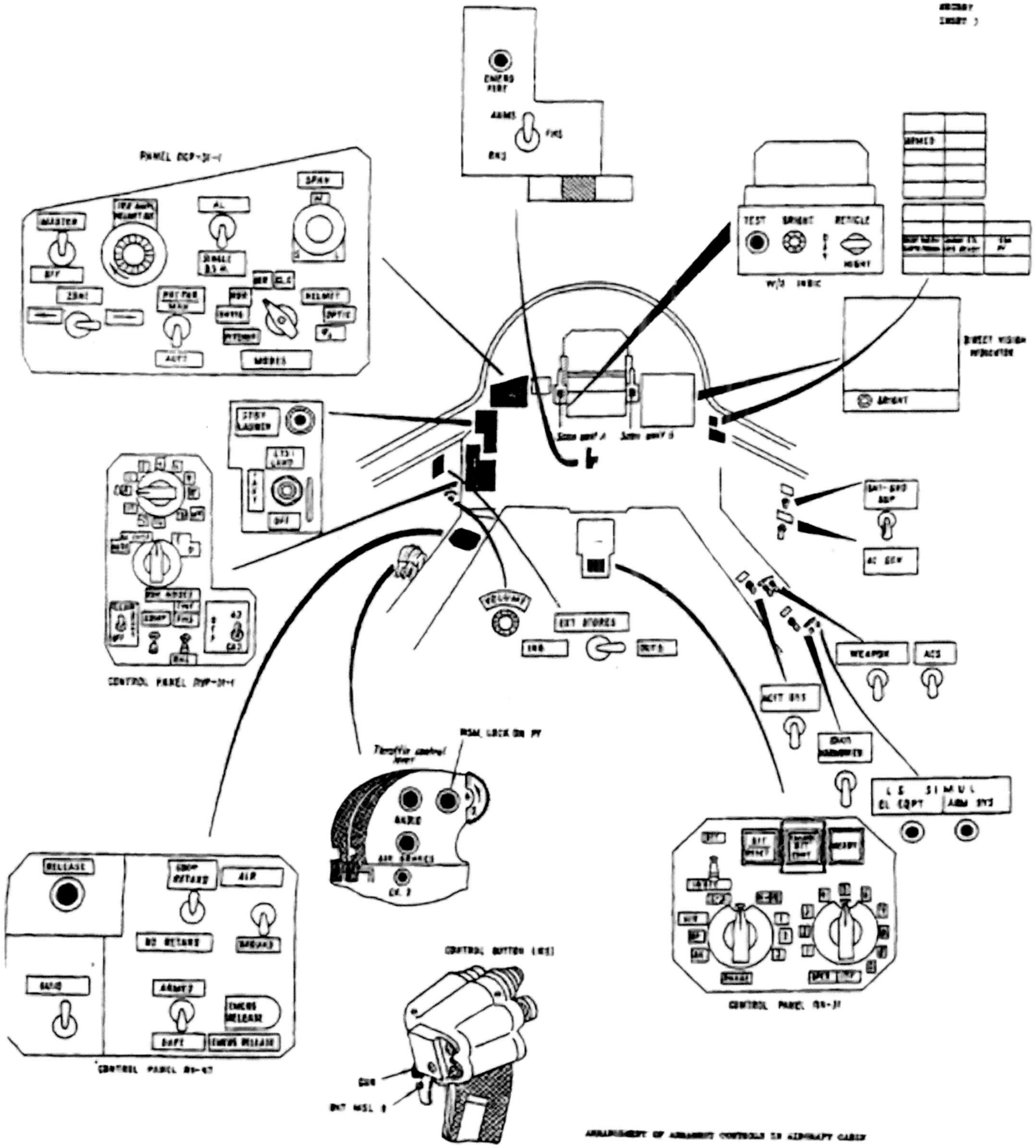

ARRANGEMENT OF ARMAMENT CONTROLS IN AIRCRAFT CABIN

• the WSK, LOCKON PF button for delivering the permission for looking on the target by the radar and optical locator laser station, delivery of the target designation command to the missile infrared homing head an well as the execution of the preliminary fix when attacking the ground target;
• the chaffs dispensing button.
• The aircraft control stick mounts:
• the RKT MSL B trigger for launching the missiles, rockets and releasing the aerial bombs;
• the GUN trigger for delivering the command to select the aiming Mode when firing the gun (with the trigger preliminarily shifted to the intermediate position) and for delivering the command to open the fire (with the trigger fully depressed);
• the control button ensuring:
• the gate control of the optical locator laser station or radar when the aerial target locked on for automatic tracking;
• the control of the zone scanned by the radar in elevation depending on the target range when guiding the aircraft by voice;
• the changeover of the aiming information to the windshield indicator and direct vision indicator in the GUN mode, after locking on the target for automatic tracking from the visually visible target display to the visually invisible target display by depressing the respective button;
• the manual trigger of the laser range finder for irradiation when attacking the ground target, the delivery of the manual target designation (in the OPTIC mode) to the missile infrared homing head, radar and optical locator laser station when attacking the visually visible aerial target;
• the INTERR button for cutting in the target nationality interrogation circuit.

The right horizontal console mounts:

• the WCS and WEAPON switches for applying the power to the weapon control system and armament circuits, respectively;
• the LG SIMUL, ARM SYS button for canceling the fire control circuitstoextended leading gear position interlock;
• the FKP HARMONIZN switch for harmonizing the camera gun.

The right console mounts the following annunciators:

• the ARMED annunciator to indicate selection of the aerial bomb "Armed"
• Emergency release mode;
• the GROUND annunciator to indicate setting of the AIR GROUND selector switch to the GROUND position;
• the AIR annunciator to indicate setting of the AIR GROUND selector switch to the AIR position;
• the RDR ILL annunciator to indicate application of the RF signal from the radar transmitter when the ILLUM. DUMMMY OFF selector switch is set to the ILLUM or DUMMY position;
• the AB RETARD annunciator to indicate setting of the COOP RETARD NO RETARD selector switch to the COOP RETARD position.

Located in the lower section of the instrument board, under the cover in the test panel which is used for performing the ground test of the weapon control system with the aid of the builtin test system, for detecting the troubles and also for introducing the ballistic code of the aerial bomb by the ground crew before executing the bombing mission.

ACS MODES OF OPERATION

ACS Operation Against Aerial Targets

The armament control system (ACS) ensures detection and lookon or an aerial target by various methods depending on target flying conditions. Information about the target on the integrated display system screen in the scanning mode until the lookon corresponds to the selected mode of operation.

After the target lockon in the continuous tracking mode by the radar or optical laser station regardless of the mode of operation selected by the ACS MODES selector switch, the information on the integrated display system screen corresponds to modes of missile launch or gun firing.

The armament control system operates in the following modes: RDR, IRR, CLOSE COMBAT (IRR), HELMET, FLIR, OPTIC, GUN.

RDR Mode

The RDR mode is the basic mode of operation of the ACS against aerial targets as it ensures the greatest ran e for target detection and automatic tracking at all altitudes of the aircraft combat use and makes it possible to accomplish an attack against the target with every type of guided missiles in the front and rear hemispheres and aimed gun firing in the rear hemisphere of the target.

To engage RDR mode, set the modes selector switch on the special modes panel to the RDR position. When interacting with the ground automatic control system, also turn on the GUID switch on the special modes panel. Set the RDR modes selector switch on the missiles control panel to position E (encounter) (B) or P (pursuit) (A) in response to a command from the ground automatic control system.

Arrival of commands from the ground automatic control system over the command radio link causes the radar antenna search zone and its attitude in space to chan e, the operating range also varies with the flight altitude of the fighter and the attack hemisphere of the target. The magnitude of stepdown (stepup) vertical separation relative to the target is selected by the pilot manually using the selector switch an missiles control panel.

In the process of closing with the target at a probable detection range the radar irradiation is activated automatically in response to a command from the ground automatic control system or manually on the missiles control panel. After detection, identification and selection of the menacing target, the lockon gate an the windshield indicator in manually aimed at the target and the target is. automatically or manually looked on for automatic target tracking (continuous direction finding mode).

A changeover to the continuous direction finding mode may take place manually by depressing the WSM, LOCKON PF button. Here. if the pilot disagrees with the selection of the menacing target as performed by the radar airborne digital computer, he may perform gating of the target selected by him using the control button on the aircraft control stick.

A changeover to the continuous direction finding mode my be accomplished manually, without preliminary gating, with the RDR MODES selector switch set to CLOSE COMBAT position. This mode is employed in close air combat under target visual visibility conditions.

To lock on the target by an aircraft maneuver, superimpose the lockon zone an the windshield indicator against the visually visible target and depress the WSK, LOCKON PF button on the throttle control lever, which results in nearest target lockon and changeover to the continuous direction finding mode.

IRR(GATE) Mode

When the ACS MODES selector switch is placed in the IRR position, the ACS is operating for search (scanning), lockon and automatic tracking of the aerial target in response to the angular coordinates with the aid of the optical locator laser station.

Choice of scanned area (transition to smaller zone) is accomplished by depressing button to the right or button to the left ensuring its indication on the integrated display system screen.

Target discrimination under interference conditions on the integrated display system screen is performed by adjusting the sensitivity of infrared radar scanner receiver with the aid of IRR AMPL, BRIGHT HELM knob on the special modes panel.

Locking on the target detected in the scanning mode is accomplished by superimposing the gate on the target blip using control knob KI on the aircraft control stick. As soon as the gate centre is brought into alignment with the target blip, the WSM,LOCKON PF button is depressed and kept. The optical locator laser station finalizes the search and locks on the target. In the process of finalizing the search and locking on the target it is essential that no change in the aircraft attitude (in yaw, roll and pitch) occurs to avoid the target going beyond the limits of the gate. When the target has been locked on for automatic tracking, the scan indication on the integrated display system screen is replaced with the sighting indication, then the WSM,LOCKON PF button is released.

If no target lockon occurs within 3 seconds, target blips on the integrated display system screen disappear. Then it is necessary to release the WSM, LOCKON PF button, and target blips will appear again on the integrated display system screen.

Adjust the position of the gate on the target and repeat the lockon.

In the lockon mode, the aircraft digital computer generates target designation angles for the infrared homing heads of missiles to ensure a target lockon and a set of commands required for the missile launch.

If infrared homing heads of the missiles have locked on the target, and the target is within the permitted launch range, the LP command is shaped. If no present target range is available, flashing LP command is generated, in which case the missile launch is possible with visual determination of the target range and manual output of the PREPARATION command.

Aiming is accomplished by aligning the sighting circle with the fixed electronic crosshairs.

In the course of closing with the target and leaving the permitted launch zone, the BKWY command appears on the integrated display system screen (LP command is cancelled). The preferable direction for breaking off attack is shown by the movement of the circle on the integrated display system screen.

If reaiming is necessary, a reset of the target locked on by the optical locator laser station is accomplished by depressing the RESET button an the PI47 control panel.

CLOSE COMBAT (IRR) Mode

In the CLOSE COMBAT (IRR) mode, the optical locator laser station is scanning a narrow zone of +2° in azimuth and +16 to 14° in elevation (a zone of +2° in azimuth and +12° in elevation is indicated on the windshield indicator). As a visually visible target enters this zone and the WSM, LOCKON PF button in depressed, an automatic target lockon by the optical locator laser station occurs without preliminary gating. Operation of the optical locator laser station after lockon with target tracking and missile launch is similar to the IRR mode.

HELMET Mode

In the HELMET mode, visual detection and sighting of the target are accomplished by means of the helmetmounted sighting device (HMSD). The aircraft digital computer of the integrated optoelectronic navigation/sighting system performs recalculation of the primary coordinates coming from the helmetmounted target designation system (HMTDS) and delivery of target designation to the optical locator laser station, radar and infrared homing heads of missiles. With the aiming mark of the HNSD collimator aligned with the target, depress the WSM,LOCKON PF button, which results in delivery of the "Lockon permitted" command to the infrared homing heads of missiles, optical locator laser station and radar at the same time. Loop the WSM, LOCKON PF button depressed and the aiming mark on target until the lockon moment.

After bringing the HMSD collimator into the field of vision of the pilot's eye (by depressing a key on the collimating night), an aiming mark (AM) is shown on the collimator as two concentric circles. When the target is locked on by the infrared radar or radar, the aiming mark on the collimator starts flickering at a rate of 2 Hz. After the target is locked on by the infrared homing heads of missiles, an audio signal is applied.

Upon arrival of the LP command, a crosshairs signal mark (SM) appears against the background of two circles the aiming mark. In the event of an inadmissible aiming error, the aiming and signal marks start flickering in opposite phase. If the helmetmounted sighting device leaves the zone illuminated by the scanning unit, the aiming and signal marks go out.

After the target is locked on by the infrared homing heads of missiles, if the target is not locked on by the optical locator laser station, keep the WSM,LOCKON PF button precisely on target until the missile lifts off.

To reaim the infrared homing heads of missiles, superimpose aiming mark on another target and depress the WSM,LOCKON PF button. Here, the infrared homing heads of the missiles again follow up target designations supplied by the helmetmounted target designation system.

The LP command is shaped when the target is looked on by the infrared homing heads.

With the target locked on by the optical locator laser station, after the WSM, LOCKON PF button has been released, all the required commands and signals to the missile infrared homing heads are supplied by the optical locator laser station. If prior to target lockon by the optical locator laser station a LP command has been supplied in response to target lockon by the missile infrared homing heads, the LP command is not cancelled after the target in looked on by the optical locator laser station or radar.

With the target locked on by the optical locator laser station, reaiming is accomplished by aligning the aiming mark with the now target and depressing the WSM,LOCKOK PF button, whereupon the infrared homing heads and the optical locator laser station again receive target designations from the helmetmounted target designation system. The aircraft digital computer compares the target sighting angles fed by the optical locator laser station and the helmetsounted target designation system. If these angles disagree, the optical locator laser station lockon is reset and the station is reaimed.

OPTIC Mode

The OPTIC mode is employed in case of failures of the main modes of operation of the radar, optical laser locator station or helmetmounted target designation system against a visually visible target. The windshield indicator shows an aiming mark which should be superimposed, by manipulating the joystick on the aircraft control stick, on the visually visible target and then the WSM, LOCK-ON PF button should be depressed. This results in a target lockon by the missile infrared homing heads and by the optical locator laser station.

Operation of the armament control system, similarly to the HELMET mode, may be accompanied by target designation to the infrared homing heads for the angles determined by the position of the sighting mark (the WSM LOCK-ON PF button is depressed) or for the angles determined by the optical locator laser station after the lookon and release of the WSM LOCK-ON PF button.

GUN Mode

The engagement of 'the GUN mode against aerial targets is achieved by placing the AIR GROUND selector switch on control panel in the AIR position and by throwing the GUN trigger to the intermediate position. In this case, aiming is achieved by:

- the asynchronous method with automatic tracking of the target by the optical locator laser station or radar to a distance of less then 1200 m;
- the "predicted lead track" method if target range measurements are not available;
- the visual estimation of firing corrections making use of the fixed collimating reticle.

On the basis of information about the target arriving from the optical locator laser station or the radar (target sighting angles in azimuth and elevation, components of absolute angular speed of the sighting line and range), and from the fighter flight conditions transmitters (altitude, speed, roll, pitch, angles of attack and sideslip, components of aircraft angular rates), the aircraft digital computer and the integrated optoelectronic navigation/sighting system compute the selected aiming correction (moving crosshairs with the present range scale) and the sighting mark (circle), as well as other commands required for aiming. The present range scale is the portion of the circle round the moving crosshairs. As the present range decreases, the circle length decreases and its readings in hours over the clock dial correspond to the range in hundreds of meters. A full circle corresponds to a range of 1200 m.

Aiming is accomplished by aligning the moving crosshairs with the sighting mark by piloting the aircraft, and as the effective firing range in reached, fire is opened by depressing the GUN trigger.

When launching an attack against the visually invisible target or when during attack the followup angles of the moving crosshairs exceed the maximum indication angle of the windshield indicator, aiming by asynchronous method is employed, using nullindication of aiming data on the integrated display system screen.

Aiming is achieved by aligning the coarse aiming circle with the fixed crosshairs in the centre of the integrated display system screen with subsequent alignment of the fine aiming point with the centre of the fixed crosshairs.

The problem of aiming by the "predicted lead track" method is solved by the aircraft digital computer of the optoelectronic navigation/sighting system on the basis of the data arriving from fighter flight conditions transmitters (altitude, speed, roll, pitch, angles of attack and sideslip, components of angular rates and linear accelerations of the aircraft) and the value for the span of the target to be attacked entered from the special modes panel. Here, for every point of time the values of aiming corrections are calculated for the whole effective firing range, which are indicated an the integrated display screen as a track. The aiming is achieved by inscribing the target wing into the inner outlines of the track through aircraft piloting.

Integration of Integrated Radar/Sighting System and Integrated Optoelectronic Navigation/Sighting System of Armament Control System in Combat Use

For better accomplishment of combat missions, concealed attack against an aerial target better immunity against interference, provision is made in the ACS for interaction between the integrated radar/sighting system and the integrated optoelectronic navigation/ sighting System.

For this purpose, the armament control system performs automatic selection in ACS of the main information channel with initial manual engagement of one of the information channels by the ACS MODES selector switch on the special modes panel.

In the interaction mode, the engagement of the integrated radar/sighting system for irradiation into space is provided automatically when placing the ILLUM WPM OFF selector switch in the DUMMY position. Changing of information channels is indicated by the illumination of symbols RL or TP (IRR).

The main channel performs shaping of the integrated display system screen indication, delivery of target designation to the missile homing heads, solves the problems of aiming and fighter control at the stage of guidance.

When setting the ACS MODES selector switch to the RDR position, an aerial target is detected, locked on and automatically tracked on the headon and followon courses similarly to the PURSUIT mode regardless of the RDR MODES selector switch positions, as the radar in interaction modes is operating on medium pulse repetition frequencies only.

After a target lockon for automatic tracking, the integrated radar/sighting system ensures delivery of target designation angles and generation of the "Lockon permitted" command to the optical locator laser station. After the target lockon by the optical locator laser station, the integrated radar/ sighting system is operating in the support mode, ready to assume the functions of the main channel it the radar fails or the target tracking is reset by the radar. In this case, the optical locator laser station will be the main channel until information flow from the integrated radar/sighting system is restored.

When laser range readings appear, the target is locked on by the integrated radar/ sighting system, which from now on operates in the support mode ready to assume the functions of the main channel in the event of an abortive infrared radar lockon or a failure of the aircraft digital computer of the integrated optoelectronic navigation/sighting system.

In the IRR mode, it is possible to effect a forced selection of the integrated radar/sighting system as the main channel by a repeated depression of the WSM,LOCKON PF button only before the laser range readings appear.

When the ACS MODES selector switch is placed in the CLOSE COMBAT (IRR) position, the ACS main channel is the optical locator laser station. In response to its target designations, the integrated radar/sighting system performs a target lockon for automatic tracking. After the target lockon, the integrated radar/sighting system operates in the support mode ready to assume the functions of the main channel if laser range readings disappear, if target tracking by the infrared radar is abortive, or if the aircraft digital computer of the integrated optoelectronic navigation/sighting system fails.

When the ACS MODES selector switch is placed in the HELMET or OPTIC positions, the lockon and changeover to the target tracking are accomplished almost simultaneously by the integrated radar/sighting system and the optical locator laser station in response to the target designations from the helmetmounted target designation system or control button KI following a depression of the WSM,LOCKON PF button. After the WSM,LOCKON PF button is released, the information channel which is the first to accomplish the lockon becomes the main channel and provides range computations. The second information channel after the target lockon operates in the support mode.

In all ACS modes of operation and with the target locked on by one of the channels, a changeover to the GUN mode (shifting the GUN trigger to the intermediate position) with laser range readings available always makes the opticallocator laser station the main channel, with laser range readings unavailable it is the integrated radar/sighting system.

Friend-or-foe identification of the target in the IRR, CLOSE COMBAT (IRR), HELMET and OPTIC modes is not provided.

ACS Operation Against Ground Targets

The modes for aimed firing and precision bombing on ground targets are shaped in the armament control system with the ACS MODES selector switch placed in the OPTIC or PITCHUp position and when the "Aerial bombs available" or "Rockets available" and "Rocket type" commands are applied from the weapons control system (WCS), or in case of gun firing with commands "Ground" (the AIR GROUND selector switch is in the GROUND position) and "Gun" (the GUN trigger is thrown aside) applied in addition to those.

Computation of aiming data is accomplished by the aircraft digital computer of the integrated optoelectronic navigation/sighting system on the basis of data about the slant range to the aiming point and data about the aircraft flight parameters (altitude, speed, roll, pitch, heading, angles of attack and sideslip, components of aircraft angular rates).

As in the case with an attack against an aerial target, if the integrated display system sighting complex or the aircraft digital computer fail, firing and bombing on ground targets may be accomplished with the aid of the collimating reticle. In this case, select the mode on the missiles control panel, place the DAY NIGHT RETICLE selector switch on the windshield indicator in the RETICLE position and depress MAN PREPAR button to engage the bombs drop manual mode. Angular corrections and range are determined by comparing the visible target dimensions and the anglemeasuring lines on the collimating reticle.

Modes for Gun and Rocket Firing Against Ground Targets

For solving the problems of aiming at ground targets when firing the gun and rockets, two methods of aiming are realized in the system: an asynchronous one and a preliminary f ix (PF) method.

The asynchronous aiming method ensures calculation of the aiming correction (displayed on the integrated display system screen as the aiming mark) without allowance made for the wind speed and target movement.

When using the preliminary fix method, the aiming problem is solved with allowance made for the total wind and target speed for the time interval of flight (drop) of ammunition to a target. This method is realized for firing at ground targets and dive or dive toss bombing.

After a depression of the WSM,LOCKON PF button the aiming mark moves over to the position of fixation and the line of bursts disappears. Maneuver the aircraft to align the aiming mark with the target and release the WSM, LOCKON PF button. From now on calculation of an aiming correction starts with allowance for the total wind and target speed. After a second alignment of the aiming mark withthe target the firing button is depressed. If within 10 s following the preliminary fix the firing button is not depressed, the preliminary fix aiming method is automatically replaced by the asynchronous aiming method.

When firing the gun, the aiming mark on depression of the button travels to the zero position.

When launching an attack (after releasing the WSM,LOCKON PF button), it is possible to achieve reaiming (refixing) by a repeated depression of the WSM, LOCKON PF button, fixation is achieved by subsequent repeated alignment and depression of the firing button. The target range and the effective firing range are indicated on the windshield indicator scale. If the aircraft is within the effective firing range, the PR command appears and an audio signal is supplied to the pilot's headphones. When the minimum firing range is reached, the OTB. command appears on the windshield indicator.

Precision Bombing Modes

The integrated optoelectronic navigation/sighting system ensures precision level bombing, dive bombing, divetoss bombing with the ACS MODES selector switch set to the OPTIC position, and toss bombing with this selector switch not to the PITCHUP position.

If in the OPTIC mode under the attack conditions the aiming angle calculated in the integrated optoelectronic navigation/sighting system does not exceed 14°, the aiming mark shows the bomb impact point and the release of aerial bombs occurs at the moment when the firing button is depressed. If this condition is not fulfilled (bombing in an invisible zone), the aiming mark follows up the aiming angle of 14° (relative to the longitudinal axis of the aircraft) and after it is aligned with the target and the firing button in depressed, a change of indication takes place on the windshield indicator: the selected gload circle and present gload vector appear, the range scale is replaced by the time scale showing the time remaining before aerial bombs are dropped. From here on it is necessary (until the aerial bombs are dropped) to keep the present gload vector end aligned with the selected gload circle centre, this will ensure stabilization of heading and flight altitude. Two or three seconds prior to dropping an audio signal is supplied to the headphones. The bombs are released automatically with the firing button (a) depressed.

When the minimum permissible diving altitude according to safe conditions of flying over the splinter scattering area is reached, the OTB. command is indicated on the windshield indicator. This command is also delivered after the time remaining till bomb release has elapsed with the RKT MSL B trigger not depressed.

Note. After depression of the MAN PREPAR lightbutton, bombs are released on depression of the firing button regardless of flight conditions and aiming.

In dive bombing and in divetoss bombing the aiming may be achieved by the preliminary fix method. The operation of the armament control system in this case is similar to launching rockets against a ground target. The position of the aiming mark with the WSM, LOCKON PF button depressed represents 1/2 of the current correction value for the correction that has been earlier calculated by the asynchronous method for the given bombing conditions.

In toss bombing, after the aiming mark has been aligned with the target and the firing button has been depressed, a change of indication on the windshield indicator takes place too.

With the current gload vector end aligned with the selected gload circle centre, the gload will remain within 4g.

Bombs drop occurs at pitchingup angles of 110 to 130° automatically with the firing button depressed. Two or three seconds before drop an audio signal is supplied to the headphones.

If at the moment of the aerial bombs drop command passage the RKT MSL B trigger is not depressed, the OTB. command is indicated on the windshield indicator. On depression of the trigger, after the drop command has passed, the bombs will be dropped without aiming.

NAVIGATION SYSTEM

The navigation system is part of the integrated optoelectronic navigation/sighting system and is intended for continuous automatic determination and output to users (to integrated radar/sighting system, integrated optoelectronic navigation/sighting system computer, AFCS, transponder and to indicators) of flightnavigation parameters required for flying the aircraft during combat operations, anroute flight, approach to the landing airfield, execution of intermediate approach and landing approach.

The navigation system is a set of interconnected onboard flightnavigation equipment and incorporates three subsystems:

- the aircraft radio and navigation equipment (ARNE);
- the information directional/vertical gyro system (IDVGS);
- the air data computer system (ADCS).

The navigation system in conjunction with azimuthrange radio beacons and landing radio beacon group (LRBG) of the groundbased radio system ensures generation and output into the automatic flight control and integrated display systems and flightnavigation instruments of signals employed in piloting the aircraft during:

- enroute flights according to the programmed three waypoints and three airfields;
- return to the departure airfield or one of two alternate landing airfields with entering the landing radio beacon coverage area at a distance of the navigation beacon of up to 80 km from the leading airfield. The distance between the beacon and the landing airfield is limited by the steady reception of its signals at an altitude of the intermediate approach;
- landing approach down to an altitude of 50 x both on programmed and unprogrammed airfields (in cooperation with the automatic flight control system during automatic or flight director control);
- repeated landing approach on programmed airfields.

INFORMATION DIRECTIONAL/VERTICAL GYRO SYSTEM EK-VK

The EKVK system is the flight navigation data transmitter and is designed for measuring and continuous transmission to the combined course indicator (PNP), flight director indicator (KPP), windshield indicator (ELS) and other systems of values for the angles of roll, pitch, yaw, as well as of the components for the absolute linear speed along the axes of gyroplatforms used in computation of coordinates.

The EK-VK system employs two inertial directional/vertical gyro systems (EKV). Apart from these, the EK-VK system includes: control and coupling unit (BIS), monitoring unit (BK), latitude updating panel (PShK), magnetic declination setter (ZMS) and fluxgate transmitter (ED).

The inertial direct ional/vertical gyro systems are the central gyro transmitters of angles of roll, pitch and gyro heading. One of the systems is the main one, the other is the standby one.

The operating principle of the inertial directional/vertical gyro system consists in measuring the angular position of an aircraft in relation to a gyrostabilized platform held in the horizontal plane, in response to integral correction signals. The integral correction is produced by integrating the horizontal components of absolute aircraft acceleration as measured by accelerometers arranged on the gyroplatform.

The control and coupling unit provides precise digital integral correction of the main inertial directional/vertical gyro system gyroplatform, conversion of analogue angle parameters into a code and their output to users, Shaping of the gyros magnetic or corrected course, output of components for the absolute aircraft linear speed used in computation of position coordinates.

Corrected course is the course measured in relation to the meridian of the system preparation point.

The monitoring unit ensures checking the inertial directional/vertical gyro system for serviceability.

The PShK7 latitude updating panel is designed for shaping the latitude correction for apparent drift of the gyro from the daily rotation of the earth.

The magnetic declination setter is intended for input of magnetic declination during the inertial directional/vertical gyro system alignment.

The fluxgate transmitter is used for shaping the gyromagnetic heading during initial course setting in the magnetic correction (MK) mode.

The gyrostabilized platform provides the basis for readout of angles of roll and pitch which are measured by transmitters installed on the axes of suspension.

The EKVK system may perform the following operating modes:

- operating mode with integral correction by digital integrators (for the main inertial directional/vertical gyro system operating in conjunction with the control and coupling unit) and with integral correction by analogue integrators (for the standby inertial directional/vertical gyro system operating without the control and coupling unit) after normal preparation;
- operating mode with integral correction by analogue integrators (for both inertial directional/vertical gyro systems) after fast preparation;
- operating mode with radial correction during failure of digital integrators (for the main inertial directional/vertical gyro system) and after accomplishing a restart (for both inertial directional/vertical gyro systems).

The operating mode with a radial correction is an emergency mode. Here, the gyroplatform is aligned in response to the vector equal to the sum of gravity acceleration and acceleration vectors caused by a change of value and direction in the aircraft speed, i.e., in response to the apparent vertical. Therefore, the minimum values of error of the EK-VK system in determining the angles of roll and pitch will occur only during the uniform straight flight.

Depending on the correction type, the heading channel may operate in the following modes:

- directional gyro (DG);
- magnetic correction (MC);
- correction from the external oncourse information source (only in the PREPARATION mode.

The directional gyro mode is the main mode for the heading channel. In this mode the corrected course is shaped in the control and coupling unit which receives the gyro course from the main or standby directional/vertical gyro system, the initial ramp heading and correction for the vertical component of the earth rotation angular speed from the latitude updating panel. The local position latitude is selected on the PShK7 Panel prior to departure. Thus, in the directional gyro mode the users are fed with the great circle course with respect to the geographical (or magnetic, if the selected ramp heading is equal to the magnetic heading) meridian in the location of departure.

The magnetic correction mode in flight is accomplished as required in the straight-inlevel flight by depressing the M HDG SLAV button. Here, if on the magnetic declination setter value, the indicator will display the magnetic heading.

Checking the EK-VK system for serviceability is achieved by the automatic builtin test system.

The EK-VK system is energized by turning on the BAT GND SUP AC GEN NAVIGATION switches. Heating circuits are energized as soon as the AC generator is turned on. If the EK-VK system is operated on the ground for a long period of time, forced air blowing should be turned on.

Prior to flight, both inertial directional/vertical gyro systems are brought into the operating position, but signals of roll, pitch and yaw are supplied to users only by one of them depending on the position of EKV MAINSTBY selector switch on the navigation system control panel.

AIRCRAFT RADIO AND NAVIGATION EQUIPMENT

The aircraft radio and navigation equipment is designed for solving the problems of navigation during plannedroute flight, approach to the landing airfield area, intermediate approach and landing approach.

The aircraft radio and navigation equipment provides:

- shaping and output to the indicators of distance, azimuth, true heading, desired heading signals, and in the return and intermediate approach modes additionally an assigned altitude deviation signal;
- shaping and output to the automatic flight control system and indicators of signals ensuring a landing approach both on programmed and unprogrammed landing airfields in response to the landing radio beacon group signals;

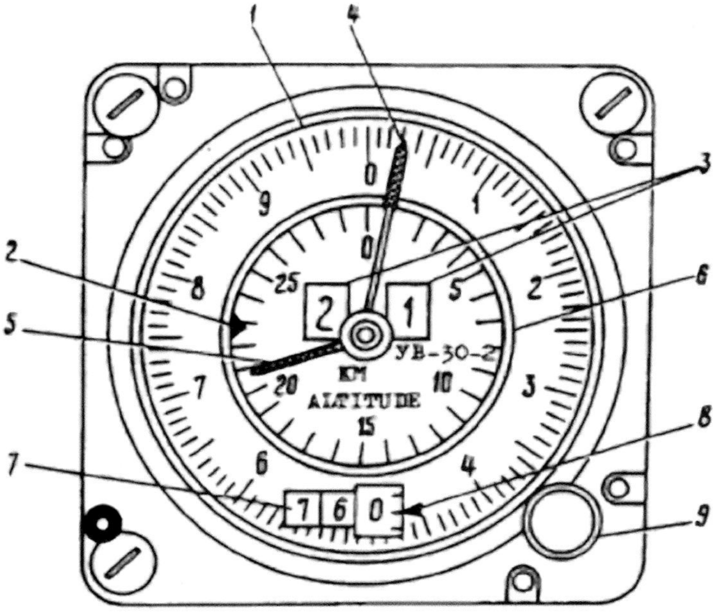

ALTITUDE INDICATOR

1 - altitude scale in meters (outer); 2 - altitude command index; 3 - altitude counter; 4 - larger pointer; 5 - smaller pointer; 6 - altitude scale in kilometers (inner); 7 - P_0 counter; 8 - tenth fraction readout index in mm Hg; 9 - P_0 input setting knob

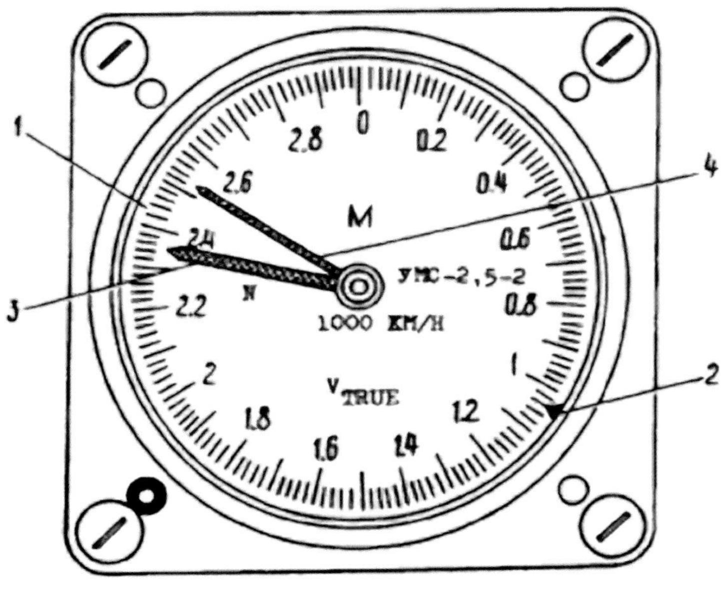

V_{TRUE} AND MACH NUMBER INDICATOR

1 - scale; 2 - speed command index; 3 - Mach number readout pointer (broad); 4 - true airspeed readout pointer (narrow)

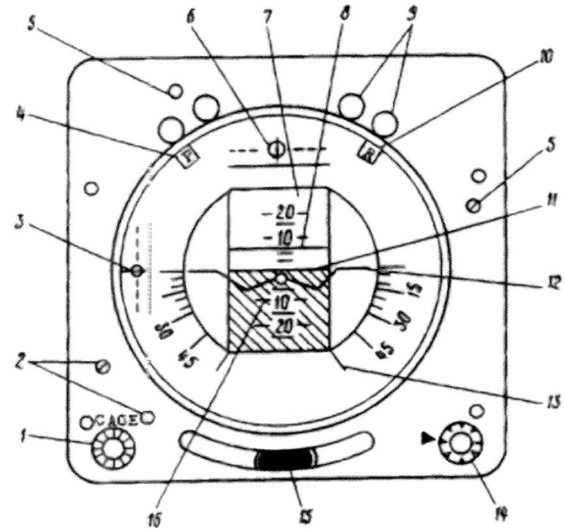

FLIGHT DIRECTOR INDICATOR

1 - CAGE (АРРЕТИР) light-button; 2, 5 - correction of deviation and command pointers; 3 - vertical deviation bar; 4 - pitch flag; 6 - course deviation bar; 7 - pitch scale; 8, 16 - command pointers-bars; 9 - illumination lamp; 10 - LOC flag; 11 - central index; 12 - present roll indicator (miniature airplane); 13 - roll scale; 14 - horizon zero setting knob; 15 - slip indicator

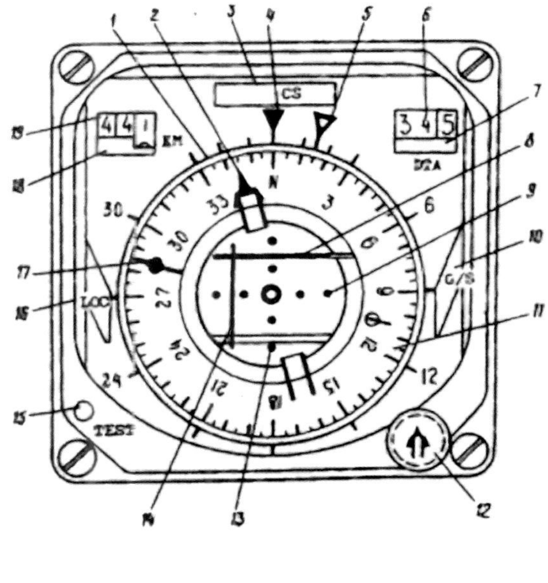

COMBINED COURSE INDICATOR

1 - combined drift angle and radio station relative bearing scale; 2 - desired track angle pointer; 3 - flag to display failure of navigation system computer or one of directional/vertical gyros; 4 - present course counter index; 5 - drift angle index (vacant); 6 - desired track angle counter; 7 - desired track angle counter flag curtain; 8 - glide slope equisignal zone deviation bar; 9 - course equisignal zone deviation scale; 10 - G/S flag; 11 - present heading scale; 12 - desired track angle setting knob; 13 - desired track or course equisignal zone deviation scale; 14 - course equisignal zone deviation bar; 15 - test-check button; 16 - LOC flag; 17 - present azimuth (tail end) and course angle (head end) pointer; 18 range flag curtain; 19 - range counter

- input and ground check of two program versions (for computers TsVI and TsVIM (2204) used for one version); each version includes the coordinates of three waypoints, three airfields, three radio beacons in the conventionalgeographical coordinate system, FCC numbers of three navigation and three landing radio beacons as well as the runway headings of the programmed airfields;
- operational selection of the waypoint (airfield) number from the navigation system control panel and the correction radio beacon from the program version selected before flight;
- selection of FCC number for navigation and landing and of navigation radio beacon types in the manual mode;
- manual actuation of the landing mode.

The aircraft radio and navigation equipment consists of the radio equipment section and digital autonomous navigation computer TsVI or TsVIM having program 2204 or 2205.

The airborne short range radio navigation equipment operates in the UHF band and through the antennafeeder system ensures the joint operation with both nondirectional ground radio beacons, type RSBN4N, and directional radio beacons, as well as with landing radio beacons, type PRMG.

The short range radio navigation equipment measures:

- aircraft polar coordinates (azimuth, and slant range) relative to the place of installation of the ground radio navigation beacon when performing a plannedroute flight and return;
- deviation of aircraft from the localizer and glide slope equisignal zones when performing a landing approach with measurements of distance to the runway threshold.

Besides, when operating with ground radio beacons, the short range radio navigation equipment ensures:

- monitoring the aircraft position and identifying the aircraft on the radio beacon planposition indicator;
- listening to the radio beacon call signs on board the aircraft.

The operating range of the equipment when operating with various types of navigation radio beacons is determined depending on the aircraft flight altitude.

The accuracy of polar coordinates determination:

- with radio beacons, types RSBN2N, RSBN-4N, "POLE is 0.25° in azimuth, 200 m + 0.03 % D (m) in distance;
- with radio beacon RSBN-6N is 1.5° in azimuth, 300 m + 0.05 % D (m) in distance.

The number of channels during operation:

- in the NAVIGATION mode:
    - with nondirectional radio beacons RSBN-2N is 40 channels, RSBN-4N is 88 channels;
    - with directional radio beacons "POLE-N" is 176 channels.

Selection of the frequencycode channels for navigation and landing and of radio beacon types for the receivingtransmitting devices is accomplished in the manual mode with the aid of control panel (the CHANNELS selector switch is in the MAN. position) and in the automatic mode from the navigational computer via the code conversion unit (BPK) depending on the selected number of the radio beacon (the CHANNELS selector switch is in the AUTO position).

In the short range radio navigation equipment the checking is effected automatically in the presence of groundradio beacons signals.

The digital navigational computer of the aircraft radio and navigation equipment is the linking element of all the subsystems incorporated in the navigation system and in designed for processing the data supplied by primary transmitters, for autonomous reckoning of the aircraft position coordinates using the inertial or airspeedandheading reckoning method and for updating the calculated coordinates in response to the data provided by the short range radio navigation equipment. On the basis of data about the waypoint or landing airfield coordinates entered prior to flight and on the basis of reckoned coordinates, the computer determines the distance, desired heading to a selected point and the height deviation. Besides, the computer is continuously calculating the true aircraft heading (relative to the geographical meridian passing through the point of reckoned aircraft position) making use of the gyro heading supplied by the information directional/ vertical gyro system. The computer also shapes discrete commands at various flight stages and performs automatic check of the whole navigation system. The information recorded by the computer reprogrammable readonly storage can be stored for 30 days, maximum (for computer TsVI, 10 days).

PREPARATION OF NAVIGATION SYSTEM FOR OPERATION

The navigation system has two preparation modes determined by the time for bringing the information directional/vertical gyro system into readiness:

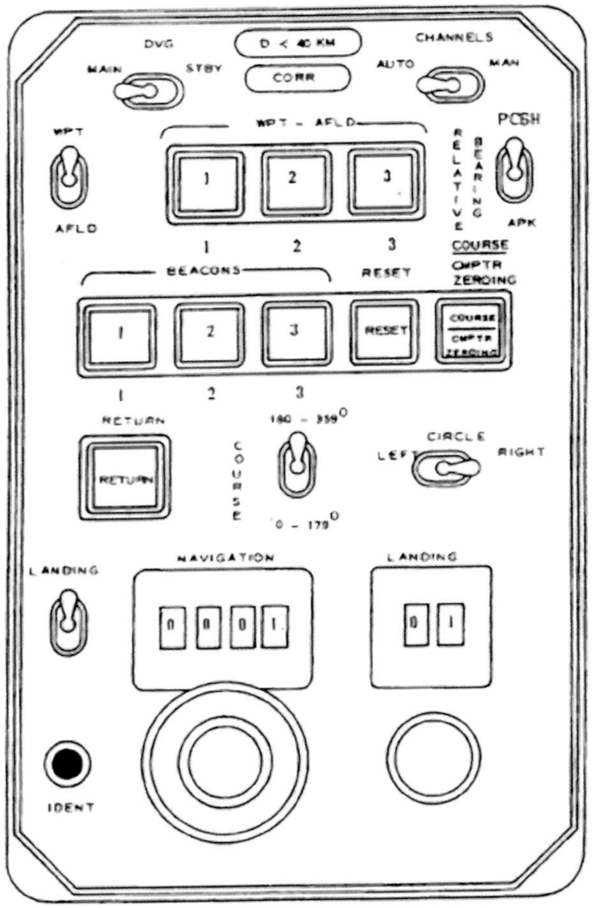

NAVIGATION SYSTEM CONTROL PANEL

- fast preparation mode (main);
- navigational preparation mode.

The fast preparation mode may be initiated both prior to engines start (with ground power supply connected) and after starting one or both engines and disconnection of the ground power supply.

The navigational preparation mode is accomplished by the technical personnel before the pilot has climbed into the cabin.

The preparation mode is required for horizon and azimuth alignment of the inertial directional/vertical gyro system, for determination of the aircraft ramp heading and puts input into the course computer and control and coupling unit, for input or adjustment of the flight program entered earlier. The mode is initiated after energizing the KV MAIN STBY NAVIGATION switches and placing the OPERN PREPAR selector switch on the starboard console in PREPAR position (should be placed before energizing).

The preparation mode includes the following stages:

- fast alignment (FA) of the gyroplatform in horizon and in azimuth in response to the zero signals produced by the sinecosine resolver and starting of gyro motors (30 s);
- coarse alignment (CA) of the gyroplatform in horizon in response to the signals supplied by the accelerometers and in azimuth in response to the signal produced by the sine-cosine resolver with a higher accuracy than at the FA stage. Gyroscopic stabilization is engaged into operation at the CA stage.

After aligning the gyroplatform in horizon plane and in azimuth with an error of not more than 10, the CA mode is terminated and the fine alignment mode is initiated. The fine alignment mode starts with analogue alignment.

Further procedure of alignment depends on the position of the OPERN PREPAR selector switch.

With the selector switch set to the OPERN position (not later than 1.5 minutes after switching on), the information directional/vertical gyro system changes over to the operating mode automatically; following 2 or 3 minutes after the navigation system is switched on the PAST PREPAR Annunciator illuminates.

The standby directional/vertical gyro in both modes of preparation ensures functioning of the system in the airspeedandheading reckoning mode only.

It allowed, if need arises, to interrupt the navigational preparation mode by placing the OPERN PREPAR selector switch in the OPERN position, but not earlier than 5 min after the navigation system is switched on. This will result in illumination of the PAST PREPAR annunciator, and the navigation system will provide the accuracy parameters corresponding to the fast preparation mode.

Placing the selector switch in the OPERN position within 15 to 5 minutes after the navigation system is switched on, is not advisable as the accuracy parameters in this case will be worse than during the fast preparation.

In the event of integrators fail in the operating mode, the main directional/vertical gyro automatically changes over to the directional/vertical gyro mode with radial correction.

Determination and input of the ramp heading (initial course setting) both during the fast and during the navigational preparation are accomplished with the OPERN PREPAR selector switch in the PREPAR position as follows:

1. In the magnetic correction (M) mode.
2. In the initial course setting (ICS) mode proper.

In the MC mode the slaving and input of the ramp heading into the computer are achieved by simultaneous depression of the M HDG SLAV and COURSE ZEROING buttons. The magnetic heading transmitters are elements sensitive to the magnetic field of the earth in fluxgate transmitters. The signal is averaged in the computer of the control and coupling unit and employed for computation of the heading. The true heading is obtained after entering into the control and coupling unit (BIS) of the present value for the magnetic declination. The magnetic declination is set on the magnetic declination setter by manipulating a knob and loaded into the control and coupling unit after the M HDG SLAV button is depressed. To eliminate instrumental errors and deviation of the magnetic corrector device, the control and coupling unit (BIS) is provided with a magnetic deviation compensator (MDC)

The mode of initial setting proper performed with the aid of the magnetic declination setter unit with the ICS (initial course setting) button depressed makes it possible to achieve the initial alignment of the system if the data determining the aircraft heading are available. The measured heading is selected an the magnetic declination set at the same time depressing the ICS button to load it into the control and coupling unit (BIS) and the COURSE/ ZEROING button to load it into the computer. The aircraft heading is determined with the aid of by measuring the aircraft position with relation to the geodetically fixed landmarks or, in the parking area, by positioning the aircraft in a rigidly fixed attitude. Besides the ramp heading which may be loaded from the program input panel (PVP) during the input of the flight information which in a more accurate method of input. In this case, the ICS and the COURSE/ZEROING buttons are not depressed and the heading is loaded automatically.

Provisions are made for adjustment of the ramp heading determined in the M mode by means of the optoelectronic sighting system of the aircraft For this purpose, depress the WPTAFLD lightbutton on control panel (with the WPTAFLD selector switch in either position) to correspond to the number of the selected landmark or depress the RESET button (for a landmark "O"), manipulate the control button (KI) on the control stick to align the moving mark an the windshield indicator with the selected landmark and depress the WSM,

LOCKON PF button. This ensures an accuracy of static ground angle determination of 0.2°.

The azimuths of four landmarks from the start of takeof f are determined by the geodetic service and loaded into the computer from the program input panel during input of all the other initial information. The ramp heading of the aircraft in the computer is determined as a difference between the azimuth of the landmark and the sighting angle towards it from the aircraft parking point. To prevent false writing of information, this method is employed only when preliminary information from other means of initial setting is available. For example, if prior to optical setting the true heading in the MC mode is stored in the computer memory, the writing of the ramp heading determined with the help of the integrated optoelectronic navigation/ sighting system in the computer memory and in the control and coupling unit (BIS) will take place only provided it differs from the heading written in the MC mode by not more than 5°.

When flying from the programmed airfield, the course updating can be effected also by reference to the runway heading (computers TsVI and TsVI-N (2204) do not ensure this updating).

Depressing the CAGE lightbutton on the flight director indicator initiates the REPEATED START mode which is an emergency mode. During repeated start the information directional/vertical gyro system automatically passes through all the stages of autonomous fast preparation, then goes over to the radial correction mode. During a repeated start it is necessary to ensure straightandlevel flight at a constant speed for 2 minutes.

After a repeated start the information output in angles with errors in roll and pitch during performance of maneuvers of up to 2-4° is ensured.

To receive information on the heading from the information directional/vertical gyro system, it is necessary to perform slaving in MC by depressing the M HDG SLAV button.

MODES OF OPERATION OF NAVIGATION SYSTEM

The navigation system has the following modes of operation: navigation; return; landing; missed approach; return to an unprogrammed airfield.

Navigation Mode

The mode is employed during a plannedroute flight or for the interception of a programmed point. To bring the aircraft to the desired point or route point, the course method is employed, i.e., the variables of this method of navigation are the desired and the present aircraft headings and the distance to the exit point.

The most precise autonomous mode of operation of the navigation system is that of inertial coordinates reckoning. In this mode the system is functioning after its navigational preparation before flight, with computer and main channel of the information directional/vertical gyro system serviceable.

Should the digital integrators of the main channel of the information directional/vertical gyro system located in the control and coupling unit (BIS) fail, the system switches over automatically to the airspeedandheading reckoning mode.

The airspeedandheading reckoning mode is used by the system also after a fast preparation before flight.

At EKV failure, as indicated by the illumination of the CAGE lightbutton on the FDI, dropout of the HDG warning flag on the CCI and the MAIN DIR VERT GYRO signal on the EKRAN system screen, and after placing the EKB MAIN STBY selector switch on the control panel in the STBY position, the navigation system is operating in the airspeedandheading reckoning mode in response to the information supplied by standby channel EKV STBY of the information directional/vertical gyro system.

When communication with the short range radio navigation system beacon in steady (the CORR annunciator is illuminated), the navigation system is operating in the inertial or airspeedandheading reckoning mode with correction from the short range radio navigation system.

With inertial reckoning mode unavailable and the air data computer system unserviceable (the AIR DATA SYS (CBC) signal is displayed on the EKRAN system screen), as well as in case of navigation system computer failure (the HAVIG COUPUTER signal is displayed on the EKRAN system screen and the HDG flag comes into view on the CCI), the aircraft coordinates reckoning does not occur and use is made of navigational data supplied by the other serviceable transmitters (short range radio navigation system, inertial directional/vertical gyro system, automatic direction finder).

In flight, the ARNE computer is continuously comparing the reckoned coordinates of aircraft position with the programmed coordinates of the exit point (WPT or airfield) and generates, with output to the combined course indicator, the selected heading and the horizontal distance to the exit point. The present heading is computed also by the computer in response to the data of the gyro heading supplied by the information directional/vertical gyro system and the ramp heading with allowance made for the latitude correction for the speed of the earth rotation and convergence of meridians.

'When approachinga waypoint at D <40 km after illumination of the D < 40 annunciator, thenext waypoint button may be depressed. When flying the distance remaining to the waypoint, allowance should be made for the aircraft drift. After flying past the waypoint, if the button of the next waypoint is not depressed, the CCI retains the fixed heading towards the first waypoint until the exit from a zone with a radius of 6 km. The accuracy of waypoint interception depends on the coordinates reckoning mode and existence of the radio correction. Transmission of the radio correction is indicated by continuous Glowing of the CORR annunciator on the navigation system control panel. Discontinuity of the radio correction transmission is indicated by extinguishing of the CORR annunciator with a delay of 60 s. For the aircraft equipped with computers TsVI and TsVIM(2204), the abovementioned delay is maintained at changeover of the radio beacon number on the navigation system control panel as well.

In the navigation mode, the following indications are displayed:

- true heading, desired heading, aircraft azimuth relative to the radio beacon, distance to waypoint on the combined course indicator;
- angle of corrective turn on the flight director indicator;
- true heading, distance and angle of corrective turn on the windshield indicator and the direct vision indicator.

Landing Mode

The navigation system is automatically changed over to the LANDING mode when the aircraft is at a distance of 8 to 37.5 km from the runway centre line and at a lateral displacement of 1 to 4 km from the runway centre line, respectively with a difference between the present and runway landing headings less than 60°, at a flight altitude below 1400 m.

For computers TsVI and TsVIM(2204), the conditions ensuring generation of the LANDING command are: a distance from the runway centre of 8 to 35 km, a lateral displacement less than 1.5 km, a difference between the present and runway landing headings less than 45°, a flight altitude below 1125 m. In response to the LANDING command, the computer generates and accomplishes changeover of the radio circuit from the channels of the navigation beacon to those of the landing radio beacon group. The mode can be also engaged manually by manipulating the LANDING switch when the aircraft is within the coverage of the localizer radio beacon. In this mode. the receiver of the shortrange radio navigation system is changed over from the azimuthrange radio beacon channel to the channel of the landing radio beacon group.

The operating principle of the short range radio navigation system equipment in the LANDING mode consists in comparing the magnitudes of signals with opposite signs from the antenna lobes and obtaining as a result of comparison the difference of signals characterizing the position of the aircraft relative to the localizer and glide slope equisignal zones. The values of an output signal identify the degree of deviation from the desired track, the signal polarity indicates the direction of the corrective turn. The operation of the rangefinding channel, as in the NAVIGATION mode, is based on the principle of active radiolocation. The distance is read out to the beginning of the runway (by a rangefinder repeater on the glide slope radio beacon).

The signals of angular deviations from the localizer and glide path zones are displayed by deviation bars on the CCI and PDJ instruments. The maximum (to the fourth point) deviation of the glide slope deviation bar is an indication of angular deviation of the aircraft from the equisignal line (radio glide slope) equal to about 0.5°. The serviceability of the system in the LANDING mode in heading and glide path are indicated by the appropriate flags on the combined course indicator. If the flags (red colour) are out of view, the system is serviceable. In this mode the computer shapes the selected course.

Since the receiver of the short range radio navigation system equipment switches over from azimuthrange beacon channels to localizer radio beacon group channels, the beacon azimuth is not indicated, but the ADF RB (KIR-ARK) in displayed regardless of the ARK RSBN RB selector switch position (this switchover should be accomplished manually on the earlier items).

In response to the signals of deviation from the localizer and glide slope equisignal zones, the APCS generates control signal. When approaching for landing from an altitude of the intermediate approach, the aircraft is always below the assigned glide slope. Therefore, if the control signal is shaped in the longitudinal channel, prior to interception of the equisignal zone stabilization of the intermediate approach altitude in accomplished.

During the interception of the glide slope equisignal zone, shaping of the signal in the longitudinal channel occurs in response to a deviation signal from the equisignal zone and to the angle of pitch.

An approach for landing with automatic control is ensured down to an altitude of 50 to 60 m. Descent along the glide slope below this altitude is forbidden owing to safety conditions due to curvature of the equisignal zone.

Missed Approach Mode

When performing an approach for landing to a programmed airfield as the need for a go around arises, the navigation system may be switched over to the missed approach mode by depressing the MISSED APPROACH button on the AFCS panel. The missed approach mode may be initiated in the zone with coordinate X less than 17 km. During a missed approach procedure the selected heading is continuously followed up into a moving point located at a distance of 2.5 km towards the turn onto base leg from the foot of a perpendicular dropped from the aircraft position point to the line removed from the runway axis by 10 km and parallel to it. In the vertical plane the SN29 navigation system shapes a deviation from the assigned altitude of 600 m.

To execute the missed approach, depress the MISSED APPROACH lightbutton mounted on the APCS panel. If the LANDING mode is to be selected manually, do not fail to preliminarily turn off the LANDING switch and depress the lightbutton with the correction radio beacon number. The MISSED APPROACH lightbutton glows up to the moment of initiating the turn to the base leg.

For computers TsVI and TsVIM(2204), the MISSED APPROACH mode can be selected only after automatic generation of the LANDING command.

A missed approach may be accomplished in the left or right traffic circuit depending on the position of the CIRCLE LH RH selector switch on the navigation system control panel.

Return to Unprogrammed Airfield Mode

To accomplish a return to an unprogrammed airfield after placing the AUTO MAN selector switch on the navigation system control panel in the MAN position and manual selection of channels for navigation and landing of the chosen airfield, the flight to the radio beacon or to the homing radio station of the given airfield in performed. In this case, the selected heading to an offset point and the descent path are not shaped. After entering the coverage zone of a selected radio beacon and twitching on the radio correction, the CORR annunciator light is illuminated on control panel W, the combined course indicator gives readings of the distance to the radio beacon, the corrected heading supplied by the information directional/vertical gyro system and the aircraft azimuth with relation to the radio beacon or the relative bearing to the homing radio station (depending on the position of the ARK RSBN selector switch). After the interception of the calculated point of the start of the base turn, manually set the runway landing heading on the combined course indicator by manipulating the heading select knob and turn on the LANDING switch.

PREPARATION FOR ENROUTE FLIGHT WITH USE OF NAVIGATION SYSTEM

Preparation of the pilot for an enroute flight with the use of the navigation system includes:

- selection of the route and preparation of the initial data for programming;
- checking the navigation system before flight.

The selection of the route is determined by the mission to be accomplished during the flight. Distinctive landmarks should be designated, if possible, as the waypoints so that they can be used for the purpose of visual correction.

To program the route, do the following:

- determine the initial geographical coordinates, for which purpose on a map, scale 1:500000 or 1:1000000, select the origin of coordinates in such a way that all the route waypoints, programmed airfields and radio beacons are located to the north and to the east of the initial latitude and the initial meridian, i.e. , within a spherical square of 40° (36° for computers TsVI) limiting the operating zone of the system
- from the map read the geographical coordinates of the waypoints, distance to WPT and airfields, headings for waypoints and airfields;
- write from special lists the coordinates of the runway centreline and of the radio beacons with an accuracy of 0.05;
- convert the angular values of coordinates read into degrees with fractions of hundredth;
- calculate the conventionalgeographic coordinates of waypoints, centrelines of runways and radio beacons in terms of difference between the coordinates read and the initial geographic coordinates by subtracting the latter from the former;
- measure the azimuths of four landmarks from the point of start napped earlier with an accuracy of at least 1;
- measure the aircraft ramp heading;
- fill in the flight mission form for flight programming by entering:

  (1) conventionalgeographic coordinates of waypoints airfields, radio beacon (within the range of 0 to 39.99° (up to 35.999°, for computers TsVI);

  (2) azimuths of four landmarks (within the range of 0 to 39.999° (UP to 35.999° for computers TsVI);

  (3) coordinates of the reference zero point latitude;

  (4) true headings of airfield runways (within the range of 0 to 179.99°);

  (5) true aircraft ramp heading (within the range of 0 to 35.999°);

  (6) numbers corresponding to the frequencycode channel numbers of radio beacons;

- distances and headings to the waypoints and airfields, distances and azimuths of landmarks, true headings of runways and magnetic declinations of airfields; frequencycode channel numbers of radio beacons should be recorded in a convenient place on the map or tabulated.

The aircraft armament system includes:

- guided missile armament;
- rocket armament;
- bombing armament;
- gun armament;
- chaff dispensers

The aircraft armament is arranged on six wing external store stations numbered in the direction of flight away from the longitudinal axis of the aircraft (1, 3, 5 on the left side, 2, 4, 6 on the right side), the gun is built into the left wing extension, and the chaff dispensers are located forward of dorsal fins on the wing.

Provisions are made on the aircraft for the use of similartype and combined armament versions. In the latter case, symmetrical stations should carry weapons of the same modification. Indication of availability type and readiness for use of armament is ensured by means of the integrated display system, information reporting system and aural warning system.

In the presence of the target lockon by the integrated radar/sighting system or the optical locator laser station or when the ACS MODES selector switch is placed in the OPTIC HELMET and PITCHUP positions, the integrated display system indicates the type of armament selected for use by symbols 27R, 60, S-24B, S8, AV and its availability by horizontal strokes corresponding to the external store stations. If the missile is ready for launch, the corresponding stroke indicating availability is marked by the illuminated digit of the external store number. In case of a missile failure to lift off and deenergizing of the missile after a depression of the trigger button, the station number of the missile failed to lift off is framed in a rectangle. Permission for the use of weapons is indicated by the PR (LP) index and for missiles and rockets it is regularly dubbed with a "Launch permitted" voice command.

The audio signal in the pilot's headphones during the use of weapons is supplied in the following cases.

AIRCRAFT ARMAMENT

- when at least one missile is ready for use (target lockon by the infrared homing head);
- when the aircraft is within the effective firing range for the gun or rockets against ground targets;
- 2 or 3 seconds prior to aerial bombs drop during divetoss bombing, toss bombing, level bombing in the invisible zone;
- 2 or 3 seconds prior to pitchingup entry during toss bombing.

MISSILE ARMAMENT

The guided missile armament of the aircraft includes "airtoair" missiles, types R27, R60.

The R27 missiles are suspended from inboard stations only, the R-60 missiles from any stations. The longitudinal axes of missiles suspended from launchers are deflected by 2° downwards relative to the longitudinal axis of the aircraft and the centre of the fixed aiming reticle.

Selection of the missile type to be used is done by the INB OUTB selector switch. In the INB position of the selector switch the launching sequence will be 1, 2, 3, 4, 5, 6, in the OUTB position of the selector switch 5. 6, 3, 4, 1, 2. Missiles may be launched singly or in series of two. With the AL SINGLE 0.5 AL selector switch set to the SINGLE 0.5 AL position every depression of the RKT MSL B trigger will cause one missile to be launched. In the AL position of the selector switch every depression and holding of the trigger will result in a launch of a chaff dispenser.

For a combat launch of a missile the following requirements have to be fulfilled:

- the LP (launch permitted) command should be present;
- the aircraft landing gear should be in a retracted position;
- the MASTEROFF selector switch should be turned on.

The LP (PR) command is shaped as the sum of the missile readiness signal (target lockon for the infrared homing head) and launch permitted range D signal, which is calculated by the armament control system in the RDR and IFF modes or is selected by the pilot manually using the MAN PREPAR selector switch. In the armament control system modes: IRR, HELMET, OPTIC and RETICLE, the "Preparation" command is supplied automatically when selecting any of these modes and can be duplicated by the MAN PREPAR selector switch.

In the process of automatic tracking of the target by the missile infrared homing heads, when a certain angular rate of tracking is exceeded by the target sighting line, the flickering of the LP signal is an indication of the impermissible aiming error (IAE). If missiles are launched under these conditions, the probability of target hit is reduced. In the presence of the IAE signal it is necessary to maneuver the aircraft until the tracking rate decreases.

The LP command is also flickering during a target attack by missiles with infrared homing head in the IRR, HELMET and OPTIC modes when no information about the target range is available.

All missiles are fitted with the target designation system.

Target designation is provided most accurately for the missiles selected for use by the INB OUTB selector switch.

For other missiles with infrared homing heads, the target lockon by the homing heads, under the identical conditions of launching, may not be available due to disregard of external store conditions.

Upon launching the guided missiles the weapon control system deenergizes missiles powered by an aircraft source following 0.8 s after a depression of the RKT MSL B trigger. Should a missile fail to liftoff within 2 s from the moment of trigger depression, it is deenergized and as the trigger is held depressed the signal is generated to launch the next missile ready for use. If the missile at the selected station is not ready, the ready missile at the next station is launched. Uncontrollable emergency launching of missiles is initiated by depressing the STBY LNCH button, whereby pairs of missiles lift off at 1s intervals in the following order: I and 4, 3 and 2, 5 and 6.

For R60 missiles the launches through combat and emergency circuit differ only in that during an emergency launch the missile infrared homing heads are not functioning (the missiles follow the ballistic trajectory). Therefore, if on depression of the RKT MSL B trigger the solidpropellant gas generator has been activated in R60 missiles, these missiles will not lift off even though the emergency circuit is used.

With R27 missiles the emergency launch signal is applied directly to the engine contacts. In case of an aborted liftoff in the combat circuit the rudders may be found in arbitrary positions, which does not ensure a safe liftoff in the emergency circuit. Therefore, an emergency launch of missiles with at least one stuck R27 missile on board IS PROHIBITED.

In spite of some difference in the design of the missiles used on the aircraft have similar composition, purpose of components and modes of preparation for launch, launch, control in flight to the target and destruction of the warhead.

The missiles employ the "canard" configuration with cruciform wings, rudders and ribs forward of rudders (or destabilizers).

The missiles comprise 5 or 6 compartments hermetically interconnected by means of joining assemblies and including the following elements:

- homing head;
- operational equipment (warhead, contactsensing and proximity fuses, safetyand-actuating mechanisms);
- power supply system;
- solid-fuel rocket engine.

The nose section of missiles is covered with fairings whose shape and material depend on the type of the homing head.

The R27 missile is fitted with a semiactive radar homing head using the method of quasicontinuous target illumination by means of an integrated radar/sighting system. Provisions are made in the radar homing head for the missile guidance by the inertial method with radio correction without tracking the target by the homing head an up to 70% of the flight path. Correcting signals are transmitted on side lobes of radar emissions. On the final leg of the missile flight path the target lockon is achieved by the missile radar homing head and semiactive homing is accomplished.

The infrared homing heads installed on the R60 missiles are fitted with photodetector thermoelectric cooling system.

All missiles employ an active, pulsed proximity radio fuse. For R27 missiles, the radio fuse actuation range depends on the signal about the

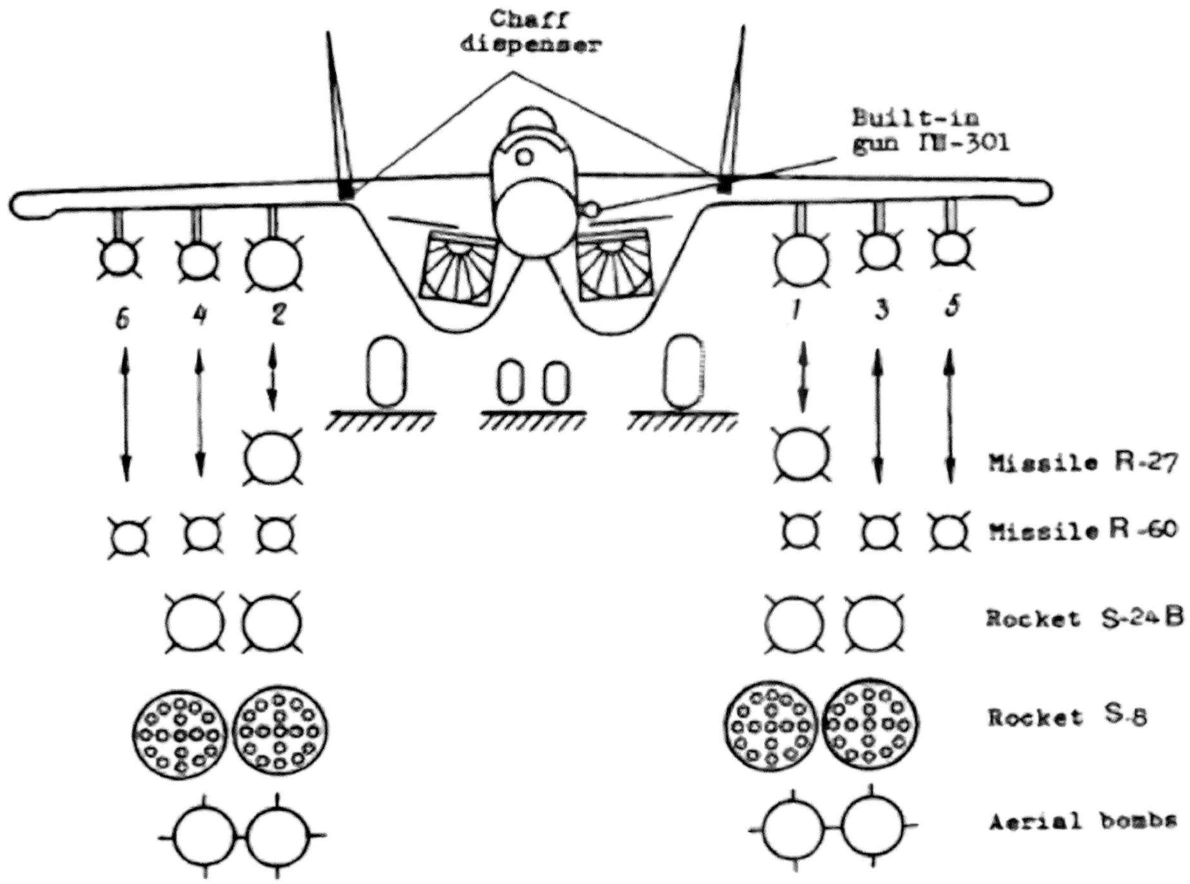

No.	Armament	Qnty, max	External store stations (gun)	Armament mounts
1	Missile R-27 (item 470)	2	1, 2	Launcher APl -27 (item APl -470)
2	Missile R-60 (item 62)	6	1, 2, 3, 4, 5, 6	Launcher APl -60-1DB1 (item P-62-1DB1)
3	Rocket S-24B (item 240)	4	1, 2, 3, 4	Launcher APl -68УМ2(3) on rack DB3-УМК2B
4	Rocket S-8	80 (4x20)	1, 2, 3, 4	Rocket pod B-8M1 on rack BD3-УМК2B
5	Aerial bombs	4	1, 2, 3, 4	Bomb rack BD3-УМК2B
6	30-mm cartridges	150	1 xGSh-301 (item 9A-4071L)	Built-in gun
7	Chaff dispenser	60		

AIRCRAFT ARMAMENT

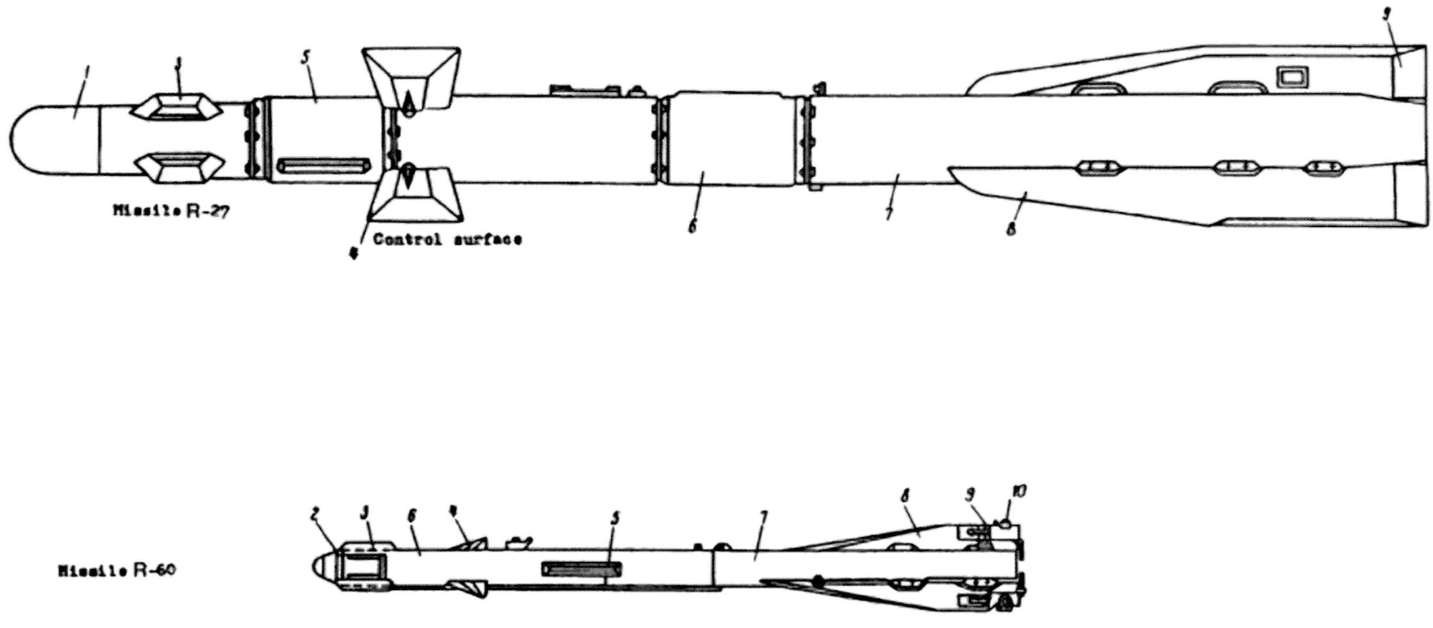

MISSILE COMPONENTS

1 - radar homing head; 2 - infra-red homing head; 3 - destabilizer; 4 - control surface; 5 - radio fuse; 6 - warhead; 7 - solid-propellant rocket engine; 8 - wing; 9 - aileron; 10 - rolleron

target type (large, medium, small) supplied to the missile prior to its launch by the armament control system of the aircraft.

To explode the missile warhead in a direct target hit or in case of an impact against the ground, all missiles are fitted with contactsensing fuses. An action of the radio fuse or contactsensing device initiates the electric igniter of the safetyandactuating mechanism to explode the missile warhead.

The R60 (R60MK) missiles are equipped with an active proximity fuses.

The safety and actuating mechanism is designed for preventing the output of a detonating pulse to explode the missile warhead under all conditions of storage, operational use and flight of the missile to target until the expiration of time for the outer arming and for generating a detonating pulse to explode the missile warhead after the expiration of time for the outer arming upon receiving a command from the radio fuse or from the contact sensing device. The safety and actuating mechanism has two electrical and mechanical stages of safety:

- the first stage is cancelled after the missile switchover to its own sources of power supply;
- the second stage is cancelled during the missile liftoff from the launcher;
- the third stage is cancelled upon the termination of outer arming mechanism operation, whose arming time depends on the gload value at the moment of missile launch and amounts to 1.5 3 seconds. In the R60 missile provision is made for self destruction along the flight path.

If the contact sensing device or the fuse fails to operate within 23 to 25 s after the missile liftoff, the safety-and-actuating mechanism sends a command to explode the missile warhead. Destruction of the R27 missiles after a miss is triggered by the contact sensing device only after a ground impact.

When missiles are launched against a ground target, the proximity fuse is not switched off. The missiles are equipped with a coretype warhead.

After the detonation of an explosive charge the core spring opens up into a ring, which having great kinetic energy hits the target.

To visually monitor the missile flight path, a tracer producing a visible fire trace of red or orange colour is installed on the wing of a R27 missile or in the tail section of R60 missiles.

During the autonomous flight, power supply of R27 and R60 missile equipment is accomplished by a power unit consisting of a solid propellant gas generator, turbogenerator and a stabilizer-rectifier unit shaping the required parameters of power supply.

Stabilization of missiles on the flight path and flight control in response to the signals fed by the homing equipment is done by an autopilot.

The R27 missile is equipped with a hydraulically operated rudder steering system.

Rudder actuators of R60 missiles are also driven by the gas generator (a solid propellant gas generator).

When using R60 missiles to destroy a target in the front hemisphere (against afterburning targets), for maintaining the homing accuracy as with the target attack in the rear hemisphere, it is necessary to alter the gain of the aircraft control circuit. A command for altering this gain factor is generated by the armament control system with the FHS RHS selector switch set to the FHS position on the missile control panel.

After starting the aircraft engines and switching on the armament control system, in response to a command from the armament control system a standby mode is engaged in the missile. In this mode the homing heads and the autopilot gyros of the missile are energized.

To avoid overloading the aircraft onboard power supply sources and to ensure the required time and level of the missile equipment serviceability, a special "Preparation" command is introduced and after its output the target designation signals may be fed into the homing head. Simultaneously, data on launch and guidance conditions are supplied by the integrated radar/sighting system and the optical locator laser station (or manually for the missiles with the infrared homing heads). Following a lockon of the target heat emission by the infrared homing head or a followup of the target designations by the radar homing head, the WR (warhead ready) (GG) signal is generated and if the permitted launch range is available, the armament control system shapes the approval for a missile launch (the LP signal).

After a depression of the RKT MSL B trigger its power supply unit is switched over to the autonomous supply source mode. The solidpropellant gas generator is activated on R27 and R60 missiles to ensure the operation of the turbogenerator. As the solidpropellant gas generator gains the normal rating, the aircraft power sources are cut off, the missile equipment is fully turned on, the first safety stage of the safetyandactuating device is cancelled, a command for the missile engine start is supplied. After the mine is started, the missile leaves the launcher and in the process the electrical connectors coupling the missile and the launcher are detached (or sheared off), the second safety stage of the safetyandactuating device is cancelled. The gloads affecting the missile determine the moment of canceling the third stage of safety and the moment of arming the safetyandactuating device.

After launch, the missile control system shapes the launching log of flight path ensuring the missile stabilization regarding gloads, angular rates and pitch angle, counter action of liftoff disturbances. deflection of control surfaces to angles related to the carrier gloads at the moment of launching. Further guidance of the missile to the target is accomplished by a missile control system in response to signals fed by the homing head and the sensors.

ROCKET ARMAMENT

The rocket armament comprises rockets, types S8 in pods B8 and S24B. Pods and rockets S24B are suspended at the inboard and middle stations from BDZ beamtype racks. When suspending the S24B rockets, use is also made of API launchers.

With the AL SINGLE 0.5 AL selector switch in the AL position and the RKT MSL B trigger depressed, rockets S24B are launched in succession from stations 1, 2, 3, 4 at 0.16s intervals or rockets S8 of all the four pods are launched until the entire ammunition load is used. Here, the unloading of pods occurs in the following sequence: one rocket from each pod at stations 1, 2 in a salvo, one rocket from each pod at stations 3, 4 in a salvo, then one rocket from pods 1, 2 each and so on. The time of launching the entire ammunition load is 2 S for S8 rockets and 0.5 s for S24B rockets.

With the AL SINGLE 0.5 AL selector switch set to the SINGLE 0.5 AL position, rockets S24B are launched in series of two at 0.16s intervals from stations 1, 2 after the first depression of the RKT MSL B trigger and from stations 3, 4 after the second depression.

In this position of the selector switch the liftoff of S8 rockets from the pods occurs as in the AL position but with a outoff after the liftoff of ten S8 rockets from every pod.

In all the cases the rocket liftoff is discontinued as the RKT MSL B trigger is released.

In the course of firing the S8 rockets, aiming corrections applicable to ballistic characteristics of rockets are calculated in the integrated optoelectronic navigation/ sighting system.

Emergency jettisoning of the loaded and empty pods and S24B rockets together with the launchers is initiated by a depression of the EMERG RELEASE button. In this case, the pods (rockets) are dropped in pairs from stations 1, 2 and, then at 0.6s intervals from stations 3, 4. Drop of launchers without S24B rockets is interlocked by the weapon control system.

Provisions are made in the weapon control system for single launching of S24B rockets at every depression of the trigger regardless of the AL SINGLE 0.5 AL selector switch position. For this purposes turn on, prior to flight, the WEAPONTRAINING MODE switch located in the cabin power system. The rocket liftoff in this case occurs in turn from stations 1, 2, 3, 4.

The remainder of S8 rockets in pods is not indicated.

BOMBING ARMAMENT

The bombing armament consists of 250kg and 500kg calibre bombs.

The use of bombs is ensured from the inboard and middle external store stations by singlelook beamtype racks (BD). The racks ensure both free dropping of bombs and dropping of bombs with forced repulsing.

The dropping of aerial bombs by the RKT MSL B trigger in possible only In the "armed" condition. In this came, with the AL SINGLE 0.5 AL selector switch set to the AL position, the aerial bombs are released in series at 0.16s intervals from stations 1, 2, 3 and 4 after one depression of the firing button. With the AL SINGLE 0.5 AL selector switch set to the SINGLE 0.5 AL position, the first depression of the trigger releases aerial bombs from stations I and 2, the repeated depression from stations 3 and 4 (at a 0.16s interval).

Under the SINGLE 0.5 AL conditions, the small load packs are released in series of two bomb clusters BKF from stations 1 and 2 at a 0.16s interval when the RKT MSL B trigger is depressed for the first time and from stations 3 and 4 when it is depressed once more. In the TRAINING MODE condition, the small load packs are successively released one by one from stations 1, 2, 3, 4 at each depression of the trigger irrespective of the AL SINGLE 0.5 AL selector switch position. With the wafer switch set to the SINGLE position on the pod control panel the small load packs release their clusters BKF one by one at each depression of the trigger.

To stop releasing, set the MAIN OFF selector switch to the OFF position.

With the WEAPON TRAINING MODE switch turned on prior to flight, regardless of the AL SINGLE 0.5 AL selector switch position the aerial bombs are dropped singly at every depression of the trigger in the sequence of stations 1, 2, 3. 4.

The emergency jettisoning of the bombs is initiated by the EMERG RELEASE,button depression in the "armed" or "safe" conditions depending an the position of the ARM SAFE selector switch. This causes simultaneous release from stations 1 and 2 and after 0.6 s simultaneously from stations 3 and 4.

The input of ballistic data for the aerial bombs is carried out before flight from the monitor centre by the ground crew setting the wafer switches (coded input) to positions corresponding to the code of aerial bombs used. Depending on the conditions of combat use and the bombing mode, additional information about the target vertical separation relative to the departure airfield and the selected drop angle of aerial bombs in toss bombing is entered from the input and monitor panel.

Besides, for all aerial bombs the input and monitor panel provides also the information input of ballistic data.

GUN ARMAMENT

The gun armament of the aircraft consists of one singlebarrel builtin gun GSh301 of 30 mm calibre with a rate of 1500 rounds per minute. The gun ammunition load is 150 rounds. Firing is triggered off by the GUN trigger located an the aircraft control stick.

When the GUN trigger is shifted from its initial position to the preliminary position, the "Gun " signal for the use of aircraft gun is supplied into the weapon control system regardless of the availability of external stores on board.

Firing the gun can be effected in one of the following modes:

- automatic mode, the AL SINGLE 0.5 AL selector switch is set to the AL position. In this mode, with the GUN trigger depressed, the maximum number of shots in a burst is limited by 3/4 ammunition load. At the repeated depression of the trigger, the remaining ammunition is fired;
- cutoff mode, the AL SINGLE 0.5 AL selector switch is set to the SINGLE 0.5 AL position. Each depression of the GUN trigger limits the length of a burst by 01.0 s (25 to 27 shots).

In both the modes, when the GUN trigger is being depressed for the time less than indicated above, the number of shots in a burst is decreased. During the repeated depression of the GUN trigger the firing is resumed until 1/4 ammunition load remains In the automatic mode or within the given cutoff (1.0 s).

When the TRAINING MODE switch is turned on, the firing is limited by 0.25 s with outoff after 5 to 7 shots, regardless of a position of the AL SINGLE 0.5 AL selector switch.

In all the firing modes (except for the training mode) the interval between the bursts should be at least 3 s.

Prior to a mission the gun is loaded ready to fire. Firing of ammunition starts at the first depression of the GUN trigger.

In the event of stoppage of the "misfire" type in the process of gun firing (failure of the electric primer) with the GUN trigger depressed, after 0.4 s an alternate primer is activated and firing is resumed if the AP cartridges are available. Elimination of one stoppage only is possible.

The fired cartridge cases are ejected overboard, while the links are collected into a link collector. Cartridge cases are ejected to the right downwards as looking forward between the engine air intake nacelles, therefore this factor should be remembered in flight when flying in closely spaced formations. With a ventral tank suspended, the gun firing is interlocked by the weapon control system.

Indication of the ammunition load remainder in quarters is displayed on the integrated display system screen as the GUN trigger is moved to the preliminary position and is truthful only when the total ammunition load is present,

A change in the ammunition load remainder indication occurs after firing of the next 37 to 40 rounds.

In the process of gun firing the aircraft is turned to the left upwards (towards the gun), therefore this should be taken into consideration when aiming.

CHAFF DISPENSING SYSTEM

The chaff dispensing system is designed for producing passive jamming to the missiles having infrared homing heads.

Chaff dispensing units are installed above wing forward of the aircraft dorsal fins. The units are loaded with PPE cartridges. For controlling and monitoring the chaff dispensing system, the aircraft cabin is equipped with the following controls:

- the AAMS FHS RHS
- the EMERG DECOY THERM CARTR FIRING button;
- the JAMM CTL SYS READY annunciator light and DECOY THERM CARTR FIRING - annunciator light;
- the cartridge consumption indicator of the jamming system indication panel (PESP);
- CH D (BP) button on the throttle control lever.

The firing modes, the interval between salvos and the interval within salvo are set by the specialists of the ground crew. After the connection of power supply and outin of the ACS and WEAPON switches the JAMM CTL SYS READY annunciator light glows which will extinguish after the PPE cartridges in units have been fully extended.

The PESP indicator bears the 60 symbol changing in the process of PPE cartridge firing to full expenditure of cartridges.

In the combat operations area, in response to flickering of the DECOY THERM CARTR FIRING annunciator light depress the CH D (BP) button, the DECOY THERM CARTR FIRING annunciator light will be illuminated and extinguished upon termination of the dispensed series.

Note. In the event of continuing illumination of the aircraft, the DECOY THERM CARTR FIRING annunciator light keeps flickering.

In order to continue firing cartridges, depress again the CH D (BP) button and go on doing so until the entire load is spent, whereupon the DECOY THERM CARTR FIRING annunciator light is extinguished and the BOX symbol appears on the PESP indicator.

To discontinue firing jamming cartridges, set the AAMS FHS BHS selector switch to the AAMS (ZRK) position.

When operating against the ground, set the AAMS FHS BHS selector switch to the AAMS position.

Firing of jamming cartridges should be accomplished by depressing the firing button BK or by depressing the CH D (Bn) button.

Section Nine

AMMENDMENTS

PECULIARITIES OF DETECTING, LOCKING ON AND TRACKING TARGETS UNDER PASSIVE JAMMING CONDITIONS

1. In the presence of passive noise, set the SPAN knob to the M (C) position to prevent operation of the missile radio fuze because of the window cloud. Thus, when attacking the target and on getting information from the command post concerning the use of passive jamming by the enemy, turn the SPAN knob to the M position. The other actions are as those performed when attacking the target in the absence of jamming.

The tracking of a low-flying target, when attacked in the rear hemisphere, may happen to become abortive. In this case, lock on-the target once more.

The passive jamming of different density against the earth background in the "Seeking" and "Tracking" modes of operation of the integrated radar/sighting system an well as against the free space background in the "Scanning" and "Tracking" modes of operation at aspect angles up to 2/4 exerts no influence.

The ranges of the target detection and lock-on do not decrease.

PECULIARITIES OF DETECTING, LOCKING-ON, AND TRACKING OF ELECTRONIC COUNTERMEASURES AIRCRAFT

2. When attacking an electronic countermeasures aircraft, check to see that the AT-OFF-CAJ selector switch in met to the AJ position. When guided by voice, set the radar modes-of-operation selector switch to the ENCOUNTER or PURSUIT position on the basis of the information from the direction control post. The settings of other controls resemble those used for attacking the target in the absence of jamming.

The radar can be influenced by the jamming actions both in its scanning mode and in the lock-on mode.

3. The detection of the electronic countermeasures aircraft is indicated by the appearance of the AJ index and vertical dash lines, the azimuth positions of which correspond to the target bearing, on the windshield indicator screen.

The noise gating widths depend on the jammer power as well as the target range, the source of jamming. The maximum width of noise gating can amount to 10° to 15°.

Note: When flying the aircraft at altitudes below 2500 m during the target attack against the earth background as well as after the radiation and selecting the ENCOUNTER, PURSUIT, and AUTO modes at any altitudes, the AJ index can illuminate on the integrated display system screen for a short time even in the absence of the man-made jamming.

The ECM aircraft target azimuth is determined by the azimuth which corresponds to the noise gating centre.

The information concerning the target range can be obtained only from the command radio link in the event of the instrumental guidance or may be given by voice from the direction control post in this case.

To lock-on the ECM aircraft target operate the control button to superimpose the azimuth lock-on gate on the noise gating centre at the range which equals the present range provided by the command radio link or at the range informed from the direction control post by voice and press the WSM, LOCKON PF button.

If no information is available from the direction control post, position the gate at the range not exceeding 50 km when attacking in the front hemisphere or at that not exceeding 30 km when attacking in the rear hemisphere.

After the target is looked on, the integrated display system displays information on its aiming. The presence of index AP is indicative of the ECM aircraft lock-on.

Besides, the guidance by voice causes a sudden change of the present range by the range corresponding to the control button (KY) position. In

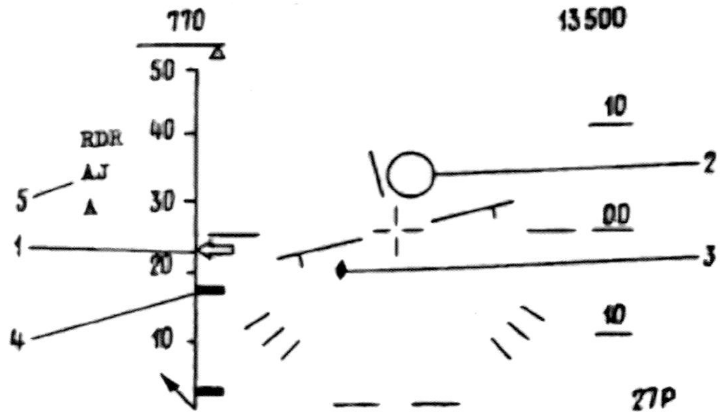

"LOCK-ON" RADAR MODE. LOCK-ON OF ECM TARGET AIRCRAFT IN REAR HEMISPHERE
1 - D_{pres} present target range mark;
2 - aiming circle; 3 - radar antenna position mark; 4 - $D_{perm.\ max}$ mark;
5 - "Active jamming" command

case of the integrated guidance, the radar in provided with the present range from the command radio link. In this case, continue to attack the target an in the case of the jamming absence.

If the target range information is not provided by the command radio link, proceed as follows:

- pilot the aircraft with the aid of the aiming circle as in the case of the target attack in the absence of jamming;
- position the target present range mark by means of the control button at the range specified by the direction control post;
- launch the missiles after the target present range mark enters the zone of the missile launches and after delivery of the LP command.

4. The indication of the active jamming signal affect on the radar during lock-on is the appearance of index AP on the integrated display system screen. Besides, the guidance by voice causes a sudden change in the present range by the range corresponding to the control button position. In this case start to control the present range a mark by means of the control button as directed in Step 3 above.

Notes: (a) To increase the target kill probability under both active and passive jamming conditions, it is recommended to launch the missiles in salvo.

(b) If the range information cannot absolutely be obtained from the grounds use can be made of the kinematic ranging method as a means of information to determine the target range.

(c) When the target is tracked only in the angular coordinates and the range data are entered in the integrated radar/sighting system from the control button or from the command radio link, the integrated display system displays only one mark.

KINEMATIC RANGING METHOD (KRM) MODE

5. When noise is induced during lock-on, the moment of inducing noise is determined by the appearance of indices AP and KD on the integrated display system screen.

In this case, the range is read on the basis of the memorized speed of closure with the target (the "extrapolation" mode).

The "extrapolation" mode functions at the aircraft roll angles below 30° and is followed by flickering of index KD.

The attack of a maneuvering target in this mode may result in the buildup of the range error. In this case, change over to the kinematic method mode by rolling the aircraft in excess of 30°.

The kinematic ranging method mode is set up automatically and, after 7 or 8 s, index KD is illuminating constantly. Then make the S-turn, executing the, turn until the rhomb reaches the end of the aircraft reference symbol wing tip in its horizontal position and then rolling the aircraft in the opposite direction. Perform such maneuvers causing the rhomb to move to the aircraft reference symbol wing tip until the present range mark is displayed.

After its appearance (simultaneously, the permitted missile launch zone also appears), fly closer to the target keeping the rhomb close to the running gear of the aircraft reference symbol.

The target range data are provided not earlier than 12 s after index KD starts illuminating constantly.

To adopt a decision to attack the target on the basis of the information obtained by the use of the kinematic ranging method for the specification of the missile launch permitted zone after delivery of the LP command. it is recommended to:

- set the radar modes-of-operation selector switch to the ENCOUNTER or PURSUIT position depending of the hemisphere used for attack;
- remember the present target range obtained by the kinematic ranging method;
- turn off the kinematic ranging mode by pressing the control button;
- use the control button (without pressing it) to set the present target range which is equal to the range obtained by the kinematic ranging method;
- launch the missiles upon delivery of the LP command.

In the absence of the LP command (the present range mark locates beyond the newly computed zone of permitted launches), shift the present target range ma into the permitted launch zone with the aid of the control button (without pressing it).

Launch the missiles upon delivery of the LP command.

As soon as the noise action ceases, the radar automatically relocks on the target. The indication of its re-locking is the disappearance of index KD and absence of the present range – movement when the control button is operated.

The operating time of the kinematic ranging method mode from the moment of its switching-on or after the change-over from the "extrapolation" mode amounts to 3 min.

The kinematic ranging method mode cannot be employed when the gun is used and in all cases when the target is not tracked in angle during lock-on.

The kinematic ranging method mode is automatically turned off after the target range is obtained from the radar station or command radio link or manually by pressing the control button.

6. When the target mark is duplicated in azimuth during the attack of the ECM target aircraft but its size is less than that of the gate, perform the attack as in the case of noise-free situation.

If the size of the target mark duplicated in azimuth is larger than that of the gate, turn on the COMPENS switch, ensure that the burst of the target marks became narrower and carry out the attack as in the case of noise-free situation.

7. After the target lock-on during attack of the ECM target aircraft, the permitted launch zone oscillates within 5 to 7 km in range and the LP command flickers. In this case, continue to approach the target until the LP command is delivered steadily and then launch the missiles.

In the event of the lock-on reset, detect and lock-on the target once more.

8. In the AUTO mode, the integrated radar/sighting system provides the detection of the ECM target aircraft in a way similar to that in the ENCOUNTER and PURSUIT modes.

To lock on the ECM target aircraft and to use the weapons against It after the noise detection, set the RDR MODES selector switch to the ENCOUNTER or PURSUIT position depending on the data obtained from the command post or at the pilot's discretion.

The integrated radar/sighting system should be switched over to the ENCOUNTER or PURSUIT mode to permit the computation of the missile launch zones.

It in not recommended. to employ the TRACK-WHILE-FLYBY mode against the ECM target aircraft under the AUTO mode conditions.

ACTIONS PERFORMED WHEN ATTACKING ECM TARGET AIRCRAFT WITH NOISE BEING TURNED ON AND OFF DURING TRACKING

9. The indication of the ECM target aircraft detection is the display of the vertical dash lines, AJ command and target mark against the background of the vertical dash lines an the integrated display system screen.

Note: At a low frequency of the noise cancellation, index all and the noise indication can periodically disappear from the screen.

To lock on the target in this case, superimpose the lock-on gate an the target mark and depress the WSM, LOCKON PF button.

After the target is locked on, the scanning information on the integrated display system screen is replaced with the aiming one.

Depending on the frequency of the noise signal cancellations three cases of locking on and tracking the target are possible when the available aiming information is similar to the information obtained after the noise lock-on. In this came, attack the target in the same way as the ECM target aircraft

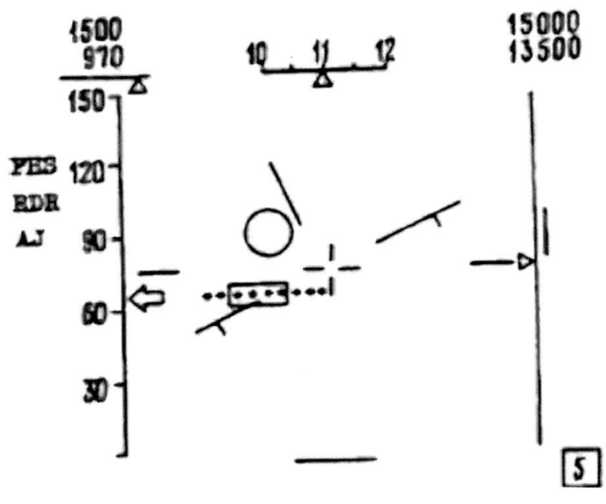

"ENCOUNTER" RADAR MODE DURING SCANNING OF ECM TARGET AIRCRAFT.ATTACK IN FRONT HEMISPHERE DURING INSTRUMENT GUIDANCE (WITH "AJ - OFF - CAJ" SELECTOR SWITCH SET TO "AJ" POSITION)

(b) when the available aiming information is similar to the information obtained after the target lock-on under noise-free conditions (the range information is available, index AP is absent). Make sure that the range from the radar to the target equals the range to the target received from the direction control post and carry out the attack as in the case of noise-free situation;

(c) when the ECM target aircraft turns on and off its jammer during tracking. In this case the radar automatically, when the jammer is turned off, changes over to tracking in range without switching over to the scanning mode. The indication of the radar change-over to tracking In range is the disappearance of index all and automatic recovery of the range information.

In the event of the repeated turning-on of the jammer, the radar automatically changes over to tracking of noise only in angle without switching over to the scanning mode. The integrated display system displays index AP, the target range shifts stepwise to the range corresponding to the control button position (as in the case of noise action in the lock-on mode).

Note. When the radar is switched over to the target tracking only in angle after launching the missile equipped with the radar homing head, continue to illuminate the target until it is destroyed without aligning the aiming circle with the crosshairs to prevent attainment of the target altitude. Turn away 15 to 20 s after the missile liftoff or in case of visual determination of the moment the missile passes by the target.

PECULIARITIES OF ATTACKING TARGET OBSCURED BY JAMMER

10. The jammer-obscured target can be attacked if they are separated in angle. The target and ECM aircraft marks can be separated by maneuvering the attacking aircraft or decreasing the target range (approaching it).

To lock on the target, superimpose the gate on the target so that it cannot contact the noise-gated area and depress the WSM, LOCKON PF button. The integrated display system will present the aiming information.

In tracking the jammer-obscured target, the radar can be changed over to tracking mode jammer. The Indication of its change-over to tracking the jammer is the appearance of index AP on the integrated display system screen, stepwise displacement of the present target range mark to the range corresponding to the control button position. In this case, reset the tracking by depressing the INTERROGATION - RESET button on the control stick, detect the target once more and re-lock on it. Attack the target as in the case of noise-free attack.

11. The non-standard indications can be seen on the integrated display system screen during the serial target attack in the "lock-on" mode:

- the target aspect angle index sharply changes its position and size or oscillates, including its circular rotation;
- the present range does not correspond to the range obtained during the range scanning from the command radio link (from the direction control post commands) or starts energetically increasing (decreasing):
- index AP is absent;
- the permitted launch zones oscillate within several kilometers.

In such cases, without waiting for the automatic look-on reset, reset the look-on manually and set the PREP: MANUAL - AUTO selector switch to the MANUAL position. Then continue to approach the target In the scanning mode, keeping the target mark (the scanning zone mark in the TRACK-WHILE-FLYBY mode) in the centre of the screen. Control the lock-on gate depending on the target mark position, range obtained from the command radio link or orders received from the command post.

On reaching the confident combat employment range, lock on the target and launch the missiles immediately after appearance of the LP command.

PECULIARITIES OF CONDUCT OP CLOSE COMBAT UNDER NOISE CONDITIONS

12. There is a peculiarity in locking on the target and launching the missiles during attack of a visible aerial target (ECM aircraft) with the CLOSE COMBAT radar mode used.

As soon as the ECM aircraft is locked on, index AP appears on the integrated display system screen and the range mark gets "frozen" there.

It the LP command is not furnished (the present range mark locates beyond the permitted launch zone), bring the present target range - within the missile launch zone with the aid of the control button. Launch the missiles as soon as the LP command is furnished.

Section Ten

COLOR PHOTOGRAPHS

Ejection handles are flexible, easy to reach.

Cockpit is comfortable, roomy, and functional.

Instrument layout is straight forward, practical.

Overall of Front Fanel.

Above: HUD controls include TEST, Day/Night, and Brightness Control. Right: HUD is large and easy to use. Backup compass can be seen in upper right.

Center of Front Instrument Panel contains; Top Left Clockwise: Flight Director Indicator, Gyro Horizon Duplicator, Radio Altimeter, Tachometer, Exhaust Gas Temperature, Combined Hydrogas, Combined Life Support, Clock (missing), Mach Meter (above Clock), Combined Course Indicator. Yellow Handle is Nose Wheel Brake Control.

Control Panel (center) for IRR, Attack Modes, SPAN, Master. Upper Round Gauge is Combined Angle-Of-Attack, G-Load Indicator. These indicators are red-lined at 15° and 9-g, respectively.

Fuel Quantity & Flow Indicator Vertical Indicator (Top Left), EKRAN (Aircraft test equipment) Panel (Top Right), Intake Ramp Position Indicator (Bottom Left), Passive Radar Warning Station Indicator in the shape of an aircraft. The RWR is accurate to 12°. (Bottom Right).

Left: Lower Left Instrument Panel: Airspeed Indicator (Top Left), Altimeter (Center Left), Landing Gear and Flaps Position Indicator (Bottom Left), Emergency Landing Gear Control (Red Handle, Bottom Left).

Above: Primary Instruments are easy to read.

Right: Flight Computer Control

Upper Right Instrument Panel showing sun shields with radar display in center. Radar can also be projected onto the HUD.

Center Bottom of Instrument Panel Contains Cabin Pressure Indicator (Top Right), Voltmeter (Top Left), Climate Control Knob (Center).

The HOTAS (Hands On Throttle And Stick) Control Stick Contains Numerous Switches: AFCS Cutoff Switch (Top Left), Trim Control (Center Top), Auto Leveling Button (Top Right), Sight and Radar Control Selector (Bottom Left), INTERR Release (Bottom Center).

Grip Top Contains Clear Labels.

Left Side Console is dominated by dual throttle control. Throttle control is unique in that it is mounted on rails, sliding straight rear to forward as opposed to the usual rotating levers found on most aircraft.

AFCS Mode Selector (Main Grip Lever), Brake Lever (Extended), Main Gun and Launch Trigger.

Throttle control is moved following the finger-controlled releases in front.

Front cf Left Console with Throttle Controls in Full-Afterburner position.

The two throttles contain controls for the Radio (Top Left), WSM LOCKON PF Switch (Top Right), Air Brakes (Center), Radio Channel Selector (Bottom Center) and Flaps.

Missile Control Panel.

Front of Left Console showing Taxi/Landing Light Switch (Top Right), Landing Gear Control (Center Right), RDR Modes (Center Left), Auto Pilot Flight Control Modes Indicators (Bottom Center), Rudder Trim (Lower Right)

Radio Channel Selector (Top Right), Oxygen Control Panel (Center Right), Oxygen Control Valve (Lower Right), Flightsuit Ventilation Controls (Center Left),Pilot Hookups- G-suit, Microclimate, Radio, Oxygen, Pressure Suit (Center Far Right).

Engine/Fuel Control Panel (Top Center), Flaps Control (Bottom Left)

Drogue Chute Release (Left), Illustrated Canopy Control Lever (Center Right).

The covered nylon cord surrounding the left side leg area is the ejection seat leg restraint in its "stored" position.

Extensive Indicator Light Panel (Top Center), Navigation Selection Panel (Lower Center)

Right Console

Right: Navigation System Control Panel

Radio Controls, IFF, Distress, Etc.

Cabin Heat Controls

Emergency Canopy Release (Red Handle, Top Left), External Lights Controls (Below Red Handle), Internal Lights Controls (Center, Top), Electric/Battery Control Panel (Top Right).

Cabin Vent Control Lever (Bottom, Center), Radar Control (Center).

Engine (s) Start Control Panel

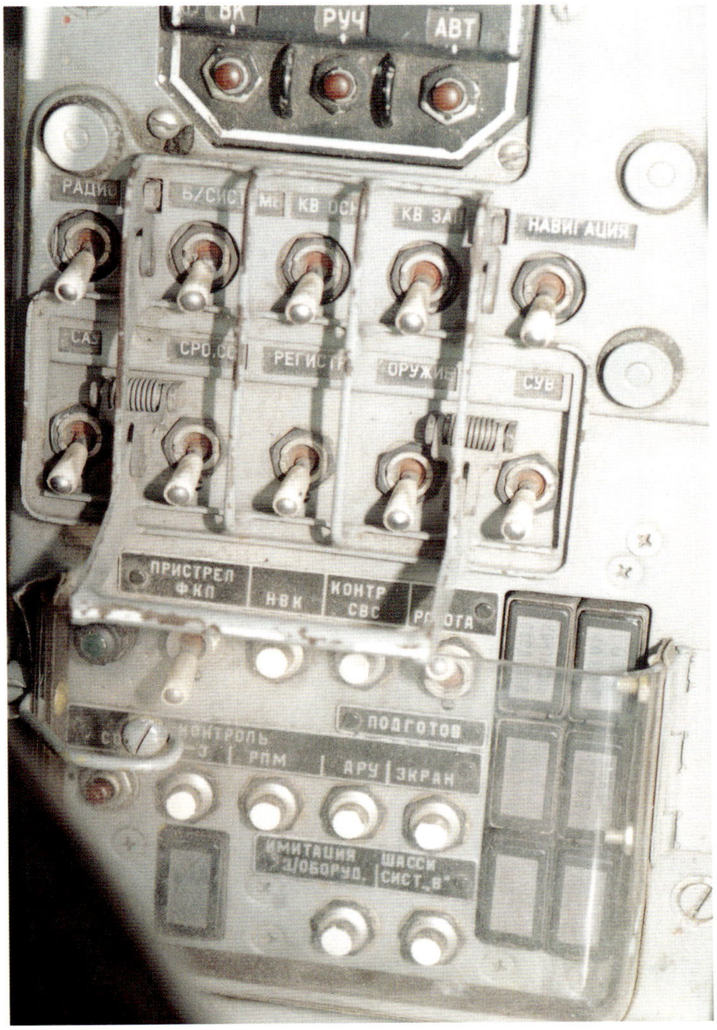

Power Supply Control Panel (Top), Primary Circuit Breakers (Bottom)

Cabin Power System Circuit Breakers

Transponcer Controls

Right: Electrical Equipment Mounted Behind Ejection Seat.

The Zvesda K-36DM Series II Ejection Seat is used in most Russian combat aircraft. Seen here minus black leather seat cushion.

Red ejection handles are flexible and easy to reach. To operate, the pilot squeezes, then pulls the handles up. Black leather leg restraints are well padded.

Common connector for pilot systems.

The K-36 is World-renowned and is proven as one of the very best.

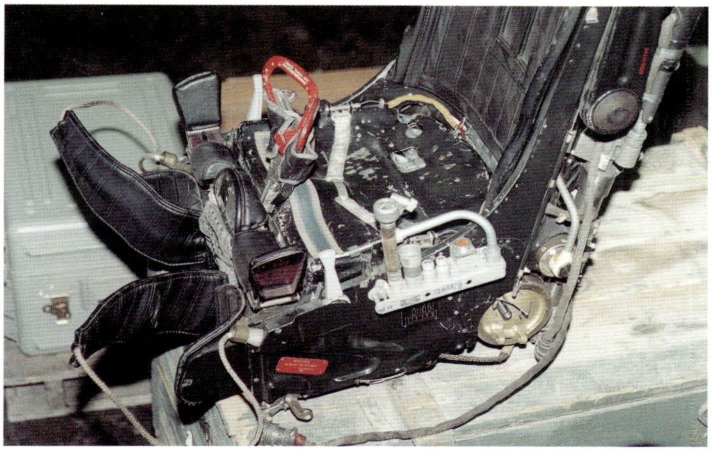

Leg restraints require no "Leg Gaiters". Stabilizer can be seen in rear.

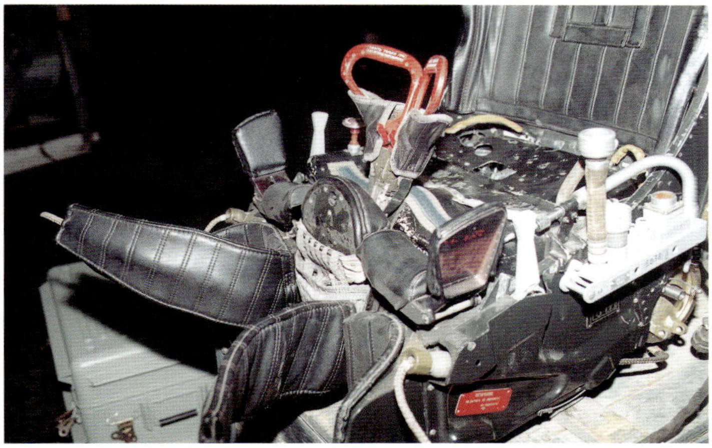

Emergency/bailout oxygen activation knob is seen behind ejection handles.

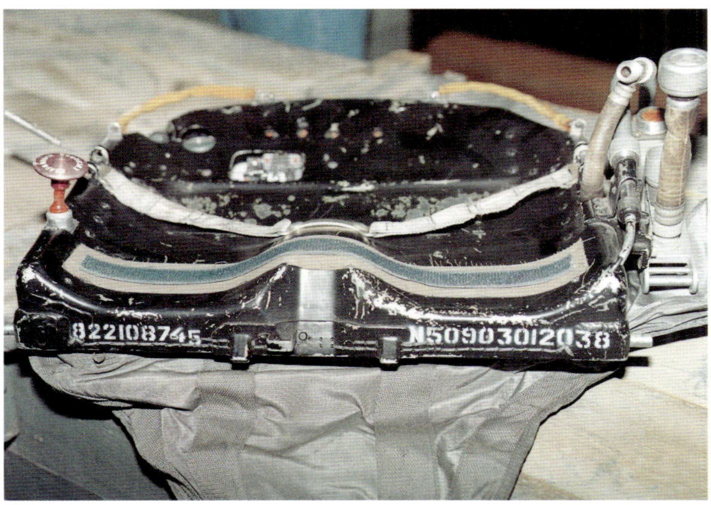

Seat base, which separates to expose life raft and survival kit, is serial numbered for aircraft/pilot identification.

Seat base compartment for NAZ-7M survival kit and life raft.

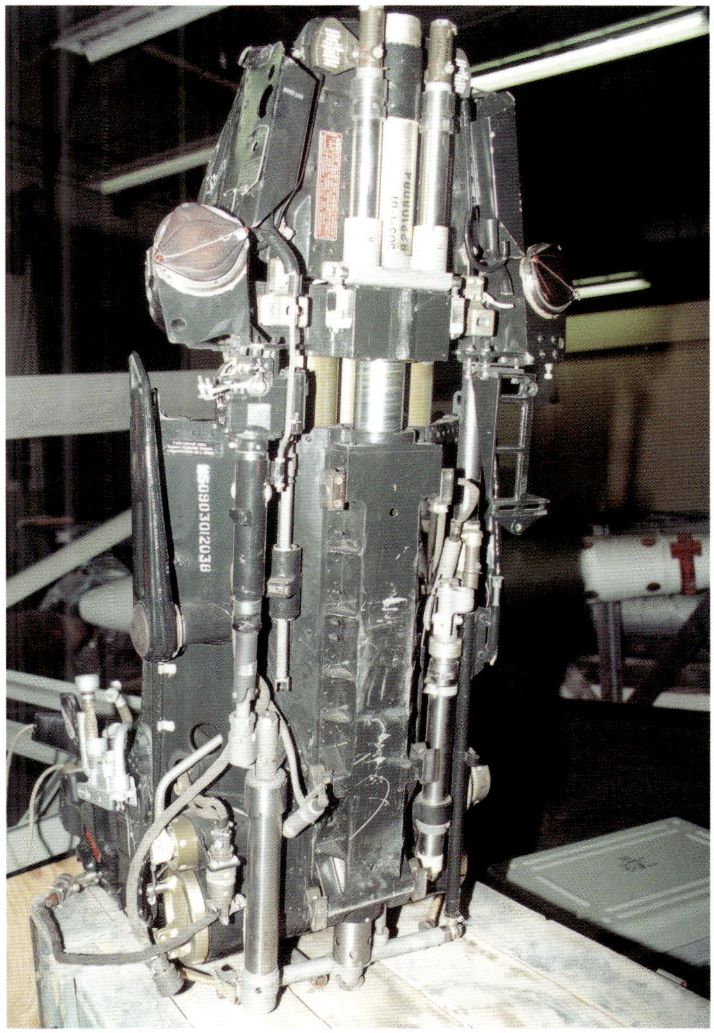

Rear of seat shows complex gas generators, stabilizers, and time delay mechanism.

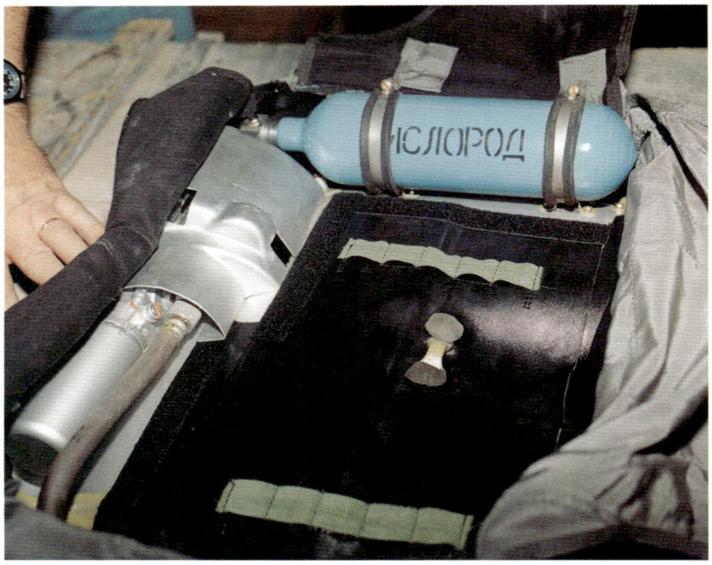

Emergency/bailout oxygen system, including bottle and regulator are mounted under the seat base.

The headrest is designed to be compatible with the ZSh-7 series helmet but is also suitable with the earlier ZSh-5 series and high-altitude GSh-6 series. The pilot's parachute is stored behind the headrest, inside the head box.

In service with the German Luftwaffe, this MiG-29 was spotted in Canada. Photo by David Johnston.

R-60 missile minus control fins is mounted on left pylon. Photos by David Johnston.

Photos by David Johnston.

Photos by David Johnston.

Above: Photo by David Johnston.

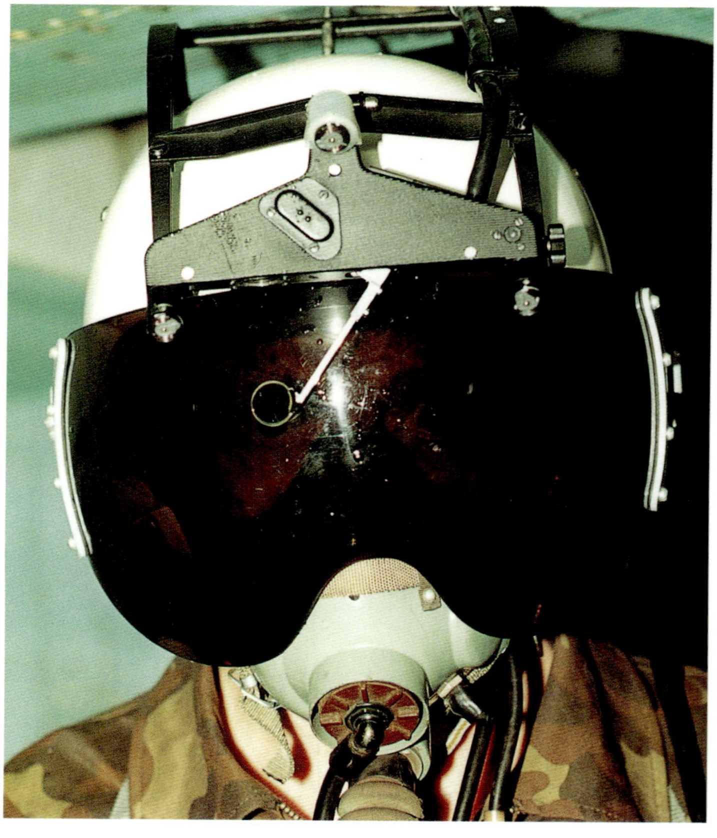

Opposite: The helmet mounted sighting device mounted on a ZSh-5MKV helmet and modeled by Rob Young.

Right: The sighting monocle can be used with or without the visor and is easily adjusted.

Above: R-60M (NATO designation AA-8 "Aphid") air-to-air missile.

Right: Close-up of control vanes of the Vympel R-73 (NATO designation AA-11 "Archer") close-range air-to-air missile. It has a seeker capable of locking onto a target up to 60° off-boresight, using the helmet-mounted sight system. The newest R-73M version is capable of 90° off-boresight seeking.

Also from the Publisher

Northrop F-89 Scorpion: A Photo Chronicle. Marty J. Isham & David R. McLaren. Detailed developmental and operational history of the first USAF dedicated all-weather, jet-powered interceptor. It also features the colorful paint schemes applied to F-89s, which should appeal to both modelers and historians.
Size: 8 1/2" x 11" • over 200 b/w and color photos • 128 pp.
ISBN: 0-7643-0065-2 • soft cover • $24.95

Republic F-84: Thunderjet, Thunderstreak, & Thunderflash - A Photo Chronicle. David R. McLaren. The F-84 performed its unheralded role in a true yeoman fashion. It, and its pilots and groundcrews, fought the ait-to-mud role as a fighter bomber in Korea. It served as an interceptor, and in photo reconnaissance. It was the first jet fighter to be operationally capable of air refueling, and it was the first to be able to deliver a nuclear weapon. 4,300 of the straight-wing fighters were built, along with 2,713 of the swept-wing F-84Fs, and 715 of the reconnaissance RF-84Fs.
Size: 8 1/2" x 11" • over 450 b/w and color photos • 208 pp.
ISBN: 0-7643-0444-5 • soft cover • $29.95

Lockheed F-94 Starfire: A Photo Chronicle. David R. McLaren & Marty Isham. The first U.S. night/all-weather fighter aircraft is chronicled, as is its use by Air Defense Command, Continental Air Command, Alaska and others.
Size: 8 1/2" x 11" • over 220 b/w and color photos • 128 pp.
ISBN: 0-88740-451-0 • soft cover • $19.95

The 4th Fighter Wing in the Korean War. Larry Davis. This book covers the history of the 4th Fighter Wing, from re-activation in 1946, through the end of the "short TDY" to Korea in 1957. Photo coverage includes most of the aces and their aircraft, maintenance, and airfield scenes. A complete list of every victory, and all the losses, is also contained. Interviews with pilots, crew chiefs, and factory tech reps tell the complete story of the "Fourth But First" before, during, and after the Korean War.
Size: 8 1/2"x11" • over 390 color and b/w photos • 200 pp.
ISBN: 0-7643-1315-0 • hard cover • $45.00

Convair F-102 Delta Dagger. Wayne Mutza. With vivid detail and many exclusive photographs, Wayne Mutza chronicles the Convair F-102 Delta Dagger's unsteady course through history, from its inception to present-day survivors. Presented for the first time are the fascinating details surrounding the F-102 in the air defense role, special projects, its unconventional use during the war in Southeast Asia, service with foreign air arms, the Air National Guard and its extensive involvement with the drone program. Includes a detailed listing of each aircraft's assignment history and many of the insignia associated with the F-102.
Size: 8 1/2" x 11" • over 530 b/w and color photos • 192 pp. •
ISBN: 0-7643-1062-3 • soft cover • $29.95

Roll Call: THUD - A Photographic Record of the Republic F-105 Thunderchief. John M. Campbell & Michael Hill. Information is included as to where each plane served, who flew it and the ultimate fate of each THUD. Contains photographs of different serial numbers, including the two F-105s that were flown on Medal of Honor missions.
Size: 9" x 12" • over 700 color and b/w photographs • 176 pp.
ISBN: 0-7643-0062-8 • hard cover • $59.95

Gray Ghosts: U.S. Navy and Marine Corps F-4 Phantoms. Peter E. Davies. This new study of the sea-going Phantoms includes contributions from over a hundred aircrew and maintainers who were involved in the Phantom's success. These voices are blended with a detailed technical and developmental description of the aircraft, supported by comprehensive appendices and over 450 photos in color and black and white.
Size: 8 1/2" x 11" • over 450 color and b/w photos • 192 pp.
ISBN: 0-7643-1021-6 • hard cover • 49.95

Crusader! Last of the Gunfighters. Paul T. Gillcrist. Once known as the "Last of the Gunfighters", the Vought F-8 Crusader has since become a legend in the histories of the U.S. and other countries. Colorful, and sometimes humorous, accounts of events involving the F-8 and "Crusader Drivers" abound in this chronicle of carrier aviation covering the three decades when this remarkable airplane was an important element of the U.S. Navy's carrier strike forces.
Size: 8 1/2" x 11" • over 200 color and b/w photos • 288 pp.
ISBN: 0-88740-766-8 • hard cover • $39.95

Phantom in Combat. Walter J. Boyne. Phantom in Combat puts you in the cockpit with the missile-age aces as they fight for their lives in the skies of Vietnam and the Middle East. Leading USAF ace Steve Ritchie speaks more in sorrow than anger of the politically inspired rules that so frustrated him and his comrades in Vietnam. Here is the human face of modern air warfare, described by the commanders and crews who earned for the Phantom its reputation as the world's finest fighting aircraft.
Size: 8 1/2" x 11" • over 300 b/w and color photos • 192 pp.
ISBN: 0-88740-599-1 • hard cover • $35.00

Tomcat!: The Grumman F-14 Story. Paul T. Gillcrist. Told in an anecdotal format, this new book is richly marbled with the salt air of fleet experience. Perhaps the F-14 program's greatest success is its overwhelming acceptance by the youngsters in the fleet. Anecdotes about the Tomcat legend abound...especially in the chapters devoted to its employment by the fleet in the oceans of the world. This is not a garden variety history about an airplane. It is a sometimes heart-stopping story of how a controversial airplane finally made it into the hearts and minds of the fleet...after stumbling at the starting gates of an outmoded defense acquisition system.
Size: 8 1/2" x 11" • over 150 color & b/w photographs • 208 pp.
ISBN: 0-88740-664-5 • hard cover • $39.95

Tomcat Alley: A Photographic Roll Call of the Grumman F-14 Tomcat. David F. Brown. Contains images depicting the F-14 Tomcat, bureau number by bureau number. Informative captions provide details concerning every Tomcat model, modification, and the fate of each aircraft, including operational accidents, the "boneyard," or "gate guards" at military installations. The majority are shown in full-color. Also includes every squadron commander's aircraft, nose art, tail colors, over 150 patches and insignia, bicentennial schemes, and Tomcats which participated in Operation Desert Shield/Desert Storm.
Size: 9" x 12" • over 875 color photographs • 256 pp.
ISBN: 0-7643-0477-1 • hardcover • $59.95

McDonnell-Douglas F-15 Eagle: A Photo Chronicle. Bill Holder & Mike Wallace. This new photo chronicle covers the F-15 Eagle from its planning and development, to its success in Operation Desert Storm and post-Desert Storm operations in over 170 photographs, most in color. All types are covered, including foreign - Israel, Japan and Saudi Arabia - and Saudi Arabia - and the Strike Eagle.
Size: 8 1/2" x 11" • over 150 color and b/w photos • 88 pp.
ISBN: 0-88740-662-9 • soft cover • $19.95

USAF F-15 Eagles: Units, Colors & Markings. Don Logan. This book chronicles the story of the F-15A, B, C, D, and E. The book includes F-15 program history, prototype and operational aircraft paint schemes, and unit markings of all USAF and Air National Guard units operating the F-15. A section on USAF F-15 MiG killers in both Operation Desert Storm and Operation Allied Force, with photos of all USAF F-15 MiG killers, is included. Over 130 colorful official and unofficial unit patches involving the F-15 are included. The book also includes a description of the aircraft, its systems, and weapons.
Size: 8 1/2" x 11" • over 750 color photographs • 304 pp.
ISBN: 0-7643-1060-7 • hard cover • 59.95

Republic's A-10 Thunderbolt II: A Pictorial History. Don Logan. This book chronicles the story of the A-10, including program history, aircraft paint schemes, all units operating the aircraft and their markings, and selected nose art and ladder door art. Over 100 colorful official and unofficial patches of the units and combat operations involving the A-10 are included. The book also includes a description of the aircraft, its systems, and weapons.
Size: 8 1/2" x 11" • 550+ color photos, line drawings • 240 pp.
ISBN: 0-7643-0147-0 • hard cover • $49.95

Lockheed C-130 Hercules and Its Variants. Chris Reed. Lockheed C-130 Hercules and its Variants chronicles the development and career of the world's predominant military airlifter, from its origins in the depths of the Cold War through the present day. Over 170 photographs, many never before published, show the "Herk" in a variety of roles and paint schemes; coverage includes gunships, electronic warfare and reconniassance models, testbed and special use aircraft, and Navy/Models. Close-up detail photos of several variants will be of interest to scale modellers.
Size: 8 1/2" x 11" • over 200 color and b/w photos • 136 pp.
ISBN: 0-7643-0722-3 • soft cover • $29.95

The Boeing C-135 Series: Stratotanker, Stratolifter and other Variants. Don Logan. This book tells the story of the Boeing C-135 series including: tankers, reconnaissance, airborne command post, weather, test, and special purpose models. All C-135 aircraft types, along with their operating units are covered. Tables and serial number lists are included listing all C-135 configurations by serial number. Also included: a listing of all C-135 losses, including date and reason for loss; three views of C-135 major configurations; selected aircraft nose art; and all USAF, ARFES, and Air National Guard unit markings.
Size: 8 1/2" x 11 • 600+ color photos, line drawings • 256 pp.
ISBN: 0-7643-0286-8 • hard cover • $49.95

Lockheed C-5 Galaxy. Chris Reed. Still the largest U.S. military aircraft more than three decades after its first flight, the Lockheed C-5 Galaxy remains an integral part of Air Mobility Command into the 21st century. Chronicles the development and service career of this behemoth of the airways, starting with the early 1960s requirement for a larger airlifter to supplement the C-141A, through Galaxy's troubled early history to the type's service during the Nickel Grass, Desert Shield, and other airlift operations. Several Galaxy counterparts, such as the Antonov An-124, are also depicted.
Size: 8 1/2"x11" • over 140 color photographs • 80 pp.
ISBN: 0-7643-1205-7 • soft cover • $19.95

Mcdonnell-Douglas F/A-18 Hornet: A Photo Chronicle. Bill Holder & Mike Wallace. The McDonnell-Douglas F/A-18 Hornet is presented here showing its development, production and use by the United States Navy, Marine Corps and by various foreign air forces. Use by the Blue Angels, and the very latest model updates are included.
Size: 8 1/2" x 11" • over 120 color and b/w photos • 80 pp.
ISBN: 0-7643-0243-4 • soft cover • $19.95

Vulture's Row: Thirty Years in Naval Aviation. Paul T. Gillcrist. Told in anecdotal form, Vulture's Row tells a fascinating story about an important period covering nearly one half of the entire history of U.S. naval aviation. "Vulture's Row" is an area dubbed by naval flyers, on the island structure of an aircraft carrier where pilots who aren't flying can overlook carrier launchings and recoveries on the deck below. This new book by acclaimed author Paul Gillcrist is a series of true stories about the U.S. Navy carrier aviation from the perspective of a Navy pilot who spent thirty-three years directly involved in that exciting profession.
Size: 6" x 9" • over 100 color and b/w photographs • 256 pp.
ISBN: 0-7643-0047-4 • hard cover • $29.95

Feet Wet: Reflections of a Carrier Pilot. Paul T. Gillcrist. Paul Gillcrist was a navy carrier pilot for almost thirty years, from the early days of flying propeller planes from straight deck carriers, to the days of high-tech, lethal "teen" jets and supercarriers. In his remarkable career - from "nugget", to competent jet aviator, to test pilot, to Vietnam fighter pilot, to air wing commander, to head of "Fightertown, USA," Gillcrist flew the F-8 Crusader, F-4 Phantom, F-14 Tomcat, and a myriad of other tactical aircraft. As part of his absorbing stories. Gillcrist shares technical information on carrier aviation; the reader sees, step-by-step, how an airplane is launched, and then trapped again on a small deck aboard a moving ship - this, the most dificult feat in aviation.
Size: 6" x 9" • 27 b/w photographs • 368 pp.
ISBN: 0-7643-0284-1 • hard cover • $29.95

Northrop's YF-17 Cobra: A Pictorial History. Don Logan. The Northrop YF-17 holds a special place in aircraft history. The YF-17 was one of two prototypes tested in the U.S. Air Force Air Combat Fighter competition, a program which attempted to reverse the trend of increasing cost and complexity of new fighter aircraft, and which resulted in the selection and manufacture of the F-16 as the next generation free world fighter. Even though the YF-17 lost the USAF competition, it was the prototype for the U.S. Navy's F/A-18 aircraft.
Size: 8 1/2" x 11" • over 100 color and b/w photographs, line drawings • 64 pp.
ISBN: 0-88740-910-5 • soft cover • $19.95

Lockheed-Martin F-22 RAPTOR: An Illustrated History. Mike Wallace & Bill Holder. It's called the Raptor. The F-22 will be the U.S. Air Force's premier fighter for the first decades of the 21st century. Like other expensive military programs, the F-22 endured continuous scrutiny throughout its development – the aircraft that evolved was worth the wait. The Raptor is planned to start operational service with the U.S. Air Force in the year 2004, and when on-line, will provide a capability far beyond that of any other fighter aircraft.
Size 8 1/2" x 11" • over 140 color photographs • 72 pp.
ISBN: 0-7643-0558-1 • softcover • $19.95

General Dynamics F-111 AARDVARK. Don Logan. Covers all models from the F-111A through the F-111K. In addition to a "Roll Call" including histories of each individual aircraft, photos of over 90% of the aircraft are included. Nose art of the FB-111A is covered with 48 photos. One Hundred and twenty Unit and Campaign patches are included, along with appendices on External Differences, Avionics Systems, F-111 Specifications, a Unit and Tail Code Summary, Attrited Aircraft, and a list of the 113 Aircrew members who died flying the F-111.
Size: 8 1/2" x 11" • 860+ color photos, line drawings • 320 pp.
ISBN: 0-7643-0587-5 • hardcover • $59.95

Lockheed F-117 Nighthawk: An Illustrated History of the Stealth Fighter. Bill Holder & Mike Wallace. The F-117 was probably the most secret aircraft ever developed. The Stealth technology upon which the plane was based made it unique in its appearance. With its stark black appearence and wispy configuration, the plane was a strange departure from the sleek supersonic fighters of the period. Covers the technical and operational aspects of the Nighthawk from its initial use over Panama through its shining moment during Operation Desert Storm.
Size: 8 1/2" x 11" • over 120 color and b/w photos • 64 pp.
ISBN: 0-7643-0067-9 • soft cover • $19.95

Lockheed T-33: A Photo Chronicle. David McLaren. The T-33 jet trainer was derived from the Lockheed F-80C Shooting Star, the USAF's first operational jet fighter, in 1948. In the fifty years since its introduction the T-33 has trained more jet pilots than any other training type.It was also a testbed in many development programs including tests on ejection seats and missile guidance systems, and continues yet today in various military and civilian roles.
Size: 8 1/2" x 11" • over 360 b/w and color photos • 128 pp.
ISBN: 0-7643-0646-4 • soft cover • $24.95

Northrop's T-38 TALON: A Pictorial History. Don Logan. This is the story of the most successful pilot training jet ever produced: Northrop T-38 Talon. The history of the aircraft is broken down by the roles it has played in over thirty years of service. All units flying the T-38, their markings, and paint schemes are covered in over 300 color photographs-including a chart of colors used listing Federal Standard (RS) color numbers.
Size: 8 1/2" x 11" • over 300 color photographs • 152 pp.
ISBN: 0-88740-800-1 • soft cover • $24.95

Northrop Grumman B-2 Spirit: An Illustrated History. Bill Holder. It has been called the most amazing and unbelievable aircraft of the 20th century. One look at the bat-like lines of the USAF's stealthy Spirit bomber quickly confirms this. But that futuristic shape has lineage back to the 1940s when the Northrop Company designed and built the B-35 and B-49 models. Granted, the development program has had to endure a rocky road, but it is now here, and even though in smaller numbers than many desired, it will still present a formidable presence in the 21st century.
Size: 8 1/2" x 11" • over 150 color and b/w photos • 80 pp.
ISBN: 0-7643-0591-3 • softcover • $19.95

Convair B-36: A Comprehensive History of America's "Big Stick". Meyers K. Jacobsen. From the origins of the use of six engines on bombers from around the world and the history and development of SAC and the use of its numbers of B-36s, this book has a greater amount of information than has ever before been published on the Convair B-36. Covered are the electronic countermeasures and various armament configurations applied in the B-36, B-36 attrition, and detailed accounts of the survivors that exist today.
Size: 9" x 12" • over 730 b/w and color photographs • 400 pp.
ISBN: 0-7643-0530-1 • hardcover • $69.95

Convair B-36 Peacemaker: A Photo Chronicle. Meyers K. Jacobsen. Explores the history of the Strategic Air Command's biggest bomber that helped keep the peace during the early years of the Cold War. This book will give the reader a concise overview of the story of the Peacemaker in the 1940s-1950s. A serial number listing is included, as well as a list of all ten B-36 bomb wings.
Size: 11" x 8 1/2" • over 130 b/w photographs • 96 pp.
ISBN: 0-7643-0974-9 • soft cover • 9.95

Martin B-57 Canberra: The Complete Record. Robert C. Mikesh. No story about one type of aircraft could be more complete than this coverage about the B-57 Canberra. A brief history of its British inception sets the stage for the conversion that took place to American standards for production in the United States. The Canberra was needed to fill the night intruder role in the USAF that was identified during the Korean War and later Vietnam. The B-57 did that, and far more.
Size: 8 1/2" x 11" • over 420 color and b/w photos • 208 pp.
ISBN: 0-88740-661-0 • hard cover • $45.00

North American XB-70 Valkyrie: A Photo Chronicle. John M. Campbell & Garry R. Pape. The North American XB-70 is one of the most unusual looking aircraft in aviation history, and only two were constructed. It was originally designed as a Mach 3 high-altitude bomber, but was later used as a research aircraft. This book gives a short, detailed history of the XB-70, including production, flight tests, and the fatal crash of Aircraft #2 in 1965.
Size: 11" x 8 1/4" • 100+ b/w photos and line schemes • 48 pp.
ISBN: 0-88740-906-7 • soft cover • $9.95

North American XB-70 in Color. John M. Campbell. This is an all-color photo study of the XB-70, including production, flight tests, and the fatal crash of Aircraft #2. Aircraft #1 is now at the U.S. Air Force Museum.
Size: 11" x 8 1/2" • over 110 color photos • 96 pp.
ISBN: 0-7643-0507-7 • soft cover • $19.95

The U-2 Spyplane - Toward the Unknown: A New History of the Early Years. Chris Pocock. The full story of the development and early use of the U-2 has never been properly told – until now. This book describes in vivid detail how the high-flying spyplane was conceived, designed, built, and deployed in record time. Will appeal to students of aviation and intelligence history, and to anyone wishing to learn more about a key episode in the Cold War.
Size: 6" x 9" • over 110 b/w and color photographs, line drawings • 288 pp.
ISBN: 0-7643-1113-1 • hard cover • $29.95

Boeing B-52: A Documentary History. Walter J. Boyne. Conceived in 1948, first flown in 1952 and projected still to be in front-line service in the 21st century, the Boeing B-52 Stratofortress is one of the most extraordinary aircraft in history. This book features a comprehensive history of the development of the U.S. heavy bomber, and intensive discussion of the Boeing B-47 and its effect upon the B-52 design, and perhaps more important than either of these today, the enormous number of modifications and changes which have kept the aircraft viable.
Size: 8 1/2" x 11" • over 200 b/w photographs, cut-away drawings, charts, diagrams • 160 pp.
ISBN: 0-88740-600-9 • hard cover • $29.95